Springer Complexity

Springer Complexity is a publication program, cutting across all traditional disciplines of sciences as well as engineering, economics, medicine, psychology and computer sciences, which is aimed at researchers, students and practitioners working in the field of complex systems. Complex Systems are systems that comprise many interacting parts with the ability to generate a new quality of macroscopic collective behavior through self-organization, e.g., the spontaneous formation of temporal, spatial or functional structures. This recognition, that the collective behavior of the whole system cannot be simply inferred from the understanding of the behavior of the individual components, has led to various new concepts and sophisticated tools of complexity. The main concepts and tools - with sometimes overlapping contents and methodologies - are the theories of self-organization, complex systems, synergetics, dynamical systems, turbulence, catastrophes, instabilities, nonlinearity, stochastic processes, chaos, neural networks, cellular automata, adaptive systems, and genetic algorithms.

The topics treated within Springer Complexity are as diverse as lasers or fluids in physics, machine cutting phenomena of workpieces or electric circuits with feedback in engineering, growth of crystals or pattern formation in chemistry, morphogenesis in biology, brain function in neurology, behavior of stock exchange rates in economics, or the formation of public opinion in sociology. All these seemingly quite different kinds of structure formation have a number of important features and underlying structures in common. These deep structural similarities can be exploited to transfer analytical methods and understanding from one field to another. The Springer Complexity program therefore seeks to foster cross-fertilization between the disciplines and a dialogue between theoreticians and experimentalists for a deeper understanding of the general structure and behavior of complex systems.

The program consists of individual books, books series such as "Springer Series in Synergetics", "Institute of Nonlinear Science", "Physics of Neural Networks", and "Understanding Complex Systems", as well as various journals.

Understanding Complex Systems

Series Editor

J.A. Scott Kelso
Florida Atlantic University
Center for Complex Systems
Glades Road 777
Boca Raton, FL 33431-0991, USA

Understanding Complex Systems

Future scientific and technological developments in many fields will necessarily depend upon coming to grips with complex systems. Such systems are complex in both their composition (typically many different kinds of components interacting with each other and their environments on multiple levels) and in the rich diversity of behavior of which they are capable. The Springer Series in Understanding Complex Systems series (UCS) promotes new strategies and paradigms for understanding and realizing applications of complex systems research in a wide variety of fields and endeavors. UCS is explicitly transdisciplinary. It has three main goals: First, to elaborate the concepts, methods and tools of self-organizing dynamical systems at all levels of description and in all scientific fields, especially newly emerging areas within the Life, Social, Behavioral, Economic, Neuro- and Cognitive Sciences (and derivatives thereof); second, to encourage novel applications of these ideas in various fields of Engineering and Computation such as robotics, nanotechnology and informatics; third, to provide a single forum within which commonalities and differences in the workings of complex systems may be discerned, hence leading to deeper insight and understanding. UCS will publish monographs and selected edited contributions from specialized conferences and workshops aimed at communicating new findings to a large multidisciplinary audience.

L. Kocarev G. Vattay (Eds.)

Complex Dynamics in Communication Networks

With 168 Figures and 13 Tables

Springer

Ljupco Kocarev
UCSD
Inst. Nonlinear Science
Gilman Drive 9500
La Jolla, CA, 92093-0402
USA

Gabor Vattay
Eötvös University
Dept. of Physics/Complex Systems
Pázmány sétány 1
1117 Budapest
Hungary

ISBN 978-3-642-06356-5 e-ISBN 978-3-540-32381-5
ISSN 1860-0832

Springer is a part of Springer Science+Business Media
springeronline.com

Printed in Germany

Cover-Design: Erich Kirchner, Heidelberg

Printed on acid-free paper 54/3141/Yu – 5 4 3 2 1 0

Preface

The study of complex systems pervades all of science, from cell biology to ecology, from computer science to meteorology. A paradigm of a complex system is a network, where complexity may come from different sources: topological structure, network evolution, connection and node diversity, and/or dynamical evolution. Network structures are being found pervasively throughout natural and engineered systems, and the modern study of networks is requiring ideas from computer science, physics, dynamical systems, statistics and biology. Networks consist of nodes which are interconnected by a mesh of links. The macroscopic behavior of a network is determined by both the dynamical rules governing the nodes and the flow occurring along the links.

Computer and communications networks are among society's most important infrastructures. The emergence of a global Communication Network during the past decade changed the way we live and work. The physical infrastructure and the logical network of the Internet grew like a living creature producing one of the most sophisticated systems humanity ever built. At the same time the components of the networking technology went trough enormous engineering assisted evolution. Routers and communication lines are orders of magnitude faster and more complicated than a decade ago, while previously separate technologies such as telephone and mobile communication systems got integrated under the umbrella of the Internet Protocol. While each and every part of the system evolves toward an enormous complexity, in a paradoxical way, all these details are hiding beneath the Internet Protocol allowing us to disregard most of the underlying details from a modeling point of view.

The Internet is a giant (global) network of networks without central control or administration: its design is guided by the end-to-end principle. Transmission Control Protocol/Internet Protocol (TCP/IP) is the predominant transport protocol used in today's Internet. TCP/IP is a protocol used by source computers to inject packets into the Internet, and used by Internet routers to store-and-forward packets among multiple routers along a path, and then finally to forward the packet to its destination computer. TCP/IP is an end-to-end protocol operating on logical connections between pairs of computers. A better understanding of the TCP/IP network structure, its topology and dynamics, is essential for optimal design, efficient protection, robustness, and modeling the Intetnet. This book is a first collection on the new and emerging filed of nonlinear dynamics of TCP/IP networks. It has fourteen chapters. The book opens with the chapter "Nonlinear Dynamics of TCP and its Implications to Network Performance," written by András Veres and Miklós Boda. The authors demonstrate that the TCP protocol, while competing for networking resources, generates complex non-linear dynamics. An important message of this work is that random traffic behavior is not exclusively due to "random" effects, but also due to complex behavior of the TCP protocol. The TCP protocol, although driven by deterministic rules, may produce time-series seemingly indistinguishable from stochastic processes.

In the second chapter "Dynamics of Congestion Control" A. Gilbert analyzes synchronous behavior in communication networks. Many experiments and the intuitive explanations of these experiments show that TCP sources competing for bandwidth on a congested link will synchronize through the weak coupling inherent in congestion control. Intuitively speaking, a population of sources will synchronize because they all experience loss at roughly the same time, they all scale back their transmission rate in the presence of these losses, and then they increase their transmission rate until the next bout of congestion. In this paper, the author explores: (1) the conditions under which periodic aggregate behavior can occur, (2) whether there is evidence that it is occurring in the Internet today, and if so, (3) whether there are simple ways to prevent it.

In the third chapter "Statistical Properties of Chaos in Communication Networks" written by G. Vattay, K. Diriczi, A. Fekete, L. Kocarev, M. Maródi and J. Stéger, the authors demonstrate that the packet sending dynamics in computer networks can be chaotic. In particular, chaotic properties of the TCP congestion avoidance mechanism are investigated. The analysis focuses on the origin of the complex behavior appearing in deterministic TCP/IP networks. From the traffic modeling point of view the understanding of the mechanism generating chaos is essential, since present models are unable to cope with this phenomena.

N. S. V. Rao, J. Gao and L. O. Chua in the fourth chapter "On Dynamics of Transport Protocols Over Wide-Area Internet Connections" present a state-space model of TCP and show that its dynamics embed a tent-like map which generates chaotic trajectories under certain conditions. Moreover, in this chapter the authors show that Internet dynamics are equally dominated by the stochastically of Internet traffic and the deterministic chaos due to the non-linear TCP dynamics, and are best characterized by anomalous diffusions defined by a large diffusion exponent.

G. Simon, P. Pollner, P. Hága and I. Csabai in "Dynamical Properties of Externally Driven TCP traffic" study the effect of periodically sent user datagram protocol (UDP) packets on a single persistent TCP connection, and on an aggregate flow consisting of two persistent TCP connections. The authors also introduce the problem of measuring a system whose characteristics are changing as a result of the measuring process, and present a comprehensive study of the influence of active probing on realistic time-varying TCP traffic.

In the sixth chapter "Data Traffic, Topology and Congestion" D. K. Arrowsmith, R. J. Mondrag, and M. Woolf use of intermittency in iterated maps to provide various relevant statistical types of binary data will be described. The dynamical modeling of packet traffic using intermittency maps is introduced together with the dynamics of Transmission Control Protocols. Regular and scale-free network topologies are used for Internet packet traffic modeling and the congestion behavior of packet lifetimes on these networks under increasing load is investigated.

The treatment of non-Poisson fractal-like time-series describing packet traffic featuring a bursty behavior and a high variability over a range of time scales is discussed by G. Setti, R. Rovatti and G. Mazzini in "Chaos-Based Generation of Artificial Self-Similar Traffic". One-dimensional chaotic maps have been shown to be able to reproduce such intermittent processes and the construction of simple models for realistic traffic sources can be considered a substantial contribution of the theory of complex dynamics to the growing field of implementation of new network control units. In this chapter, the authors establish theoretical ground for the formal development behind the chaos-based modeling of network traffic and other similar phenomena.

The eight chapter "Macroscopic Dynamics in Large-Scale Data Networks" by J. Yuan and K. Mills, the authors study space-time characteristics of congestion in large networks, and analyze system behavior as a coherent whole. This chapter also demonstrates the macroscopic effect of distributed-denial-of-service flooding attacks, and shows how the technique developed by authors could provide significant information to detect and defend against such attacks.

G. Chen, Z. Fan, and X. Li in the ninth chapter "Modeling the Complex Internet Topology" develop a novel multi-local-world (MLW) model with a localization property for better description of the Internet. Clearly, the Internet can be considered as a collection of many interconnected subnetworks. If a subnetwork in the Internet is viewed as a "local-world," then the Internet consists of several interconnected "local-worlds." This new mode is based on a carefully study of the Internet AS graphs, with a comparison to the BA model.

In the tenth chapter "Evolution of the Internet Topology and Traffic Dynamics of Data Packets" K. Goh, B. Kahng, and D. Kim investigate how the connection profile of the Internet at the autonomous systems (ASes) level evolves, finding that it is extremely dynamic and can be described in the framework of the multiplicative stochastic process. Extracting relevant parameters for the growth dynamics of the Internet topology, the authors are able to predict the connectivity (degree) exponent of the Internet AS map successfully.

H. Zhang, M. Liu, V. Vukadinović, and L. Trajković in "Modeling TCP/RED: a Dynamical Approach" study interaction between Transmission Control Protocol (TCP) and Random Early Detection (RED) gateways using dynamical models. The communication network is viewed as a discrete-time feedback control system where TCP adjusts its window size depending on whether or not it has detected a packet loss during the previous round trip time (RTT) interval. In this chapter, the authors describe a discrete-time nonlinear dynamical model for interaction between TCP and RED gateways. The model, constructed using an iterative map, captures a detailed dynamical behavior of TCP/RED, including slow start, fast retransmit, and timeout events common in TCP.

In the twelfth chapter "Nonlinear Instabilities in TCP-RED" the authors P. Ranjan, E. H. Abed and R. J. La present a novel modeling paradigm of dynamical negotiation between clients running TCP (Transmission Control Protocol) and routers with RED (Random Early Detection) active queue management scheme. Basic aim of this modeling is to understand the inherent nonlinearity in their interaction and how it manifests itself in the form of parametric sensitivities observed in practise. The model proposed here is used to study network dynamics over large parameter variations. Both smooth bifurcation like period doubling and non-smooth bifurcation like border collision are shown to occur as system parameters are varied.

In the next chapter "Synchronization in Complex Networks" the authors L. Kocarev and G. Vattay study synchronization in complex networks topologies; in particular they analyze synchronization properties of classical and power-law random graph models.

In the fourteenth chapter "Dynamic Complexity in the Internet Traffic" M. Takayasu reports on observation and simulation results showing the evidence of the phase transition in the Internet between congested and non-congested phases accompanying critical fluctuations at the phase transition point. The input mean flow density into the Internet is regarded as the control parameter of this phase transition. Results show power-law distribution of congestion duration time, divergence of correlation length of jams, and discontinuity in the differential coefficient of the probability of jam occurrence at the critical point.

San Diego, Budapest,
November 2004

Ljupco Kocarev
Gábor Vattay

Contents

Nonlinear Dynamics of TCP and its Implications to Network Performance

András Veres and Miklós Boda

Traffic Lab, Ericsson Research, Budapest Andras.Veres@ericsson.com, miklos.boda@nkth.gov.hu

1 Introduction

TCP flows continuously intertwine in the Internet competing with each other for service capacity and buffer space in bottleneck routers. The window based flow and congestion control algorithms implemented in end-hosts control competition between traffic flows. Previous work has modeled TCP competition from a macroscopic point of view that is the average TCP flow rate have been derived as a function of delay, loss and service rate [9][15]. In this paper, we introduce a novel approach to model the competition between multiple TCP connections sharing a common bottleneck buffer. We demonstrate that the end-to-end congestion control used by the TCP protocol, while competing for networking resources, generates complex non-linear dynamics. An important message of this work is that random traffic behavior is not exclusively due to "random" effects, but also due to complex behavior of TCP. The TCP protocol, although driven by deterministic, rules may produce time-series seemingly indistinguishable from stochastic processes. On the other hand, non-linear systems have unique properties and they are able to produce a diversity of phenomena.

In this paper, some of these properties and phenomena are demonstrated and analyzed. We have built a test network consisting of real hosts and protocol implementations where we can create several configurations representing different network scenarios in a simplified setting. The dynamics of this system can be characterized from two aspects.

First, the dynamics of the system are investigated in the time domain that is the frequency and duration of traffic bursts injected by the TCP protocol is analyzed. The efficiency of router buffering depends on short scale bursts in the order of the buffer emptying timescale, while end-users are more concerned by long timescale variations of the traffic rate.

Second, we analyze the dynamics in the phase space, where we observe co-dependence of the system variables while the time dimension is hidden. This analysis helps to analyze, for example, periodicity and non-periodicity. The phase space analysis also enables the investigation of the system sensitivity. A system of high sensitivity can be easily perturbed by small interactions, in such systems the effect of small interactions can grow to large-scale changes. High sensitivity might also make it possible to apply "smart network management", which means that the network performance is improved by precise, minute interventions.

The observed and analyzed phenomena are the product of complex protocol mechanisms as they are competing for the network resources. Competition is unavoidable, and TCP algorithm is the glue that keeps the Internet together. It is thus important to better understand the properties of the competition and the performance impacts of them. Ultimately better understanding can lead to better networks.

2 TCP Congestion Control

The TCP protocol [16][6] provides reliable data delivery between two computers using data acknowledgements and retransmissions. The protocol is rather complex and it has numerous variants. In this section, we explain the most important mechanisms with a focus on how TCP rate control works.

TCP uses the so-called window based congestion control to control its data transfer rate over the interconnecting networks. The congestion window (*cwnd*) controls the amount of data that can be outstanding unacknowledged in the network at any time. So if the window is x bytes, then at most x bytes can be delivered maximum during the end-to-end round-trip time RTT between the two computers. The rate of a TCP connection is thus controlled by $8 * x/RTT$.

TCP does not have explicit information about the optimal congestion window it should use, so it uses implicit information instead by means of detecting packet losses and estimating the round-trip delay. TCP assumes that packet losses are indications of network congestion. When a packet loss happens and the acknowledgement packet (Ack) does not arrive before the time-out kicks in, TCP reduces its window (and thus reduces its speed). In between losses, TCP gradually increases its sending rate. If the congestion window is small, TCP increases fast; it increases by one packet after it received an acknowledgement packet of each sent packet (slow-start phase). After it has reached the so-called slow-start threshold, it increases by one packet every round-trip time (congestion avoidance phase).

There are several optimizations of this basic algorithm, like fast retransmit, fast recovery, or selective acknowledgements, just to name the most important ones [19] [17] [18]. The fast retransmit/recovery modifications accelerate TCPs recovery time from packet losses. According to these mechanisms, after a packet loss, TPC does not have to wait for a time-out and then go to slow start, instead, it drops its rate by half and continues with congestion avoidance from there (Figure 1).

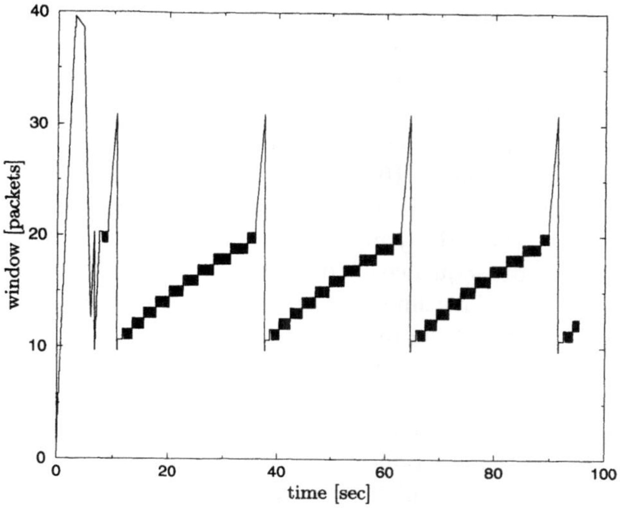

Fig. 1: Window process of a single TCP.

3 The Basic Lab Model and TCP Dynamics

Our goal is to investigate TCP download dynamics in a simplified model where we can capture the important aspects of TCP rate control. See Figure 2. We built this model using real networking components in a laboratory.

The model consists of two end-hosts, both running Linux kernel version 2.4. One of them was a TCP server and the other one was the client. In between the two hosts there is a router, which acts as a bottleneck between the two hosts. In the router we can set the service rate and the maximum buffer size. In the real world the two hosts can be far from each other: the propagation delay in our lab experiment is emulated by the *NISTNet* [11] network emulator which can delay every packet by an arbitrarily set duration.

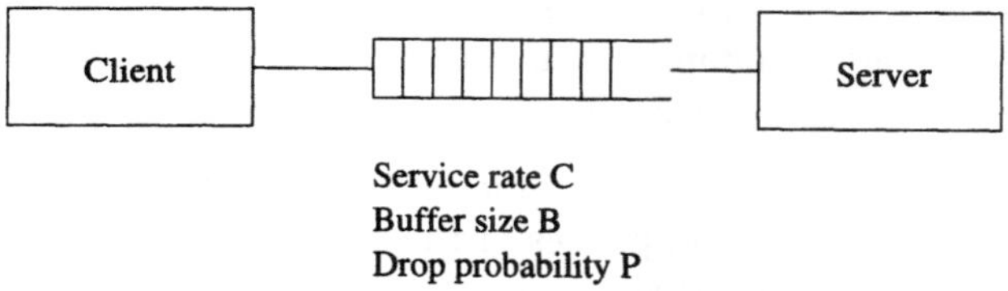

Fig. 2: Testlab configuration

To observe the TCP dynamics we monitored the packets sent and received by the server. Every packet was logged with a precise timestamp. As we discussed before, a key variable of TCP dynamics is the TCP window or the number of outstanding unacknowledged bytes. We used the *tcptrace* [20] utility which calculates the window size as a function of time from the earlier recorded packet logs.

In this first experiment we set the bottleneck rate to 80 kbps, buffer size 20 packets and propagation delay 100 ms. We start a single TCP download from the server to the client. The calculated window as a function of time is shown in Figure 1. Observe the orderly behavior, which starts with a slow-start then is followed by a periodic sequence of congestion avoidance and window halving after a buffer overflow. This simple case demonstrates how TCP adapts to a static network situation. A more interesting case is when TCP has to share the network with other traffic, e.g., with another TCP download. This case is shown in Figure 3a. We can see that the periodic behavior remains. The two TCPs start at the same time and lose packets near the same time. The reason is that when the bottleneck buffer is near full both TCPs will have a high chance to lose a packet. This phenomenon is called TCP synchronization.

One can modify several parameters in the above model. We can change the rate, the size of the buffer, the number of TCP connections, etc. Here we change the rate and see how it affects the window process. The window processes are shown in Figure 3. The periodic behavior vanishes as we reach speed of 1200 kbps where the process becomes seemingly random. What happens is that at higher speed the packets are scheduled at with smaller time between them. In every real-life systems there are inherent sources of different kinds of noises, for example, an interrupt may delay the service of a packet by a minute amount of time. As the service speed increases and the gap between served packets gets closer to the devices' inherent noise levels, the system becomes unstable. This intuitive explanation obviously deserves more discussion.

In order to investigate the sensitivity of a system we artificially perturb the system by a single packet in the middle of the usual TCP download. This small, 40 byte packet will occupy the bottleneck server by some time (e.g., at 1000 kbps this delay will be $40 * 8/1000000 = 3.2msec$. Figure 4 shows the window dynamics at two different speeds, in each case we sent in a single small disturbing packet. The result shows that the periodic system is kicked out from the orderly behavior, but some time later it turns back to order. In case of a seemingly random system we do not see any particular change. Based on the above observations, we can formulate several related questions.

- Can we characterize orderly and random behavior in some way?
- What are the statistical properties of the TCP flow rates?
- Can we measure the sensitivity of the system to external perturbations?
- Can we control a sensitive system, or at least what impact sensitivity may have on end-to-end performance?

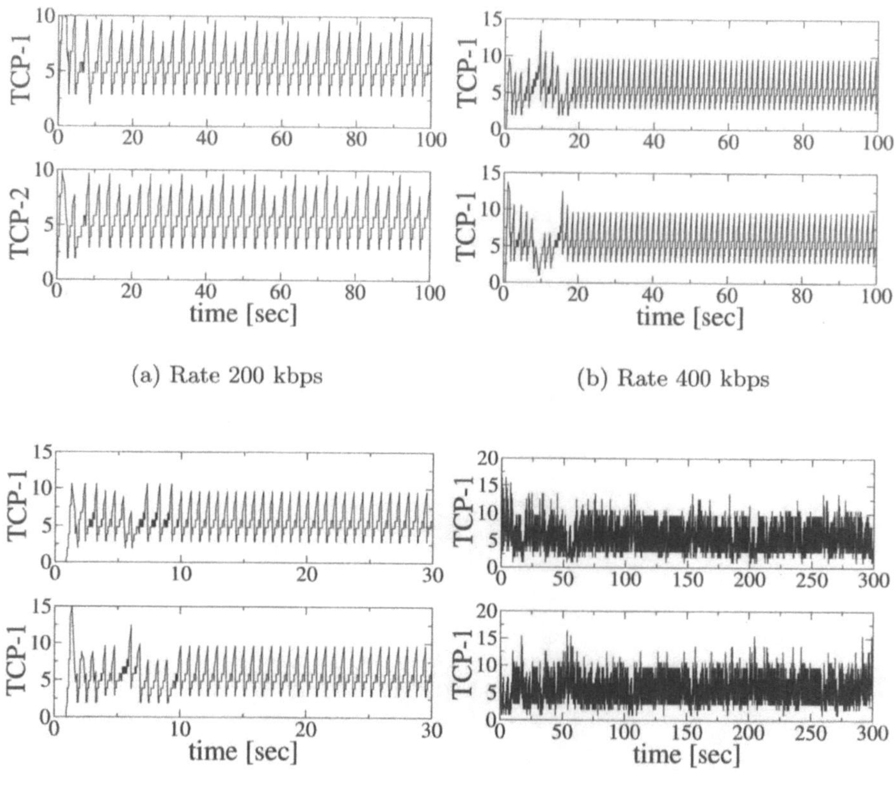

(a) Rate 200 kbps

(b) Rate 400 kbps

(c) Rate 800 kbps

(d) Rate 1200 kbps

Fig. 3: TCP congestion window dynamics at increasing speeds. Each figure shows both TCP window processes one on top of the other.

4 System Trajectories

One can observe the system's evolution not only as a function of time, but also by drawing the trajectory of the system as it moves in the phase space. If the system is periodic, then the corresponding trajectory will be a loop and *vice versa*, if the system evolution can be represented by a loop in the phase space, the system is periodic. This method is thus very appealing to examine the periodicity of a multi-TCP system.

Even in this 2-TCP system the number of state variables that completely describe the system is very large (e.g., the whereabouts of previously sent packets, internal variables of the sending and receiving TCPs), it is not possible to draw them on a single piece of paper but it is possible to properly

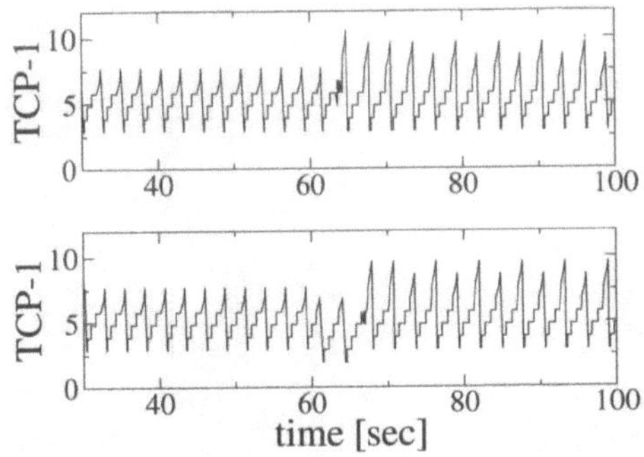

(a) Rate 200 kbps

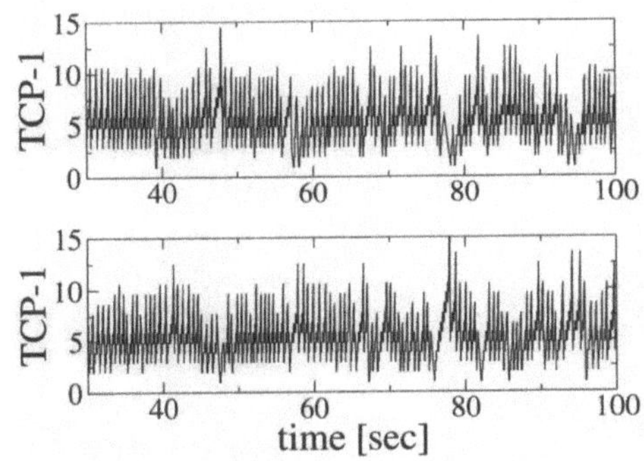

(b) Rate 1200 kbps

Fig. 4: Impact of a perturbing packet (which happens exactly at 60 sec) on TCP window dynamics at different service rates.

choose a section of the phase space. We chose the TCP congestion window (*cwnd*) size because it has a close relation with the sending rate of TCP.

In [14] the authors propose to use the time shifted past values $[x_t, x_{t-\delta t}, x_{t-2\delta t}, \ldots]$ of an easily measurable quantity for complex systems to equivalently reconstruct the underlying multidimensional trajectories if there is no access to all the state variables. The choice of δt can be nearly arbitrary in a wide range. The result is a multidimensional vector that is projected to the 2D plane simply by averaging the values $\hat{X} = 1/n(x_t + x_{t-\delta t} + \ldots)$. The method described above is used for the *cwnd* values:

$$x[i] = \frac{1}{n}\sum_{j=1}^{n} \mathrm{cwnd}_x[i-j] \quad (1)$$

$$y[i] = \frac{1}{n} \sum_{j=1}^{n} \text{cwnd}_y[i-j] \tag{2}$$

Here x and y denote the two TCPs. n controls the scale over which the congestion windows are averaged, the larger the value is, the more hidden dimensions can be reconstructed. The method has two other benefits:

- The number of possible points on the (x, y) plane is increased from W^2 where W is the number of possible *cwnd* values to $(nW - n)^2$ (if *cwnd* is counted in packets).
- Consecutive results $x[i]$ and $x[i+1]$ are placed close to each other, actually no further than $(2 * W - 2)/n$. Thus using this construction the generated graph can be made as smooth as required.

A nice property of the graph is that it preserves the periodicity property: periodic trajectories are displayed as closed loops in (x, y). (If we choose n equal to the period, then we get a single data point - a degenerate loop.)

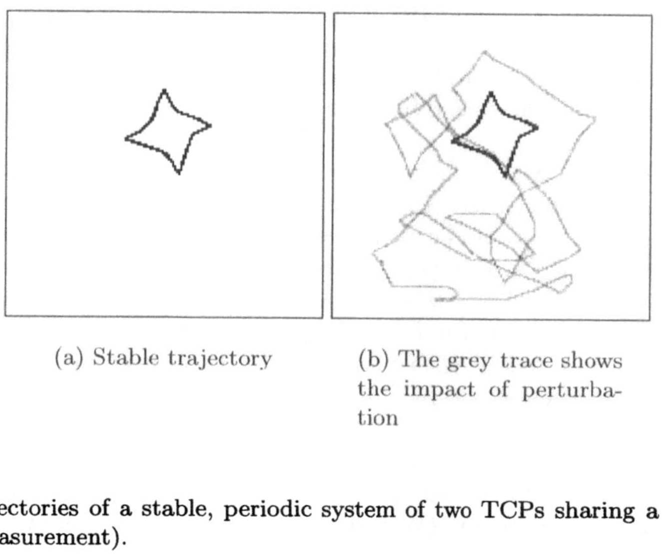

(a) Stable trajectory

(b) The grey trace shows the impact of perturbation

Fig. 5: Trajectories of a stable, periodic system of two TCPs sharing a single link (testbed measurement).

The periodicity and the effect of the perturbation on the system can be also demonstrated by displaying the trajectory of the system in the state space. The left graph of Figure 5 shows the trajectory of a stable periodic system before the perturbation. It can be seen that the periodic behavior is represented by a closed loop. On the right hand side, the trajectory of the system is shown right after the perturbation is started. The effect of the perturbation is represented by a short detour off the closed loop, but the system returns to the same pattern as in the left graph.

Figure 6 shows the result of the above method for other network configurations. In order to reduce the blur caused by transients and the impact of noise and help to differentiate orderly behavior from pure random trajectories, we applied darker shades to those points of the figure that are more frequently visited than others. With this method now we can clearly visualize stable trajectories as shapes appearing in front of the light shaded noise (right hand figures). Figure 6b is a complex loop which corresponds to a periodic pattern at rate 800 kbps, close to the edge of non-periodicity. Then at a certain speed (Figure 6c) we get a fine structured graph behind the noise, while at even higher speed we see no more than blurry shapes (Figure 6f).

The configuration seen in Figure 6d is interesting because there we do not experience periodicity or at best the period is very large. Still, this configuration has an orderly pattern of the trajectory, which is clearly distinguishable from the noise. We have seen that the clearly periodic system was immune to small perturbations, and the periodic pattern reappears after a short detour. Thus, for the complex trajectory of Figure 6d it makes sense to investigate whether the hidden order in the trajectory is robust against random perturbations. To investigate this question, we add random noise to the system by dropping randomly 0.1% and 1% of the packets on the link after the bottleneck buffer. The resulting graphs (after filtering) are shown in Figure 7. The pattern diminishes if the additional noise is high. We shall investigate how we can characterize the sensitivity of a system to perturbations later in this paper.

We have shown that by observing trajectories one can differentiate periodicity and non-periodicity as well in the same system under different settings. Does periodic or non-periodic behavior make any difference from a performance point of view? Also important to ask whether periodic or non-periodic behavior is the typical? The latter question is most difficult to answer, since they depend on many factors. Periodicity is mostly present if there is stable, constant load on a system with just a few TCP connections sharing a clear bottleneck. This scenario is typical if someone downloads long files over an access network that is slower than the backbone speeds (e.g., someone runs a file sharing application over DSL or cable modem) and the system configuration leads to a periodic pattern. Non-periodic behavior is more typical in case the traffic demands are dynamic (Web browsing) or there are many connections sharing the bottleneck (e.g., corporate/university leased line).

The first question, whether trajectories make any difference or not, has a more important implication and it leads to the problem of the sensitivity of the system.

5 Sensitivity

In this section, we demonstrate that in TCP congestion control can show large sensitivity, which means that very small perturbations in the system may

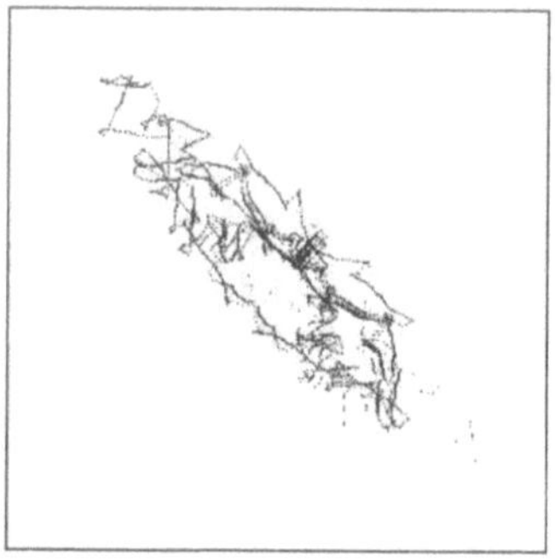

(a) Rate 800 kbps

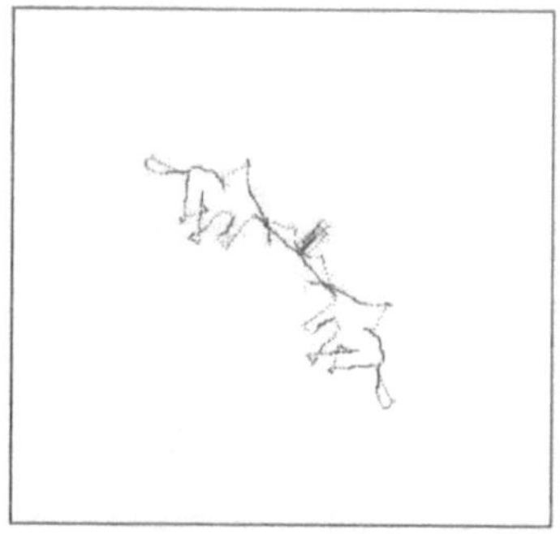

(b) Rate 800 kbps, noise reduced

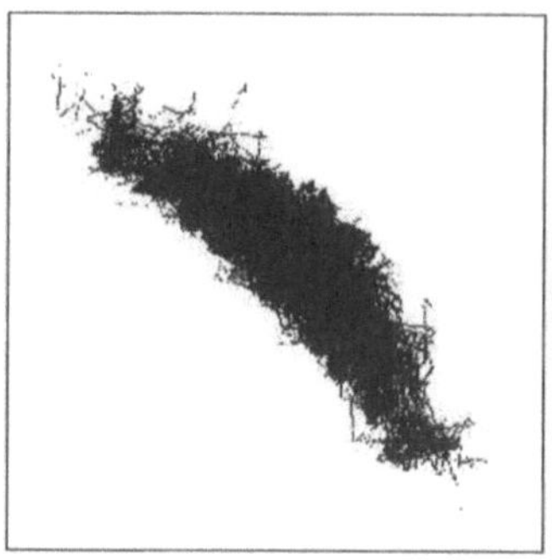

(c) Rate 960 kbps

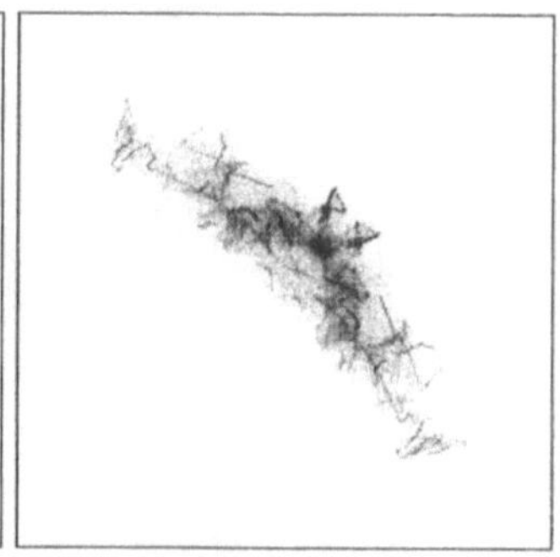

(d) Rate 960 kbps, noise reduced

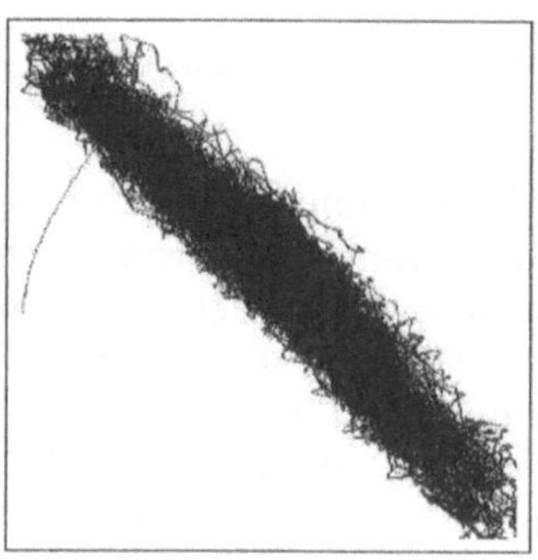

(e) Rate 1200 kbps

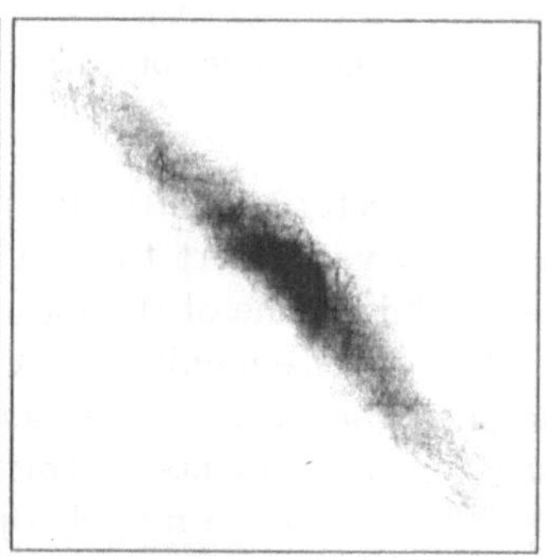

(f) Rate 1200 kbps, noise reduced

Fig. 6: TCP window trajectories at increasing speeds. Right hand figures are filtered.

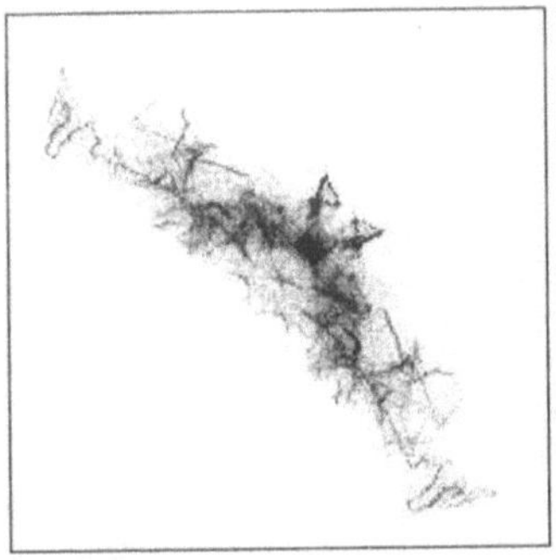

(a) No additional noise

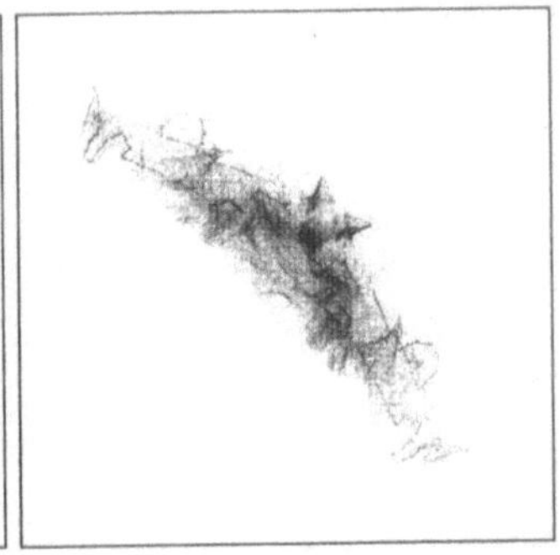

(b) Additional 0.1% random loss

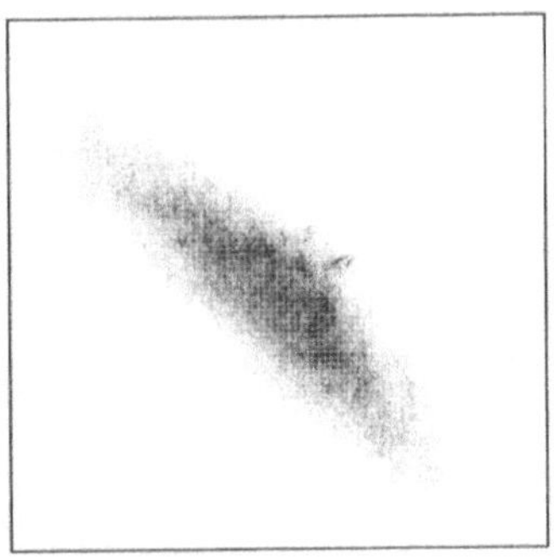

(c) Additional 1% random loss

Fig. 7: The impact of external noise on the trajectories.

cause that the trajectory departs from the original, unperturbed, system's trajectory within a very short time. The distance can grow to the range of the signal itself. This is one of the major properties of chaotic systems.

To eliminate the noise coming from operating systems so that all the perturbations in the experiment are under our control, we used simulations instead of testbed measurements in this section. We used the NS simulation software [10]. The simulation model copies the testbed setup, but simulation allows some extra freedom, so we increase the number of simultaneous TCP sessions to 30. ($C = 1$ Mbps, $d = 15$ ms, $B = 60$ packets). First, we let the system evolve for a while, then at $t = 50$ s we artificially increased the congestion window of one of the TCPs with one packet. Then we plotted the spatio-temporal graph of both systems, see the original system in Figure 8 (top) and the perturbed system in Figure 8 (middle). The length of the plot is 100s so the perturbation is done right at the middle of the plot. If we compare the two systems, first the differences are invisible, but a few seconds later the two systems look completely different. To make this more visible, we plotted the difference of the two systems in a way that each dot was colored according to the distance defined as $d(i,t) = |w^{orig}(i,t) - w^{pert}(i,t)|$, where i and t is the id. of the TCP and the time respectively, $w^{orig}(i,t)$ is the *cwnd* of ith TCP in the original system at time t and $w^{pert}(i,t)$ is the same for the perturbed system. See Figure 8 (bottom). The first part is white, which means that the two systems are identical, then a few dim dots appear and a few seconds later the difference looks like the original plots.

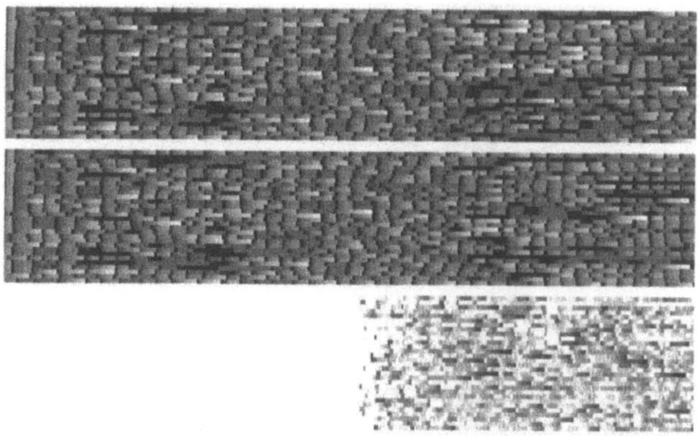

Fig. 8: Spatio-tempral graph of 30 TCP window processes sharing a single bottleneck. Time flows from left to right, light shades represent large windows, dark shaded represent low windows. Spatio-temporal graph of the original system (top). Spatio-temporal graph of the perturbed system (middle). Difference between the two systems (bottom).

To quantify how fast this divergence happens, we define the distance between the two systems at time t as the Euclidean distance in the *cwnd* space:

$$E(t) = \sqrt{\sum_{i=1}^{N}(w^{orig}(i,t) - w^{pert}(i,t))^2}. \tag{3}$$

See Figure 9.

The rate at which the systems diverge after a small perturbation of the ith TCP ϵ_i at time t_0 can be described by the so called *Lyapunov exponent*, which we approximate by measuring the time Δt it takes for the two systems to reach a given distance $E(t_0 + \Delta t) > \hat{E}$, then:

$$\lambda(t_0, i) \approx \frac{1}{\Delta t} \ln |\frac{E(t_0 + \Delta t)}{\epsilon_i}| \tag{4}$$

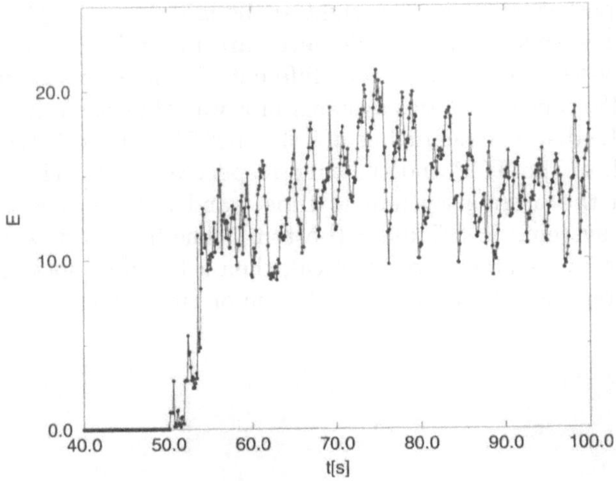

Fig. 9: Divergence of the original and the perturbed systems.

For the experiment we chose $\hat{E} = 10$ and $\epsilon_i = 1$. The motivation to calculate an *exponent* is that trajectories diverge at an exponential rate. However, as the two systems cannot get arbitrarily far from each other (as the phase space is limited) only the increasing part of Figure 9 should be considered, this explains the choice of $\hat{E} = 10$.

The Lyapunov exponent is the rate at which the two systems diverge from each other every time unit. This value of course depends on *which* TCP we perturb and *when* the interference is done. In other words, to which direction in phase space we push the system and at what part of the phase space the system is at the time of perturbation. There are cases when even an otherwise sensitive system is not affected by a small interference. A negative exponent characterizes this case. If sensitive points ($\lambda > 0$) are dense on the trajectory then the attractor is called chaotic.

To arrive at a more general numerical result for a given system we calculate the exponent at many different points of the trajectory (t_0) and for all 30 TCPs. Then for each t_0 we choose the most sensitive direction where the largest λ is measured and average these values over time to get the average maximum exponent of the trajectory (see Figure 10):

$$\lambda = E\left[\max_i \lambda(t_0, i)\right] \tag{5}$$

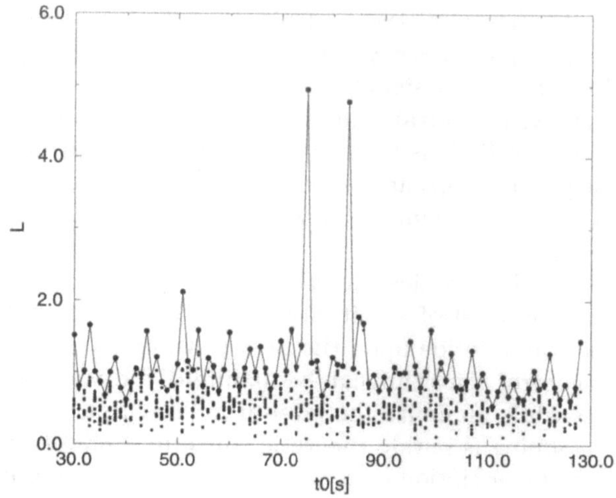

Fig. 10: Lyapunov exponents at different points of the trajectory and for all 30 TCPs, maximum exponents along the trajectory are connected with a line (average maximum $\lambda \approx 1.11$).

In the experiment we got $\lambda \approx 1.11$, which means that after a perturbation the difference between the two systems increase at an average rate of $e^{\lambda} \approx 3.03$ every second.

6 Controllability

We have shown that it is possible to perturb a network using small interventions. The question is whether it is possible to control it as well? Here we mean whether it is possible or not to improve the performance of a network using small interactions. If the answer is yes, then it could make it possible to design better networks.

There have been many proposals before to actively control the network using smart algorithms. One class of these algorithms improves the end-to-end performance by applying direct control of how the network resources are distributed among the competing flows. Generalized Processor Sharing (GPS) [13] is one of the most sophisticated of such algorithms. GPS implements separate buffers for each flow and controls the sending of packets from these buffers using a strict rule according to some shares defined by the network management. These direct methods although very effective are complex and they are usually implemented in small speed routers.

Another class of control algorithms applies heuristics based on assumptions on the usual behavior of the controlled traffic. The most important of these algorithms is Random Early Detection (RED) [4]. RED assumes that TCP flows will react to early signals of congestion and applies a control rule that drops packets with a certain small probability before the buffer reaches saturation. The idea of RED is to keep the buffer at an optimal level continuously. Since the packet drops are randomized, it is also argued that possible synchronization effects are eliminated with RED, and fairness among TCP flows is improved.

If one knows a full dynamical description of a system, e.g., the shape of the attractor and the level of sensitivity along the attractor, then, at least theoretically, one could decide upon the best possible control algorithm. The objective of this control algorithm can be manifold, for example, it may maximize the fairness while minimize the level of intervention.

The above reasoning is purely theoretic, since it is impossible to derive a complete dynamical description of a network shared by many flows, routers, short and long TCP downloads etc. Nevertheless, we may try to assess whether it is possible at all or not to effectively control a network using as small intervention as possible. In order to do this we need to have a fully repeatable model where we can try out different interventions under exactly the same circumstances. Our goal is to improve the fairness in this network. We define fairness as the standard deviation (S) of the time needed to download a certain file. If S is small then all TCPs receive almost the same service rate.

We run the same simulation experiment as before using 30 TCPs, but now we try to find the best possible intervention. First, we run the system without any action and calculate S. The result of this first experiment is denoted by S^0 Then we rerun the simulation and drop a single packet in the buffer. The result is denoted by S^1_n, where 1 means that the router dropped 1 packet and n means that exactly the nth arriving packet to the buffer was dropped. The size of the files is 100 packets, so without retransmissions altogether $30 * 100 = 3000$ packets are transmitted by the buffer, so n can take a value between 1 and 3000.

Instead of just one packet, we may try to drop several packets, in which case the fairness measure looks like $S^i_{n_1,n_2,\dots n_i}$ where $n_1, n_2, \dots n_i$ shows the list of packets to be dropped.

The objective of the router is then to find the best list of $n_1, n_2, \dots n_i$. Obviously, as i increases, the size of the problem explodes exponentially. We propose an almost greedy search algorithm to search for some local optima in this space. The algorithm has the following rules:

- Step 0. Set drop list empty. $L = \{\}$.
- Step 1. Randomly choose D new positions and temporarily append them to the list one-by-one. For each new position run a simulation and calculate S^i_L.

- Step 3. Choose the best from the D random positions and permanently append it to the list L.
- Step 4. With P probability remove one element from L.
- Go to Step 1 or exit if some limit condition is reached.

The above algorithm will try several drop positions randomly before increasing the number of drops. Remember that we would like to keep the number of drops at minimum. Also, to avoid being stuck in a local optimum, an earlier drop position is removed from the list with a certain probability, this is controlled by the parameter P.

In the actual experiment we set $P = 0.1$ and $D = 1000$, so we try approximately one-third of all possible new positions in each step. The result is shown in Figure 11. We can observe that the initial standard deviation of approximately 3 seconds can be reduced below 1 second by appropriately dropping just 4 packets. On the other hand, the wide band of variance results at each step indicates that one can do equally significant harm as one can do good at any time that is the variances range up to almost 5 seconds independently of the contents of the actual loss list. Consequently this network is rather difficult to control effectively, since a single wrongly placed packet drop can ruin the whole objective.

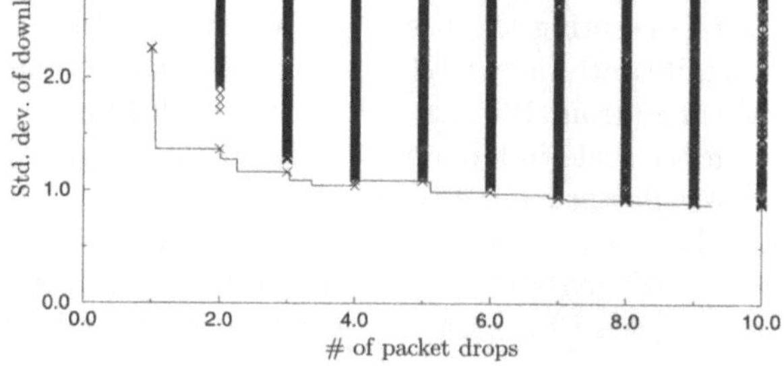

Fig. 11: Variance of download times. Each dot represents the result of a simulation run for a certain loss list L.

7 Traffic Burstiness at Different Timescales

As we have shown TCP dynamics can be quite different depending on the circumstances. The end-user, or the network operator, however, is not so much interested in whether connections follow a periodic or a random pattern. For them the time dependent properties are more important. If the TCP download rate has fast timescale fluctuations, in other words bursty, the network buffers are more stressed compared to smooth or even uncorrelated random arrivals. For the end-user the large timescale fluctuations are more important, nobody likes to see a download slow down for longer periods of times.

Common traffic modeling assumption is that TCP has only short timescale fluctuations [12], and large timescale fluctuations are due to heavy tailed file size distributions [1][2]. In contrast, real life can produce interesting phenomena and not just in extreme situations. We chose the seemingly random configuration of two TCPs (at 1200 kbps) with different buffer settings (5, 10 and 20 packets). Figure 12a-c show three test runs in the testbed, where we recorded the number of packets transmitted by one of the TCPs during consecutive 100 ms. When the buffer size is 10 packets, the traffic looks random and shows short timescale variations. On the other hand, when the buffer size is 5 and 20 packets, we see long, alternating periods of high and low transmission rates.

One way to characterize the variations of TCP rates at different timescales is the variance-time plot (VT). Let X denote the number of packets transmitted during a time period of 100 ms. Let $X^{(m)}(k) = 1/m \sum_{i=(k-1)m+1}^{km} X(i)$ denote the m aggregated series of X. The VT plot shows the variance of $X^{(m)}$ versus m on a logarithmic scale.

The slope of the VT plot summarizes the variations at different timescales in a single graph. As a comparison short-range dependent processes asymptotically drop with slope $\beta = -1$ as $m \to \infty$. The VT plots of the TCP rates are shown in Figure 13.

The VT plots representing the 5 and 20 buffer cases show that the variances drop at a significantly lower ($\beta > -1$), and almost constant rate over a significant time range from 100 msec (0 on log scale) up to 1000 sec (4 on log scale). This almost scale independent property is frequently modeled by parsimonious, self-similar processes [7][8].

We must note that we do not claim this process to be self-similar, which is anyway a purely mathematical model and should fit the process asymptotically not just in a limited time range. Nevertheless, when modeling network traffic, we are interested in a model that can capture some key statistical properties over relevant timescales. The relevant timescale in traffic theory is usually limited because traffic cannot be considered to be stationary over longer timescales.

Guo et al. [5] analyse a similar scenario using a Markov chain based model and arrives at a similar conclusion. The authors derive that self-similarity can arise in TCP traffic if the blocking probability reaches a certain value. The

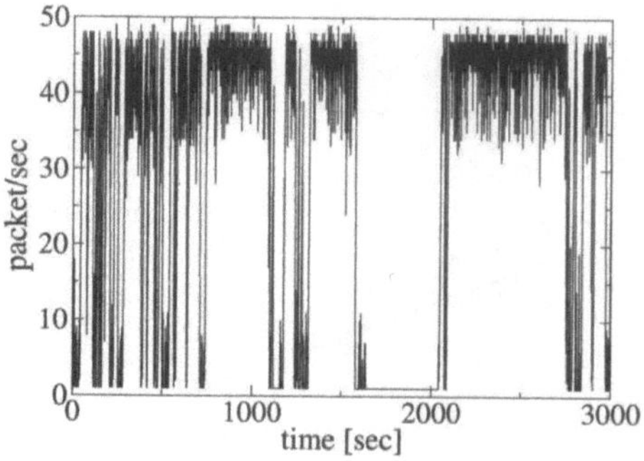

(a) Buffer size 5 packets

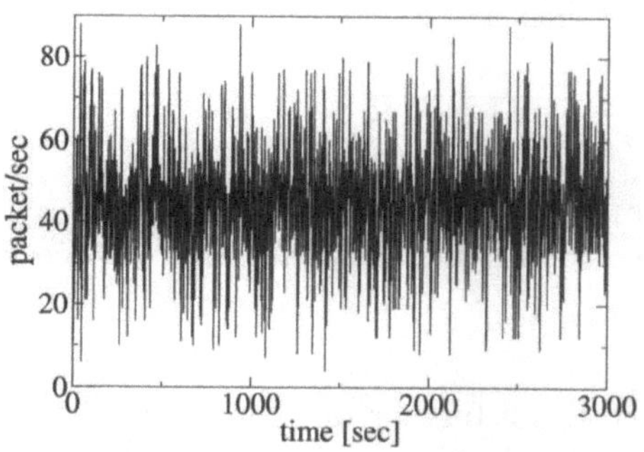

(b) Buffer size 10 packets

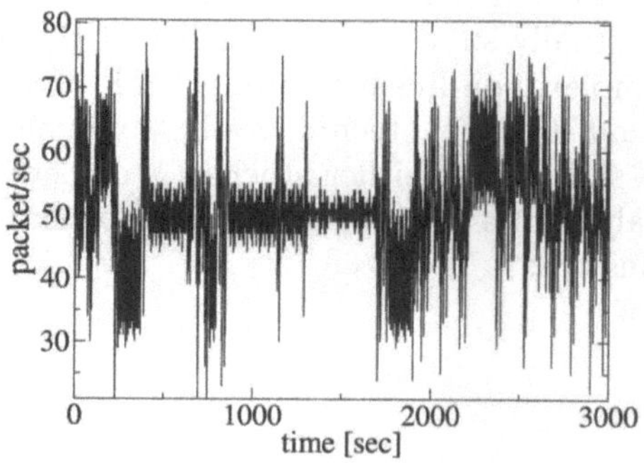

(c) Buffer size 20 packets

Fig. 12: TCP rate processes at different buffer sizes (service rate 1200 kbps, delay 100 ms).

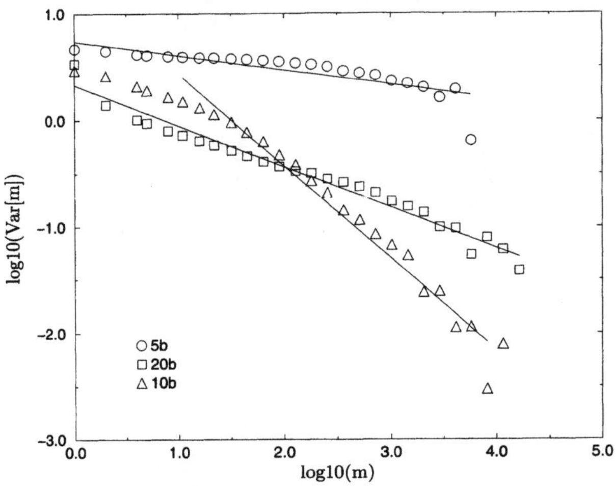

Fig. 13: Variance-time plots for different buffer sizes (service rate 1200 kbps, delay 100 ms).

work by Figueiredo et al. [3] further develops the model to include the congestion avoidance phase and not just the exponential backoff phase of TCP congestion control. The significant difference between their methodology and ours is that both papers approach the same problem from a stochastic perspective. Our work, although it takes a significantly different, deterministic modeling approach, actually supports the validity of the approach discussed in [5] and [3]. The complex, non-linear nature of TCP blurs the borderline between deterministic and stochastic techniques in traffic modeling. On the one hand, it makes it possible to use either stochastic or deterministic modeling techniques in the analysis of a system, but it calls for extra care to be taken before drawing conclusions.

8 Conclusions

In this paper, we have analyzed TCP dynamics from several aspects. We have built a testbed where we experimented with different network configurations. We have recorded periodic and non-periodic dynamics in the same system at different settings. We have defined a method to draw the system trajectory that helps to identify orderly and apparently random behavior visually. The trajectory has been characterized from an important aspect namely the sensitivity of the system to small perturbations. We have proposed a sequence of directed simulations and calculations to investigate the sensitivity of the system. We have demonstrated that TCP congestion control can create time-series with slowly decaying variances such as self-similar processes. We have analyzed the possibility of improving the end-to-end performance of such a system by using only minute control actions and have shown that although control is possible, the complexity of the system is so large that actual application is a topic for future research.

References

1. M. E. Crovella and A. Bestavros, "Self-Similarity in World Wide Web Traffic: Evidence and Possible Causes", *in Proc. ACM SIGMETRICS*, pp. 160-169, May 1996.
2. M. E. Crovella, M. S. Taqqu, and A. Bestavros, "Heavy-tailed probability distributions in the world wide web" *Preprint: A Practical Guide To Heavy Tails: Statistical Techniques and Applications*, 1996.
3. D. R. Figueiredo, B. Liu, V. Misra and D. Towsley , "On the Autocorrelation Structure of TCP Traffic", UMass CMPSCI Technical Report TR 00-55, 2000
4. Floyd, S., Jacobson, V., "Random Early Detection gateways for Congestion Avoidance ", *IEEE/ACM Transactions on Networking*, Vol.1 No.4, August 1993, p. 397-413.
5. L. Guo, M. Crovella and I. Matta, "TCP Congestion Control and Heavy Tails", Technical Report BUCS-TR-2000-017, Boston University, 2000
6. V. Jacobson, "Congestion avoidance and control", *In Proceedings of ACM SIGCOMM*, pages 314–329, Stanford, USA, 1988.
7. W. E. Leland, M. S. Taqqu, W. Willinger and D. V. Wilson, "On the Self-Similar Nature of Ethernet Traffic",*ACM SIGCOMM '93*, San Francisco, CA, USA, Sep. 1993
8. S. Molnár, T. D. Dang, and A. Vidács, "Heavy tailedness, long-range dependence and self-similarity in data traffic", *In Proc. 7th International Conference on Telecommunication Systems Modeling and Analysis*, Nashville, Tennessee, USA, March 1999.
9. M. Mathis, J. Semske, J. Mahdavi and T. Ott, "The Macroscopic Behavior of the TCP Congestion Avoidance Algorithm", *Computer Communication Review*, 27(3), July 1997
10. NS software and documentation is available at the following site: http://www-mash.CS.Berkeley.EDU/ns
11. Nistnet, http://snad.ncsl.nist.gov/itg/nistnet/
12. Nicolas Hohn, Darryl Veitch, Patrice Abry, "Does fractal scaling at the IP level depend on TCP flow arrival processes?", *Internet Measurement Workshop* Marseille, France, November 2002
13. A.K. Parekh and R.G.Gallager, "A generalized processor sharing approach to flow control in integrated services networks – the single node case", *IEEE/ACM Trans. on Networking*, 1(3):334-357, 1993
14. N. H. Packard, J. P. Crutchfield, J. D. Farmer and R. S. Shaw, "Geometry from a Time Series", *Phys. Rev. Lett.*, 45 (1980) 712-716

15. J. Padhye, V. Firoiu, D. Towsley and J. Kurose, "Modeling TCP Throughput: A Simple Model and its Empirical Validation", *Proccedings of SIGCOMM'98*, Vancouver, CA, September 1998
16. Postel, J., "Transmission Control Protocol - DARPA Internet Program Protocol Specification", RFC 793, DARPA, September 1981
17. W. Stevens, " TCP Slow Start, Congestion Avoidance, Fast Retransmit, and Fast Recovery Algorithms", *RFC2001*
18. M. Mathis et al., "TCP Selective Acknowledgement Options", *RFC2018*
19. V. Jacobson, R. Braden, "TCP Extensions for Long-Delay Paths", *RFC1072*
20. Tcptrace utility, http://jarok.cs.ohiou.edu/software/tcptrace/tcptrace.html

Dynamics of Congestion Control

Anna C. Gilbert

Dept. of Mathematics
University of Michigan
2074 East Hall
Ann Arbor, MI 48109-1109
annacg@umich.edu

This work was done in collaboration with Youngmi Joo (1970-2001). It is dedicated to her memory.

1 Introduction

One reason for the success of TCP and its widespread usage in the Internet is its ability to control network congestion. A TCP source uses implicit notification of network congestion (via packet loss) to control the rate at which it sends data. The rate is controlled by increasing or decreasing the window of outstanding, unacknowledged data. In particular, TCP uses "additive increase and multiplicative decrease" to increase the rate linearly during times of no packet loss, and to decrease the rate geometrically when loss occurs. As a consequence, the rate at which a TCP source transmits tends to be periodic over time. For example, Figure 1 shows the periodic behavior of a single TCP source in an otherwise idle network.

When many TCP sources compete for the buffers in a congested router, packet loss will cause each flow to exhibit periodic behavior. A question worth asking is: will the aggregate behavior of all of the flows also be periodic? For example, consider Figure 2(a) which shows a number of periodic TCP flows. If these flows were to pass through a single router over a shared link, will the aggregate flow be smooth (as in Figure 2(b)) or periodic (as in Figure 2(c))? The consequences of periodic aggregate behavior can be a dramatic loss of network capacity. For example, the average data rate in the output link of Figure 2(c) is significantly lower than that of Figure 2(b).

Many experiments [11] and the intuitive explanations of these experiments show that TCP sources competing for bandwidth on a congested link will synchronize through the weak coupling inherent in congestion control. Intuitively speaking, a population of sources will synchronize because they all experience loss at roughly the same time, they all scale back their transmission rate in the presence of these losses, and then they increase their transmission rate until the next bout of congestion. While much mathematical analysis has been

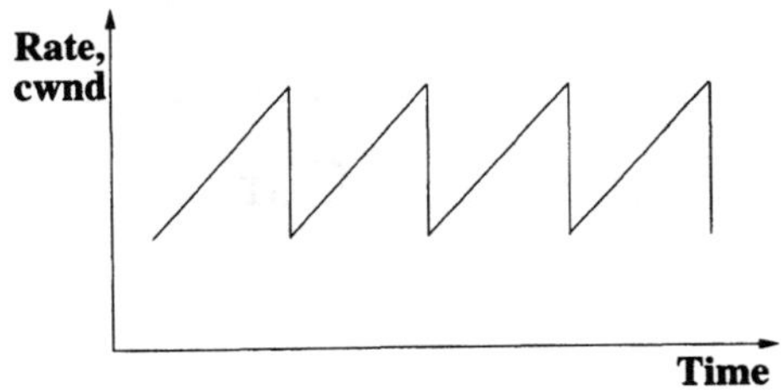

Fig. 1: Periodic oscillation shown by a single TCP connection

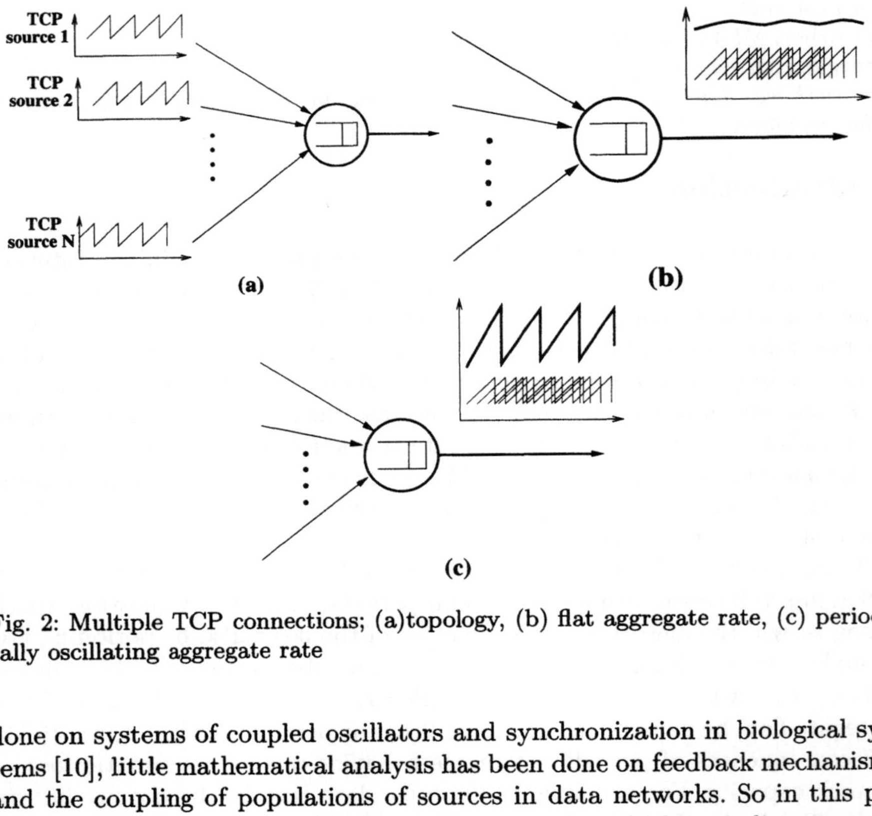

Fig. 2: Multiple TCP connections; (a)topology, (b) flat aggregate rate, (c) periodically oscillating aggregate rate

done on systems of coupled oscillators and synchronization in biological systems [10], little mathematical analysis has been done on feedback mechanisms and the coupling of populations of sources in data networks. So in this paper, we set out to explore: (1) the conditions under which periodic aggregate behavior can occur, (2) whether there is evidence that it is occurring in the Internet today, and if so, (3) whether there are simple ways to prevent it.

One mechanism that has been proposed to reduce periodic aggregate behavior is to use a drop-policy in the buffers of the routers. In particular, it is a goal of RED [6] to "desynchronize" the flows sharing a congested router buffer so that the periodic behavior of each individual flow is not synchronized with the others. RED discards packets randomly in an attempt to cause different TCP sources to reduce (and hence increase) their rates at different times. Intuition suggests that if the flows are not oscillating in phase, then their aggregate behavior will be smooth (like in Figure 2(b)) and hence link utilization will be high.

We start by understanding how periodic aggregate behavior comes about. In section 2 we present a simple analytical model of TCP that captures (in a very simplified way) the congestion control algorithm that provides feedback to the sources. We conjecture that periodic behavior, should it exist, arises primarily from the feedback control mechanism not from queueing in the network, and so our model does not model queues. Later, in section 3, we test these assumptions using simulation.

As we will see, the analytical model predicts that aggregate periodic behavior is not only likely to occur, but that a population of TCP sources will converge exponentially quickly towards an invariant periodic pattern. Furthermore, the model suggests that the periodic behavior is remarkably stable even in networks in which different sources are at different distances from the congested router, and when packets experience variable queueing delays on their way to the congested router. The simple model also leads to a prediction that periodic aggregate behavior will not be eliminated by randomized drop-policies such as RED. In section 3 we explore whether periodic behavior actually occurs in a simulated network with and without RED.

2 Mathematical Model

2.1 Description

We model a TCP connection as a source with an infinitely long file to transmit to a receiver across a bottleneck link. There are N sources or connections sharing the bottleneck link which has capacity C. In the absence of congestion or feedback from the bottleneck, each source increases its sending rate linearly over time (i.e., $dx_i(t)/dt = 1$ where $x_i(t)$ is the rate of source i at time t). The system of sources experiences congestion when the total rate $A(t) = \sum_{i=1}^{N} x_i(t)$ reaches the link capacity C. At each congestion epoch when $A(t) = C$, a fraction r of the N sources (chosen uniformly at random) experiences losses and these r sources receive a notification of congestion. Each source upon congestion notification, resets its transmission rate to zero. We assume that these events signaling congestion and reseting the transmission rate occur instantaneously when the aggregate rate reaches the link capacity. After reset, the rates of the sources continue to grow linearly until the system reaches the next congestion epoch, at which point the process repeats with another random fraction r of sources experiencing losses and adjusting their rates. This model assumes that there is an entity at the bottleneck link that can measure the instantaneous aggregate rate, signal, and reset the sources instantaneously.

2.2 Mean-field Approximation

Rather than following the possible states of all N sources, we assume that N is large enough for us to employ a mean-field approximation for the system. We describe the state of the system at time t by a density $\rho(x,t)$ that represents the percentage of sources with transmission rate x at time t. We let $\rho(x,t)$ take on any real positive value, not necessarily an integer multiple of $1/N$. Furthermore, we approximate the result of each reset upon $\rho(x,t)$ as removing a uniform percentage r of the sources at rate x and time t (i.e., that the approximate effect upon the population of sources is the mean or expected effect) and resetting the rates of that percentage r by placing a point mass of weight r at rate $x = 0$. If $\rho(x,t_n^-)$ represents the distribution of sources immediately before the nth congestion epoch, then the mean-field approximation gives us the representation

$$\rho(x,t_n^+) = r\delta_0(x) + (1-r)\rho(x,t_n^-),$$

for the distribution of source rates immediately following the reset of the sources. In addition, the linear increase of the sources' rates between congestion epochs tells us that the configuration $\rho(x,t+\delta t)$ of the system at time $t+\delta t$ is simply the translation of the distribution at time t by δt, $\rho(x,t+\delta t) = \rho(x-\delta t,t)$.

We demonstrate that:

1. there is a configuration ρ_* of the source rates that is invariant under the reset operation at each congestion epoch; i.e., the distribution of source rates immediately following a reset is the same as the distribution after the preceding congestion epoch,
2. there is a constant time interval $\Delta t = rC/N$ between congestion epochs,
3. the aggregate rate of the sources $A(t)$ is a Δt-periodic function that oscillates between a maximum value C and a minimum value $(1-r)C$ with an average of $C - rC/2$, and
4. any initial configuration of the sources will tend to this invariant exponentially fast in r.

See Figure 3 for a picture of this configuration, where the mean field approximation is plotted in black for $N = 50$ connections and $r = 1/4$.

Invariant Configurations

The invariant distribution of source rates has two important properties. First, it consists of distinct populations of sources, each population with one rate, rather than constraining all N sources to transmit at the same rate. These distinct populations point out a slight misnomer in the term "synchronization" commonly used to describe this phenomenon. The sources do not synchronize their rates with respect to one another, they conspire in groups to produce

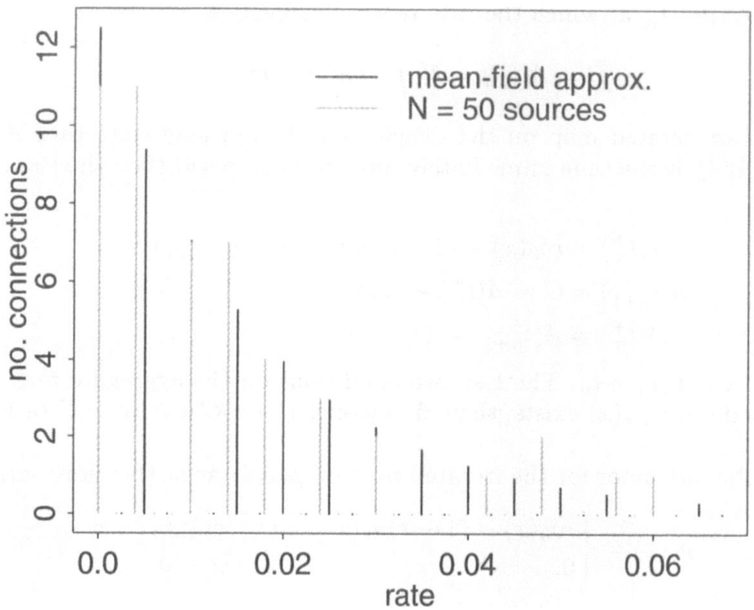

Fig. 3: The invariant configuration of sources for the mean-field approximation and a simulated instance of the model with $N = 50$ sources, the fraction $r = 1/4$ of the sources reset randomly at each epoch, the bottleneck link capacity $C = 1$, and the epoch duration $\Delta t = r/N = 0.002$. The distribution of 50 sources after 1000 cycles in a realization of the toy model is also plotted in grey.

an aggregate periodic rate, and so the term "aggregate periodic behavior" is a more accurate description of this phenomenon. The sizes of the rate populations do decay exponentially, showing that many of the sources transmit at the lowest rate while very few sources transmit at a high rate.

Theorem 1. *There is a unique density ρ_* that describes the state of N sources (for N large) immediately after resetting the rates of rN sources (chosen uniformly at random) to zero such that after the next reset, reached at time $\Delta t = rC/N$ after the first reset, ρ_* is restored. This density is given by*

$$\rho_*(x) = \begin{cases} r\sum_{l=0}^{\infty}(1-r)^l \delta_0(x - l\Delta t), & x \geq 0, \\ 0, & x < 0. \end{cases}$$

In addition, the aggregate rate $A(t)$ is a periodic function with period $\Delta t = rC/N$, oscillating between a maximum value of C and $(1-r)C$.

Proof. The aggregate rate at time t is $A(t) = N \int x\rho(x,t)\,dx$ where the density $\rho(x,t)$ describes the state the system is in at time t and the capacity constraint gives us the time t_n^- at which the rate reaches capacity C

$$C = A(t_n^-) = N \int x\rho(x,t_n)\,dx.$$

We obtain an iterated map on the density $\rho(x,t)$ and aggregate rate $A(t)$. Assume that t_n^+ is the time immediately after the nth reset, then the iterated map is

$$\begin{aligned} \rho(x,t_n^+) &= r\delta_0(x) + (1-r)\rho(x-\Delta t_{n+1}, t_{n+1}^-) \\ A(t_{n+1}^-) &= C = A(t_n^+) + N\Delta t_{n+1} \\ A(t_n^+) &= A(t_{n+1}^+) = (1-r)C \end{aligned}$$

where $\Delta t_{n+1} = t_{n+1} - t_n$. The last two conditions on the aggregate rate tell us that if a density $\rho_*(x)$ exists, then Δt satisfies $(1-r)C + N\Delta t = C$ or that $\Delta t = rC/N$.

The global attractor for the iterated map on $\rho(x,t)$ must therefore satisfy

$$\rho_*(x) = \begin{cases} r\delta_0(x) + (1-r)\rho_*(x-\Delta t), & x \geq 0, \\ 0, & x < 0. \end{cases}$$

Observe that we can translate this condition into a condition on $\bar{\rho}_*$, the Laplace transform of ρ_*, and then we can expand in a uniformly convergent power series

$$\bar{\rho}_*(s) = \frac{r}{1-(1-r)e^{-s\Delta t}} = r\sum_{l=0}^{\infty}(1-r)^l e^{-sl\Delta t}.$$

Next we compute the inverse Laplace transform of $\bar{\rho}_*(s)$ by inverting each term in the power series term-by-term, giving us

$$\rho_*(x) = r\sum_{l=0}^{\infty}(1-r)^l\delta_0(x-l\Delta t),$$

when $x \geq 0$ and $\rho(x) = 0$ for $x < 0$.

Note that if we take an iterative approach, we find the following theorem:

Theorem 2. *If a fraction r of a system of N sources whose states are initially described by the density $\rho(x,0)$ follows the dynamics outlined above, then the density $\rho(x,t)$ converges weakly to the global attractor $\rho_*(x)$, independent of the initial conditions of the system. The density at time t_n after the nth reset is*

$$\rho(x,t_n) = r\sum_{l=0}^{\infty}(1-r)^l\delta_0(x-l\Delta t) + (1-r)^n\rho(x-t_n,0)$$

and the congestion epoch length is $\Delta t = rC/N$, after the first congestion epoch (whose length depends on the initial state of the system).

The careful reader might argue that allowing even an infinitesimal percentage of the sources' rates to grow arbitrarily large is not a physically relevant result. Let us now assume that the sources' rates cannot grow larger than a maximum rate D. That is, once a source increases its transmission rate to D, its rate remains fixed there until it experiences a loss. We have to assume that $C/N < D < C$ to avoid trivial and physically meaningless situations. We express this condition upon the source rate distribution mathematically by placing a point mass at rate $x = D$ whose weight is the total percentage of source rates that at time t are larger than $x = D$ and leaving unchanged the distribution for smaller rates. Let L be the largest integer such that $L\Delta t \leq D < (L+1)\Delta t$ for $\Delta t = rC/N$. Then we may conclude using arguments similar to the above theorems that there is an invariant distribution $\tilde{\rho}_*$ for D-bounded sources.

Corollary 1. *The invariant distribution for D-bounded sources has the form*

$$\tilde{\rho}_*(x) = r\sum_{l=0}^{L}(1-r)^l\delta_0(x - l\Delta t) + m(r,L)\delta_0(x - D) \tag{1}$$

where $m(r,L) = r\sum_{l=L+1}^{\infty}(1-r)^l = (1-r)^{L+1}$ *is the percentage of sources at rate* D.

Stability of Invariant Configurations

The second important property the invariant configuration possesses is stability. To explore the stability of the invariant distribution, we define a perturbation of the system at the ith congestion epoch to be a random deviation from the time the ith reset occurs. We choose ϵ_i uniformly at random in $[-\epsilon, \epsilon]$ and jitter the duration of the ith epoch $\Delta t_i = \Delta t + \epsilon_i/N$. The time Δt between the $(i-1)$st reset and when the sources next reach link capacity C is still rC/N, we simply randomly perturb when this reset occurs as we do in a queueing policy that attempts to inject a small amount of randomness into the loss process. At the new time t_i, we reset a uniformly chosen random fraction r of the sources and let the system evolve again until the total rate exceeds capacity C. We repeat this process for each congestion epoch (drawing an independent ϵ each time) and show that the perturbed system remains close to the unperturbed system (which, we emphasize, is converging exponentially fast to the invariant distribution independent of the initial conditions). The perturbed system differs from its original trajectory with respect to its rates by a maximum value of ϵ and with respect to its congestion epoch durations by ϵ/N. On average, the perturbed system does not stray from its original course, as the perturbation ϵ is a random variable with mean 0.

We work with the more physically meaningful rate-bounded model as in Corollary 1 and assume $\epsilon/N < D - L\Delta t$ for ease of calculation (if not, then the summation in Eqn. (1) ends at $l = L+1$).

Lemma 1. *If we jitter the duration of the ith congestion epoch by ϵ_i/N for ϵ_i a uniformly distributed value in $[-\epsilon, \epsilon]$, then only the duration of the $(i+1)$st epoch is affected and $\Delta t_{i+1} = \Delta t - (1-r)\epsilon_i/N$, where $\Delta t = rC/N$.*

Proof. Because we reset the source rates after an epoch duration $\Delta t_i = \Delta t + \epsilon_i/N$, the total rate immediately before reset is allowed to grow to $A(t_i^-) = C + \epsilon_i$ and immediately following reset is $A(t_i^+) = (1-r)(C+\epsilon_i)$. We let the system of sources evolve as before until their total rate reaches capacity C. The duration of the $(i+1)$st epoch is the time interval given by $C = A(t_i^+) + N\Delta t_{i+1} = (1-r)(C+\epsilon_i) + N\Delta t_{i+1}$ or $\Delta t_{i+1} = \frac{rC-(1-r)\epsilon_i}{N}$. At this time (with no jitter), we reset a fraction r of the source rates and proceed as usual. Because at the $(i+1)$st epoch, the total rate is C and we use the same evolution procedure and reset policy at the $(i+2)$nd epoch, the duration $\Delta t_{i+2} = rC/N$ remains unaffected by the perturbation at the ith epoch.

Note that if we jitter the $(i+1)$st epoch duration by ϵ_{i+1}/N, then the total rate before reset is $A(t_{i+1}^-) = C + \epsilon_{i+1}$ (and this rate is not affected by the jitter at the ith epoch). We may then apply the lemma to the $(i+1)$st epoch. Applying this lemma inductively, we may conclude:

Theorem 3. *If we jitter each congestion epoch duration independently and let the system evolve, these perturbations do not propagate and force the system away from the invariant configuration with respect to the duration of the epochs. That is, the duration of each epoch varies only by ϵ/N, the amount by which we perturb the duration, and on average $\Delta t_i = \Delta t = rC/N$.*

These calculations do not tell us if the rates of the perturbed system remain close to those of the original system however. We return to the setting of Lemma A.1 and assume that the system has evolved long enough in time to have the form immediately after a reset

$$\rho(x,t) = r\sum_{l=0}^{L}(1-r)^l\delta_0(x - l\Delta t) + \tilde{m}_n(r,L)\delta_0(x-D)$$

where the mass $\tilde{m}_n(r,L)$ at the rate bound $x = D$ is

$$\tilde{m}_n(r,L) = r\sum_{l=L+1}^{n}(1-r)^l + (1-r)^{n+1}\int_0^{\infty}\rho(x,0)\,dx.$$

Lemma 2. *If we jitter the duration of the ith congestion epoch by ϵ_i/N for ϵ_i a uniformly distributed value in $[-\epsilon, \epsilon]$, then the source rates change by $\max\{(1-r)\epsilon_i/N, r\epsilon_i/N\}$.*

Proof. The reader is asked to verify that if $\Delta t_i = \Delta t + \epsilon_i/N$ and $\Delta t_{i+1} = \Delta t - (1-r)\epsilon_i/N$, then before the $(i+1)$st reset, the first rate population is at rate $\Delta t - (1-r)\epsilon_i/N$ while all the others are at rate $k\Delta t + \epsilon_i/N - (1-r)\epsilon_i/N =$

$k\Delta t + r\epsilon_i/N$. In addition, for $k < L - 1$ iterations after the perturbation at the ith epoch, the source rates have wandered from their original trajectory by at most $\max\{(1-r)\epsilon_i/N, r\epsilon_i/N\}$. After $L - 1$ iterations proceeding the perturbation, the system returns to its initial trajectory.

By applying this lemma inductively, we may conclude:

Theorem 4. *If we perturb each congestion epoch duration independently and let the system evolve, then the maximum deviation of the source rates from integer multiples of* $\Delta t = rC/N$ *is* $r/N \sum_{k=0}^{L} \epsilon_k$, *for* ϵ_k *i.i.d. uniform random variables on* $[-\epsilon, \epsilon]$.

We observe that if the maximum source rate D is on the order of C, then L is approximately N/r. Hence the maximum deviation is roughly $rL\epsilon/N = \epsilon$ and on average this deviation is zero since ϵ is a uniformly distributed random variable with mean zero.

Extensions and Generalizations

Several aspects of this mathematical model are only simple abstractions of more complex realistic behavior. To generate a less artificial picture of the behavior of a system of N sources, we must change both the manner in which the sources increase their rates and the rate reset policy in the face of congestion. However, neither modification changes the gross behavior of the system and so we feel reasonably confident drawing conclusions from the behavior of the simple abstract model.

The rate increase profile in this simple mathematical model does not reflect the complex behavior of the rate increase in TCP with its slow-start and congestion avoidance phases. However, if we impose a more complicated rate increase dx_i/dt, we do not change the overall behavior of the sources although we do change the details. The congestion epoch duration is different and is given by

$$A(t_{n+1}^-) = C = (1-r)C + N \int_{t_n}^{t_{n+1}} \frac{dx_i(s)}{dt}\, ds$$

but with the same reset policy, the system evolves as before. The source rate distribution moves according to the expression $\rho(x,t) = \rho(x - \int_0^t \frac{dx_i(s)}{dt}\, ds, 0)$. Similarly, if we choose a different reset policy, perhaps setting the rates to zero of those sources with higher rates, we change the congestion epoch duration slightly and the percentage of sources at a given rate no longer decays exponentially. In fact, if we adopt a "deterministic" reset policy where we reset the N/q sources with the highest rates, we will get q distinct rate populations of equal size and a constant epoch duration depending on q.

Finally, the mean-field approximation is just that, only an approximation of the true behavior of the mathematical model (itself an approximation). However, a more careful analysis of the behavior of N sources under the action of this random dynamical system, *not* assuming that N is large, reveals that the mean-field approximation is an excellent approximation of the system (see Figure 3).

2.3 Two Sources, a Special Case

Let us begin with two sources x and y. We denote the rates of the sources at time t by $x(t)$ and $y(t)$. Accordingly, the initial rates are $x_0 = x(0)$ and $y_0 = y(0)$ with $x_0 + y_0 < C$, the total capacity of the bottleneck link. The aggregate rate at the bottleneck link is $A(t) = x(t) + y(t)$. Each source increases its sending rate linearly (i.e., $x(t) = x_0 + t$) until the aggregate rate reaches the capacity rate C. At which point, one of the sources is reset; i.e., its sending rate is reset to zero. We will explore three scenarios: the source with the larger rate is always reset when the system reaches capacity, the smaller source is always reset, and one source is chosen at random and reset. Note that if both sources are always reset, the network immediately synchronizes.

Theorem 5. *If the source with the larger rate is always reset when the aggregate rate exceeds capacity C, then the aggregate rate synchronizes as time $t \to \infty$. Furthermore, this effect is independent of the initial states x_0 and y_0 of the sources. More formally, if $\Delta t_n = t_n - t_{n-1}$ denotes the time between the nth and the $(n-1)$st reset of the system, $\Delta t_n \to C/3$ as $n \to \infty$.*

Proof. Let us assume without loss of generality that $x_0 > y_0$ so that x is the larger source initially. We first show that the time Δt_n between resets n and $n-1$ satisfies the recurrence relation

$$C = 2\Delta t_n + \Delta t_{n-1} \tag{2}$$

with $\Delta t_2 = 1/4C + 1/4x_0 - 1/4y_0$. This relation imposes the form

$$\Delta t_n = \alpha_n C + \frac{(-1)^n}{2^n} x_0 - \frac{(-1)^n}{2^n} y_0 \tag{3}$$

with $\alpha_n = 1/2 - \alpha_{n-1}/2$ upon Δt_n.

We proceed by induction and observe that the first reset occurs at time t_1 when $C = A(t_1) = x_0 + t_1 + y_0 + t_1$, or when $t_1 = 1/2C - 1/2x_0 - 1/2y_0$. At which point, x is reset and the rates immediately after reset are $x(t_1^+) = 0$ and $y(t_1^+) = y_0 + t_1$. The second reset takes place at t_2 when $C = A(t_2) = t_2 - t_1 + y_0 + t_2$, or at time $t_2 = 3/4C - 1/4x_0 - 3/4y_0$. At time t_2, y is the larger source and is reset so that $x(t_2^+) = t_2 - t_1 = \Delta t_2$ and $y(t_2^+) = 0$ immediately after reset. Therefore, the difference in time between resets is $\Delta t_2 = 1/4C + 1/4x_0 - 1/4y_0$.

Next, we assume (without loss of generality) that $x(t_{n-1}^+) = 0$ and $y(t_{n-1}^+) = \Delta t_{n-1}$; i.e., x is reset at time t_{n-1} and that y was reset at time t_{n-2}. Then we reach capacity at time t_n when $C = t_n - t_{n-1} + \Delta t_{n-1} + t_n - t_{n-1} =$

$2\Delta t_n + \Delta t_{n-1}$. In other words, $\Delta t_n = 1/2C - 1/2\Delta t_{n-1}$ and source y is reseat at time t_n, two congestion epochs after its reset at t_{n-2}. Since

$$\Delta t_{n-1} = \alpha_{n-1} C + \frac{(-1)^{n-1}}{2^{n-1}} x_0 - \frac{(-1)^{n-1}}{2^{n-1}} y_0, \tag{4}$$

Δt_n satisfies

$$\Delta t_n = \left(\frac{1}{2} - \frac{\alpha_{n-1}}{2}\right) C - \frac{(-1)^{n-1}}{2^{n-1}} x_0 + \frac{(-1)^{n-1}}{2^{n-1}} y_0 \tag{5}$$

$$= \alpha_n C + \frac{(-1)^n}{2^n} x_0 - \frac{(-1)^n}{2^n} y_0 \tag{6}$$

with $\alpha_n = 1/2 - \alpha_{n-1}/2$.

In the limit as $n \to \infty$, Δt tends to $C/3$ independent of x_0 and y_0. In other words, the system resets at increasingly regular intervals with each source reset every other congestion epoch. Furthermore, the initial conditions x_0 and y_0 affect only the rate of convergence of Δt_n.

Next we show that if the source with the smaller rate is always reset, the system never synchronizes and tends to a "deadlocked" steady-state where one source sends at rate C and prevents the second from sending at all. This result is a form of chronic congestion and an extreme form of unfairness!

Theorem 6. *If the source with the smaller rate is always reset when the aggregate rate exceeds capacity C, then the aggregate rate tends to the capacity C as $t \to \infty$ with the larger source sending at rate C and the smaller source contirbuting nothing to the total rate. The nth reset t_n tends to $C - x_0$ (if x is the source which has the larger rate initially) and Δt_n , the time between resets, tends to zero.*

Proof. Assume without loss of generality that $x_0 > y_0$. We first show that source y remains the smaller source and is always reset. In general, the rates are

$$x(t) = t_0 + t \quad \text{and} \quad y(t) = t_n - t_{n-1} \tag{7}$$

for $t_{n-1} \le t \le t_n$ and the nth reset occurs at time $t_n = C/2 + t_{n-1}/2 - x_0/2$ with $t_1 = C/2 - x_0/2 - y_0/2$.

As in Theorem 5, the first reset occurs at time $t_1 = C/2 - x_0/2 - y_0/2$; at which time the smaller source y is reset. Immediately after the reset, the rates are $x(t_1^+) = x_0 + t_1$ and $y(t_1^+) = 0$. The second reset occurs at time t_2 when $C = x_0 + 2t_2 - t_1$, or $t_2 = 3/4C - 3/4x_0 - 1/4y_0$. Hence the length Δt_2 of the second congestion epoch is $\Delta t_2 = 1/4C - 1/4x_0 + 1/4y_0$.

We proceed by induction and assume that at t_{n-1}, y is reset and that x has never been reset so that the sources reach capacity at time t_n where $C = x_0 + t_n + t_n - t_{n-1}$. Thus, t_n obeys the recurrence relation

$$t_n = \frac{1}{2} C + \frac{t_{n-1}}{2} - \frac{x_0}{2} \quad \text{with} \quad t_1 = C/2 - x_0/2 - y_0/2 \tag{8}$$

and in its closed form

$$t_n = \left(1 - \frac{1}{2^n}\right)C - \left(1 - \frac{1}{2^n}\right)x_0 - \frac{1}{2^n}y_0.$$

At time t_n, source y is reset and the rates at any time t between t_n and t_{n+1} are $x(t) = x_0 + t$ and $y(t) = t - t_n$. We may now conclude that $t_n \to C - x_0$ and that $\Delta t_n \to 0$ as $n \to \infty$. In other words, regardless of the initial rates of the sources, with such an unfair penalty upon the smaller source, the system not only never synchronizes, it tends to an extreme form of chronic congestion.

Finally we exmaine the scenario in which one source is chosen at random and reset when the system reaches capacity. We show that independent of the inital conditions, the time between resets tends on average to a constant fraction of the capacity C. This fraction depends on the probability of favoring one source over another and in the uniform case (i.e., the probability of choosing the larger or the smaller source is equally likely), the fraction is less than in the first scenario where the larger source is always reset.

Theorem 7. *If the larger source is reset with probability p and the smaller source with probability $1-p$ when the aggregate rate exceeds capacity C, then in the limit as $n \to \infty$,*

$$\mathrm{e}(\Delta t_n) = \frac{p}{1+2p}C.$$

That is, the average interval length between resets is $\frac{p}{1+2p}C$ independent of the initial states of the sources. Furthermore, the limiting distribution $\Delta t = \lim_{n\to\infty} \Delta t_n$ of the congestion epoch durations is not concentrated at the mean value.

Proof. Observe that in the previous two scenarios that Δt_n has the form $\Delta t_n = \alpha_n C \pm 2^{-n}x_0 \pm 2^{-n}y_0$ where

$$\alpha_n = \begin{cases} \frac{1}{2} - \frac{\alpha_{n-1}}{2}, & \text{if reset larger source,} \\ \frac{\alpha_{n-1}}{2}, & \text{if reset smaller source.} \end{cases} \tag{9}$$

The signs decorating the initial values x_0 and y_0 do not play a role in the limiting behavior of Δt_n because the magnitude of the initial values' contributions decreases rapidly. Hence, the time between resets in the random case has the form $\Delta t_n = \alpha_n C \pm 2^{-n}x_0 \pm 2^{-n}y_0$ where

$$\alpha_n = \begin{cases} \frac{1}{2} - \frac{\alpha_{n-1}}{2}, & \text{with probability } p, \\ \frac{\alpha_{n-1}}{2}, & \text{with probability } 1-p \end{cases} \tag{10}$$

and α_n is initialized $\alpha_2 = 1/4$.

We can easily check that

$$\mathrm{e}(\alpha_n) = \frac{p}{2} - \frac{p\mathrm{e}(\alpha_{n-1})}{2} + \frac{(1-p)\mathrm{e}(\alpha_{n-1})}{2}$$

and that in the limit $\lim_{n\to\infty} \mathrm{e}(\alpha_n) = \frac{p}{1+2p}$; that is, $\mathrm{e}(\Delta t_n) \to \frac{p}{1+2p}C$.

For $p = 1/2$, the average length of the congestion epoch is $C/4$. In addition, one may check that α_n has mass 2^{-n+2} at the values $\frac{2k+1}{2^n}$ for $k = 0, \ldots, 2^{n-2} - 1$ by iterating the recurrence relation for α_n. The expected value of α_n is simply

$$\mathrm{e}(\alpha_n) = \sum_{k=0}^{2^{n-2}-1} \frac{2k+1}{2^n}\frac{1}{2^{n-2}} = \frac{\left(2^{n-2}\right)^2}{2^n 2^{n-2}} == \frac{1}{4}$$

for all n and the variance

$$\mathbf{Var}(\alpha_n) = \mathrm{e}(\alpha_n^2) - \mathrm{e}^2(\alpha_n) \tag{11}$$

$$= \sum_{k=0}^{2^{n-2}-1} \frac{1}{2^{n-2}}\left(\frac{2k+1}{2^n}\right)^2 - \left(\sum_{k=0}^{2^{n-2}-1} \frac{1}{2^{n-2}}\frac{2k+1}{2^n}\right)^2 \tag{12}$$

$$= \left(\frac{-1}{3}\right)4^{-n} + \frac{1}{48} \tag{13}$$

$$\to \frac{1}{48} \quad \text{as} \quad n \to \infty. \tag{14}$$

For $p \in (0, 1/2)$, the structure of α_n is more complicated; α_n consists of point masses at the values $\frac{2k+1}{2^n}$ for $k = 0, \ldots, 2^{n-2} - 1$ but the masses at these points are no longer equal, they are products of p and $1 - p$ depending on the value of k. More precisely, each point mass $m_{n,k}$ in α_n at the point $\frac{2k+1}{2^n}$ "spawns" two point masses $m_{n+1,2k}$ and $m_{n+1,2k+1}$ at the points $\frac{2(2k)+1}{2^{n+1}}$ and $\frac{2(2k+1)+1}{2^{n+1}}$ in α_{n+1} such that

$$m_{n+1,2k} = W m_{n,k} \quad \text{and} \quad m_{n+1,2k+1} = (1 - W)m_{n,k}$$

where $W = p$ if k is even and $W = 1-p$ is k is odd. (Note that this construction is very similar to that of Bernoulli measures, with an added twist [4].) Clearly, α_n is not concentrated at its mean in the limit as $n \to \infty$.

We pause to make several observations. We recover the result in the first scenario if we set $p = 1$ (i.e., with probability one, the larger source is reset) so that $\Delta t_n \to C/3$ with probability one. In addition, if $p = 0$ (i.e., with probability one, the smaller source is reset), we capture the second scenario with $\Delta t_n \to 0$ with probability one. If we adopt the point of view that chronic congestion is to be avoided and that the optimal network behavior is congestion epochs as long as possible, always resetting the larger source gives the longest possible congestion epochs of length $C/3$ and a (uniformly) random strategy is sub-optimal, giving average congestion epochs of length $C/4$. However, the maximum possible congestion epoch with a random strategy is $C/2$.

Let us interpret the results on the congestion epoch length for $N = 2$ sources in light of our previous section. Theorem 1 tells us that for $N = 2$, $\Delta t = rC/2$ is the congestion epoch length. In order to achieve $\Delta t = C/3$

where the larger source is always reset, using a mean field approximation, we must reset $r = 2/3$ of the sources. In other words, to achieve the "efficiency" of always resetting the larger source, we must reset two-thirds of the sources each time the link capacity is reached. Also, to reach the average congestion epoch length of $\Delta t = C/4$ where one source is reset at random, we must reset $r = 1/2$ of the sources, as one would expect. However, the density $\rho_*(x) = 1/2 \sum 2^l \delta_0(x - lC/4)$ describing the mean field approximation for the two sources is quite different from the true description of the sources' states!

3 Simulation Results

In the previous section we proposed a simple analytical model for congestion control that predicts periodic aggregate behavior; in this section we explore whether the model is relevant and its conclusions accurate.

We start with two simulation results in Figure 4. The graphs show the evolution of packet arrival rates and queue occupancies at a bottleneck link shared by 50 TCP sources sending an infinitely long file. On the top are results for a drop-tail policy; on the bottom are those for RED. In both cases there is strong aggregate periodic behavior, made more clear by the strong component in the discrete Fourier transform of the arrival rate (below each figure). The more pronounced periodic behavior caused by RED is counter to the commonly held intuition that a randomized drop-policy would prevent periodic behavior by "desynchronizing" TCP sources.

The example above represents just one scenario in which periodic aggregate behavior occurs; one in which all the sources are at the same distance from the congested link. Later in this section we demonstrate evidence of periodic behavior in a variety of settings; and then validate our model by showing that it seems to capture the essential characteristics of the TCP congestion control mechanism.

3.1 Simulation Setup

We use ns-2[2, 9] for all our simulations, with TCP Reno sources and the topology shown in Figure 5. Fifty clients (at the nodes $D_1, D_2, \ldots D_N$, with $N = 50$) on one side of a bottlenecked router retrieve an infinite-length file from a server, S, on the other side. Hence, there are 50 TCP connections, sharing a single oversubscribed 1.5 Mbps link between the nodes M_2 and M_3. Each connection starts at some random time between 0 and 300 seconds, lasting the duration of the simulation (4200 seconds). Figure 5 shows the data rate and propagation delays for each link. In our simulations, we choose the propagation delays between each endpoint $D_i, i = 1, 2, \ldots, 50$ and the node M_3 to be either a single value of 20 msec (which we call "fixed RTT"), or pick them at random and uniformly over the range 0 to 1 seconds (which we call

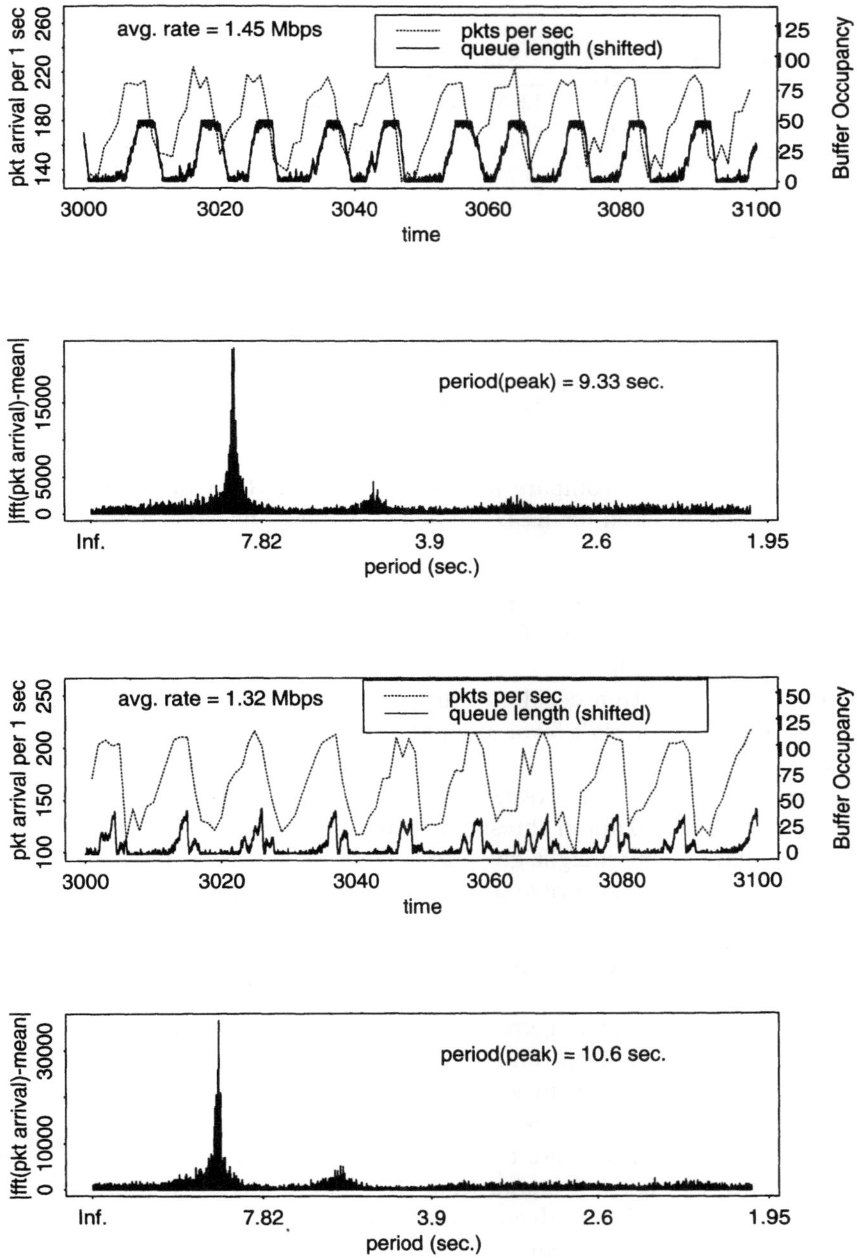

Fig. 4: Packet arrival and buffer occupancies with fixed RTT; drop-tail(top) and RED(bottom)

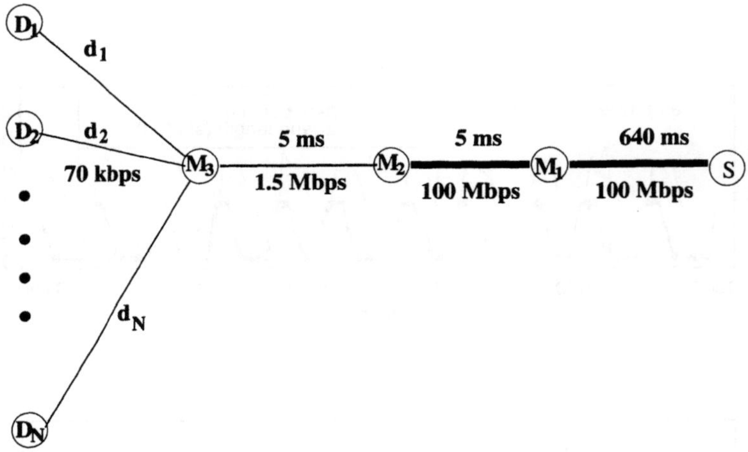

Fig. 5: Simulation setup

"variable RTT"). We compare both drop-tail and RED in our simulations; the parameters we used for RED are below. We use the same notation for the parameters as in [6].

1. Queue occupancy is measured in number of packets.
2. Dropping probability is NOT increased slowly when average queue length exceeds maxthresh.
3. When the algorithm decides to drop a packet, it drops from the tail.
4. The algorithm doesn't favor any packet based on its type(control or data) or size.
5. $min_{th} = 5$ (minimum threshold of average queue size)
6. $max_{th} = 15$ (maximum threshold of average queue size)
7. $q_w = 0.002$ (queue weight given to current queue size sample)
8. $p_b = 0.1$ (Maximum probability of dropping a packet)

3.2 Packet Arrival Rate and Queue Occupancy

We see in Figure 4 that the aggregate arrival rate shows periodic behavior with fixed RTTs with both drop-tail and RED. One might question the validity of this simulation because the fixed RTT value (1340 msec) is large, which means the time for the aggregate rate to reach its maximum will dwarf any variation caused by queueing delay and, for the case of RED, any randomness in packet drops. However, Figure 6 shows that the periodic aggregate behavior still occurs when the RTT is reduced by an order of magnitude. For an RTT of 140ms with both drop-tail and RED, the aggregate rate still fluctuates with a period of about 2 seconds, and as before the periodicity is more prominent with RED.

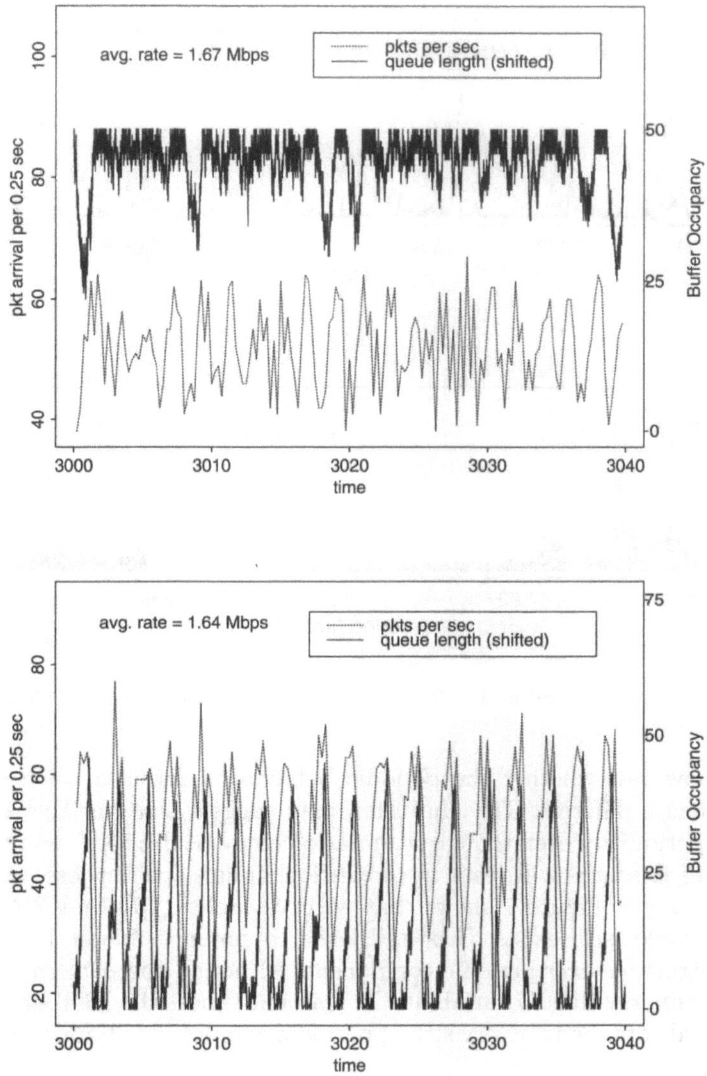

Fig. 6: Packet arrival and buffer occupancies with very short, fixed RTT; drop-tail (top) and RED (bottom)

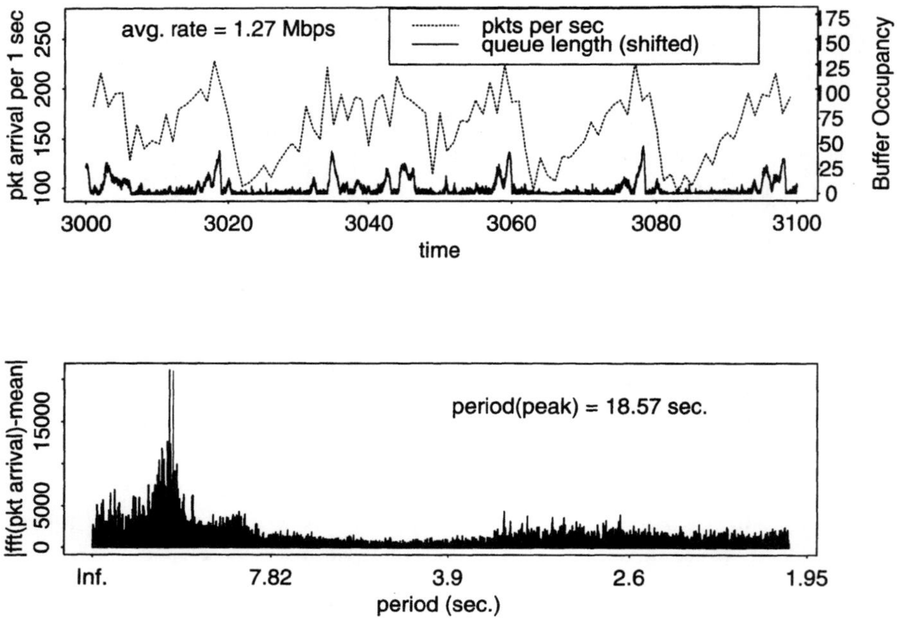

Fig. 7: Packet arrival and buffer occupancies with variable RTTs and RED

We now test whether the periodic behavior is reduced when each connection has a different RTT. Intuition may suggest that variable RTTs will "break" periodicity even though their packets may be dropped simultaneously, because each source will react to congestion signals (packet loss) at different times. However, Figure 7 shows that with variable RTTs and RED the periodic behavior still exists. Drop-tail leads to similar periodic behavior, but with a slightly *weaker* periodic component. Although the pattern looks more irregular, mainly due to variability in reaction times, the FFT of the aggregate arrival rate shows a very strong periodic component at approximately 19 seconds.

Having seen the close proximity between the behavior predicted by the model and simulation, we ask the following question: Is the similarity simply a coincidence or a consequence of valid assumptions made by the model? We try to answer this question in the next two subsections by testing two key assumptions of the model:

Assumption #1 the model assumes that the sources respond immediately to congestion, and

Assumption #2 the model assumes that an almost-constant random fraction, r, of the sources are reset to the same state at each congestion epoch.

3.3 Assumption #1: Immediate Response to Congestion

The congestion control mechanism of TCP relies on sources reacting to network congestion signaled by packet loss and timeouts. On the face of it, the assumption of an *instantaneous* feedback-response appears rather unrealistic, because it must take some amount of time for the buffer occupancy to build up and for the source to recognize the packet drop(s). However, with drop-tail, the delay due to buffering simply introduces a small fixed delay in the feedback. In RED, randomness acts over a small time interval, so we can either completely ignore it or regard it as a small random perturbation.

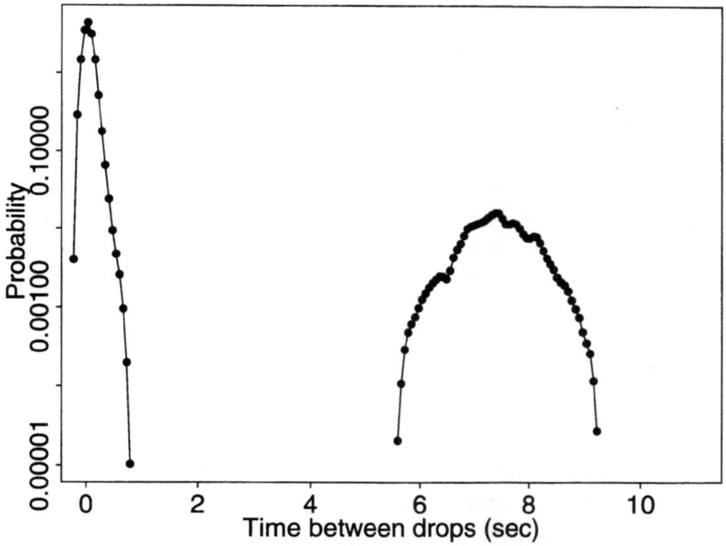

Fig. 8: Distribution of time interval between packet drops for fixed RTT and drop-tail

The time for the sources to "recognize" the packet drops and to react (or, the variability among sources with respect to this time, as in variable RTT settings) does not make the model irrelevant. The time it takes for a source to react is roughly one RTT and this time is short compared to the time for the aggregate rate to ramp up to the peak value, which is several RTTs long. We can reach the same conclusion even for variable RTTs. Among the sources to be reset, those with smaller RTT react sooner but they are more likely to have larger `cwnd` (since they build up more quickly). Hence, their reset contributes a large fraction of the decrease in the aggregate rate after the congestion. Similarly, a smaller decrease in the aggregate rate follows when the sources with the largest RTTs are reset (since those sources are likely to have the smallest `cwnds`). There is a trade-off in the size of the decrease of the aggregate rate; sources with larger RTTs effect a smaller decrease in the aggregate rate but their effect takes longer to manifest. Therefore, we can treat the variability caused by the "late" reactions as a perturbation, and the system's reaction to congestion is largely dominated by the connections with small RTT's, which occurs quickly[1].

One might think that with drops occuring one after another and not all at the same time during a congestion epoch, we can make the sources react at different times, but we find that packet drops occur in bursts of short duration, followed by longer periods in which no drops occur. For example, Figure 8 shows the empirical distribution of time between adjacent packet drops (for fixed RTT with drop-tail). We can see the large peak around very small values ($\ll 1$ second), and another set of much lower peaks around larger values ($6 \sim 9$ seconds), with a clear demarcation between the two. This indicates that the packet drops tend to occur in bursts (with short inter-drop times), separated by large time intervals. Other simulations show similar characteristics, although the demarcation between the two groups of inter-drop times become less clear with variable RTTs.

Had the burst of drops been of almost zero-length, we could safely assume that the response to feedback is essentially immediate, and so would correspond to the model. Instead we model the small (but non-zero) bursts as small perturbations from an "immediate" feedback-response, and use the perturbation analysis in section 2 which says that the invariant periodic configuration is stable.

Furthermore, from the clear demarcation of inter-drop times, we can see that the separation between bursts is unambiguous. We found that the duration of bursts is around 1 to 2 seconds for all the settings mentioned in the paper. As an aside, unlike the earlier claim[6] that RED will spread out packet drops in time by introducing randomness in selecting which packet to drop, the intra-burst inter-drop time distribution for RED is not much different from drop-tail. Analytical results [8] for batch Poisson arrival processes in an open-loop queueing system show similar results (although the model focuses on queueing details rather than on feedback to sources) and the authors conjecture that RED has little "desynchronization" effect.

Based on these observations, we conclude that packet drops at the bottleneck in the topologies under consideration tend to occur in bursts. The duration of each burst is small relative to the period between bursts. Therefore, in studying the aggregate periodic oscillation caused by TCP connections under such topologies, an immediate feedback-response mechanism augmented by the perturbation in timing appears to be a valid abstraction of the system.

---

[1]However, the sources with large RTTs can affect the ramp-up time, by slowing down the progress.

3.4 Assumption #2: Constant Fraction of Reset Sources

The model assumes that upon each congestion epoch some fraction, r, of sources set their rate back to zero. Does the feedback-reaction chain [1] in TCP Reno behave similarly?

In order to estimate r we measure the number of connections that experience *multiple* packet drops within a drop burst. Since the durations of drop bursts are comparable to round-trip times, such flows are most likely to experience timeout[5] that causes `cwnd` to be reset to 1. Figure 9 shows the histograms of these numbers for fixed RTTs, with both drop-tail and RED. The results suggest that approximately 20...40% of flows experience multiple packet drops in a congestion epoch, indicating that $r \sim 0.2 \ldots 0.4$.

Perhaps surprisingly, RED shows peaks at *larger* values of r than for drop-tail, which counters the intuition that RED prevents "global synchronization" by not resetting all the connections to the same state (`cwnd` $= 1$), at the same time. Thus, not only is the assumption inaccurate, since we show that the periodic behavior does not require the synchronization of sources, but also RED does not achieve the prevention of periodic behavior that it set out to do. Worse still, we know from our analytical model that the higher the fraction r, the faster the convergence to the invariant periodic behavior, and lower the minimum of the periodic behavior, resulting in lower average utilization of the link.

Finally, we examine the validity of the assumption that $r * N$ $(r < 1)$ TCP sources are reset at random upon each congestion epoch. In our model, we assume that the sources are chosen uniformly and randomly across all sources, but the real system may conform to some other distribution. For example, it could be close to a deterministic round-robin among the connections, possibly with overlaps between successive congestion epochs. If our assumption is correct, then the probability that a connection *avoids* being reset for k epochs is $r * (1-r)^k$. Or, if the selection follows a round-robin rule, the same probability is one for a specific value of k, and zero for all others.

To test our assumption, we look at the time between packet drops experienced by *each flow*, excluding the ones within the same burst (i.e. a slight underestimate of the "inter-reset time" for each flow). Figure 10 shows that the distributions of these values are concentrated around integer multiples of "cycle length" values, with exponential decay. This suggests that our assumption was valid.

3.5 When Does Aggregate Periodic Behavior *not* Occur?

The simulation results in the previous sections demonstrate that aggregate periodic behavior arises in a wide variety of settings: fixed or variable RTTs, drop-tail or RED queueing policies, and long or short RTTs. Because aggregate periodic behavior leads to low network utilization, it is worth exploring situations in which periodicity appears not to occur. These fall into two categories:

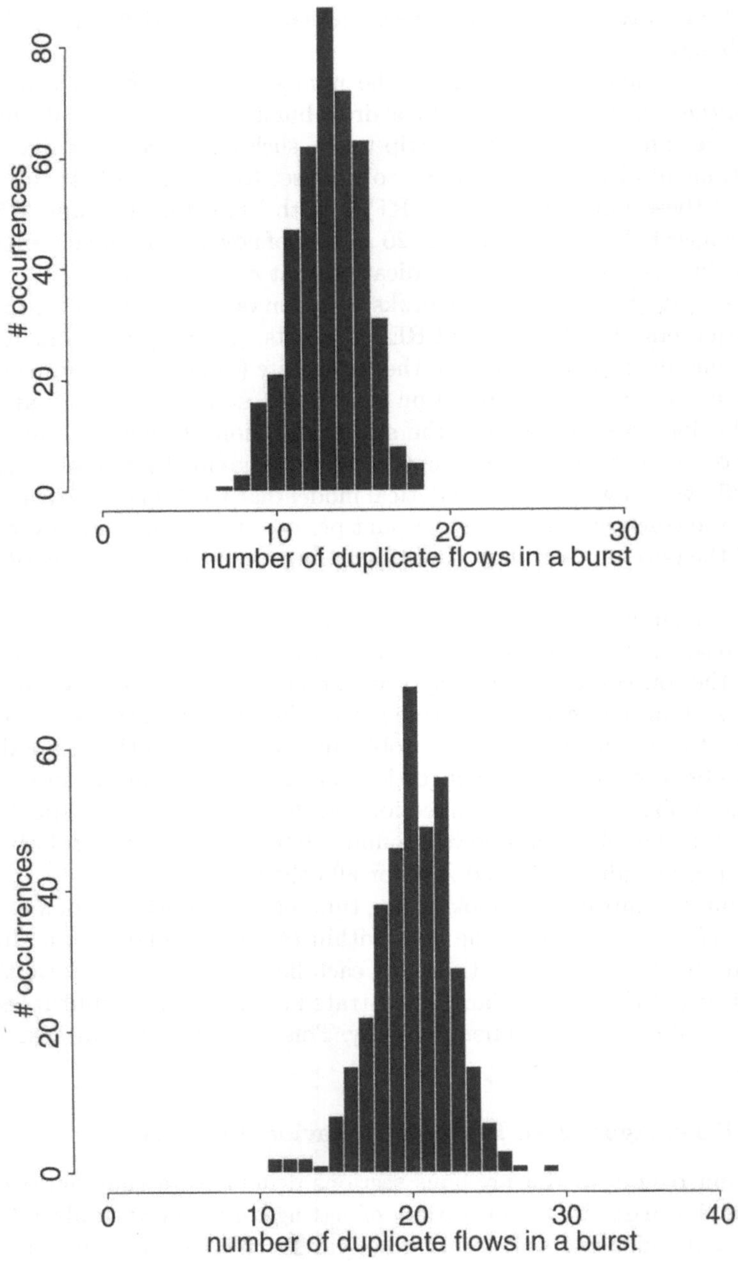

Fig. 9: Distribution of number of flows getting multiple packet drop during a drop burst for fixed RTTs; drop-tail (top) and RED (top)

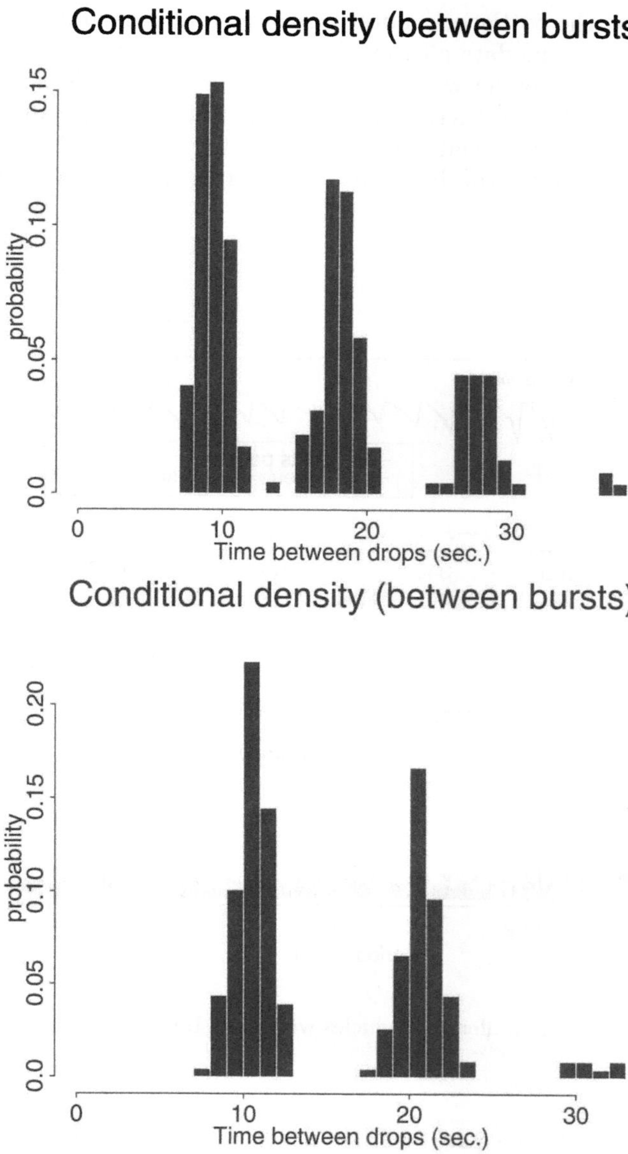

Fig. 10: Distribution of time intervals between packet drops each flow experiences under fixed RTT; drop-tail (top) and RED (bottom)

1. If the buffer at the bottleneck router is so large that it can *almost* (but not completely) accommodate all the N connections sending at their maximum rate, then the buffer will absorb (and smooth) the periodic behavior. Figure 11 shows that in this case the aggregate packet arrival rate is almost flat, except for a few deviations when the buffer overflows, and, instead of packet arrival rate, the buffer occupancies oscillate with a very long period.

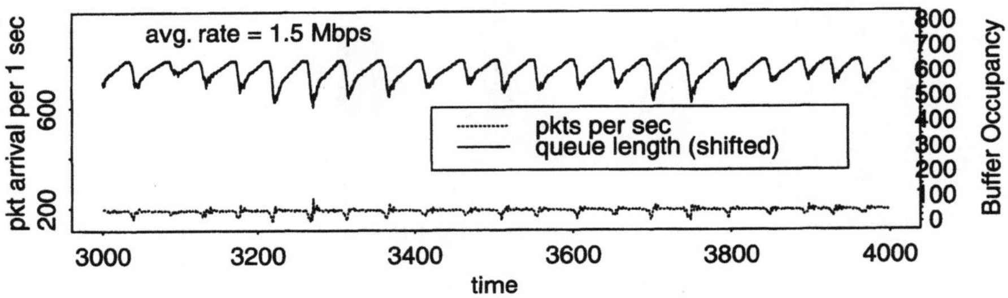

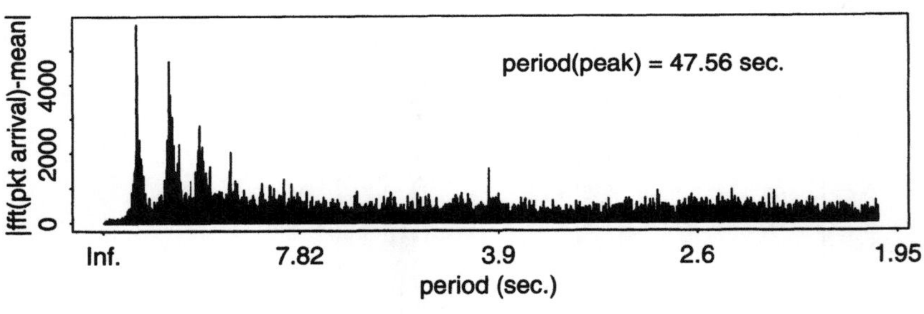

Fig. 11: Packet arrival and buffer occupancies with fixed RTT, very large buffer, and drop-tail

2. Our model and simulations both assume an infinitely long file transfer, allowing time for the feedback dynamics to be established and stabilize. But an increasing amount of traffic is sent in short connections (typically HTTP traffic). In [7], the authors demonstrate that HTTP traffic does not lead to periodic behavior, and Figure 12 shows similar results from the same single bottleneck topology (fixed RTT, drop-tail), with "Web-like" traffic. The 50 TCP clients download Web pages, with temporal and size distributions following the measurement results in [3]. Our results show no periodic behavior under either drop-tail or RED. A characteristic of the Web traffic is that the duration and inter-arrival of connections have

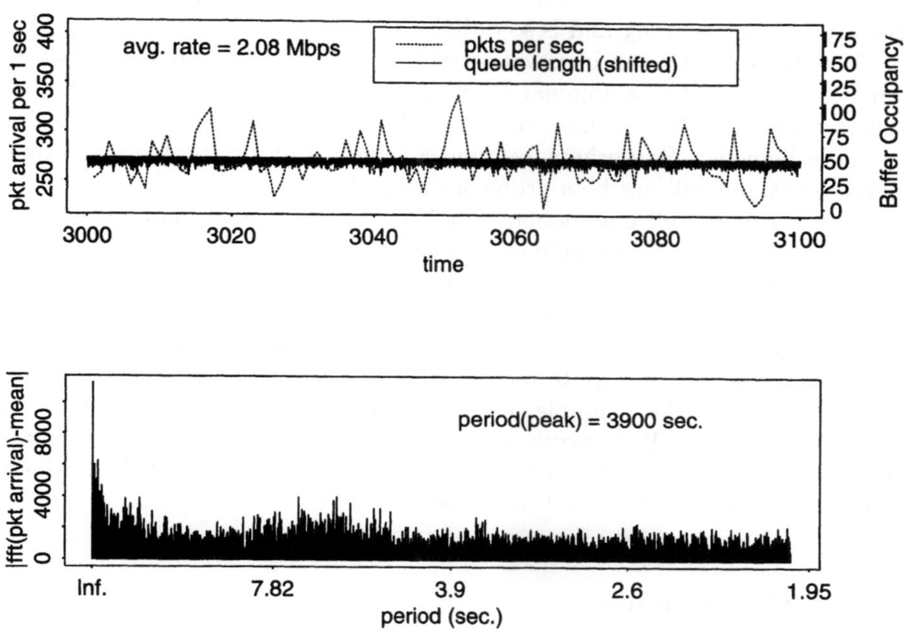

Fig. 12: Packet arrival and buffer occupancies with fixed RTT and Web-like traffic and drop-tail

heavy-tailed distribution (i.e. infinite variance), and so there is no "typical" lifetime of connections that we can study within the framework of our "toy" TCP model, even with some perturbation (in terms of number or lifetime of sources).

4 Conclusion

When network links were slower than they are today and traffic dominated by long file transfers, periodic aggregate behavior was seen as a potential problem, leading to the development of drop-policies that attempt to desynchronize TCP sources and reduce periodicity. Our simple model suggests that the concern was well-founded, and that long-lived TCP flows conspire to produce highly periodic traffic patterns. But it seems that randomized drop-policies such as RED do little to reduce the periodic behavior. On the contrary, both our model and simulations suggest that RED makes the periodic behavior more pronounced, more stable, converge faster, and cause lower overall link utilization. On one hand it is of value to know that a simple analytical model (that captures only the bare characteristics of TCP) correctly predicts periodic behavior. On the other hand, it is disconcerting to discover that periodic behavior arises so easily and is so difficult to avoid.

It is possible that this observation while interesting, may now be obsolete. Internet traffic today is dominated by sporadic HTTP flows of short duration which appear not to last long enough to become synchronized or to exhibit periodic behavior. So perhaps we can conclude that the problem identified here, is not and will not be a problem for the Internet.

However it is well understood that if the network continues to be dominated by short-lived flows, TCP cannot effectively control or avoid congestion. It seems likely that one of two steps will take place: Either flows (whether TCP or real-time UDP) will get longer, or the congestion control mechanisms of TCP will be changed to work well with short-lived flows. If longer flows prevail, our results suggest that quite radical steps are needed to prevent aggregate periodic behavior; for example, via more radical drop- and reset-policies in the network and TCP, respectively. Regardless of the paths chosen, our simple model highlights the importance of analyzing the essence of the feedback loop in any congestion control component of TCP.

Acknowledgements

The author is grateful for helpful discussions with Jonathan Mattingly, Nick McKeown, and Steven Strogatz. This work was done while the author was at AT&T Labs-Research in Florham Park, NJ and a visitor at Stanford University.

References

1. M. Allman, V. Paxson, and W. Stevens. TCP Congestion Control, April 1999. Request for Comments 2581.
2. S. Bajaj, L. Breslau, D. Estrin, K. Fall, S. Floyd, P. Haldar, M. Handley, A. Helmy, J. Heidemann, P. Huang, S. Kumar, S. McCanne, R. Rejaie, P. Sharma, S. Shenker, K. Varadhan, H. Yu, Y. Xu, and D. Zappala. Virtual InterNetwork Testbed: Status and research agenda. Technical Report 98-678, University of Southern California, July 1998.
3. P. Barford and M. Crovella. Generating representative Web workloads for network and server performance evaluation. *Performance Evaluation Review*, 26(1):151–160, 1998.
4. K. J. Falconer. *Fractal Geometry: Mathematical Foundations and Applications.* Wiley& Sons, New York, 1990.
5. K. Fall and S. Floyd. Simulation-based comparisons of Tahoe, Reno, and SACK TCP. *Computer Communication Review*, 26(3):5–21, 1996.
6. S. Floyd and V. Jacobson. Random early detection gateways for congstion avoidance. *IEEE/ACM Transactions on Networking*, 1(4):397–413, 1993.

7. Y. Joo, V. Ribeiro, A. Feldmann, A. C. Gilbert, and W. Willinger. On the impact of variability on the buffer dynamics in IP networks. In *Proc. of the 37th Annual Allerton Conference on Communication, Control, and Computing*, Allerton, IL, Sept. 1999.
8. M. May, T. Bonald, and J.C. Bolot. Analytical evaluation of RED Performance. In *Proc. of Infocom 2000*, 2000.
9. S. McCanne and S. Floyd. UCB/LBNL/VINT Network Simulator - ns (version 2). `http://www-mash.cs.berkeley.edu/ns/`.
10. S.H. Strogatz and I. Stewart. Coupled oscillators and biological synchronization. *Scientific American*, 269(6):102–109, 1993.
11. L. Zhang and D. Clark. Oscillatory Behavior of Network Traffic: A Case Study Simulation. *Internetworking: Research and Experience*, 1(2):101–112, 1990.

Statistical Properties of Chaos in Communication Networks

Gábor Vattay[1], K. Diriczi[1], A. Fekete[1], L. Kocarev[2], M. Maródi[1], and J. Stéger[1]

[1] Department of Physics of Complex Systems, Eötvös University, Budapest 1117, Hungary{vattay,diriczi,fekete,marodi,steger}@complex.elte.hu
[2] Institute for Nonlinear Science, University of California, San Diego lkocarev@ucsd.edu

1 Introduction

The behavior of communication networks, like the Internet is complex in many ways. This complexity may have many origins, including the topology of the network, the statistical properties of the traffic, and the rules governing the operation of the devices that constitute the system.

Several types of complex collective behavior have been observed in various compononents of communication networks and their models. These include self-organized criticality [1, 2], self-similar traffic [3], chaos [4], and other kinds of nonlinear phenomena (see e.g. [5]).

Recently Veres and Boda [4] have demonstrated that Transmission Control Protocol (TCP) congestion control can be chaotic in certain circumstances. Chaos in practice means that TCPs influencing each other in a computer network can produce highly complex behavior in time which is sensitive to small perturbations, yet the equations describing it are deterministic and simple. Understanding the mechanism and equations that produce chaos in a TCP/IP (Internet Protocol) network is crucial in traffic modeling. If we can identify the details of the mechanism then it is possible to build new kinds of network traffic models where the factors important from the point of view of dynamics and chaos are kept while many unimportant factors are reduced or simplified. Current TCP models are either too simplistic by assuming periodic behavior or they are entirely stochastic disregarding small details of a given network which might alter the dynamics completely and even change its statistical properties. Stochastic models can also break the temporal and spatial (topological) structure of correlations existing among TCPs[3] in an ex-

[3]Whenever it is not confusing we use the word TCP in two sense: for the set of rules that govern network dynamics, and also in the sense of devices that operate according to these algorithms (e.g. network cards).

tended network and are not able to give a correct account of possible long range dependence born between far away TCPs [6]. On the other hand, new chaos aware TCP and network models can preserve correlations and show the same level of complexity as it is observed in real network traffic. These features might turn out to be unavoidable when building reliable models of large networks where packet level simulation requires astronomical computer resources and/or simulation time. Moreover, chaos theory itself provides new tools to quantify this complexity. The complexity of a real network traffic or its packet level simulation and the complexity of a traffic model can be compared. Despite their inherent simplicity, these tools have never been used in networking and they might open new prospects in verifying TCP and network models or can even characterize the actual network state.

2 Preliminaries: TCP Chaos

We think that all these issues mentioned above are so important, that we should answer at least the basic open questions concerning the chaotic state of the TCP congestion control mechanism. For example we still do not know what causes TCP chaos exactly. Accordingly we do not know how generic its appearance is. For example, it has been argued in Ref. [7] that in the chaotic examples shown in Ref. [4] the packet loss probability is considerably higher than 1%. It has been shown [7, 8, 3] that in this case the exponential backoff mechanism plays an important role and can be responsible for the complex congestion window dynamics. In this contribution we show that chaos is not a consequence of high network congestion or loss probability and that TCPs operating in congestion avoidance mode, never entering into a backoff state show chaos. In other words chaos is the generic behavior of many TCP systems, while periodicity and synchronization is rather exceptional.

2.1 Unfairness

Also we have to ask how we can distinguish TCP chaos from stochasticity and do we gain anything by doing that. To point out the main weakness of stochastic models and to call for chaos aware models we would like to show that current stochastic TCP models are even unable to predict the traffic in a simple scenario like the one shown in Fig. 1 and investigated throughout this paper. In this setup three TCP flows sharing a common buffer that can store B packets and a common line with delay T_0 and speed C_0, then splitting into three different lines with different delays and speeds. In the actual simulations $B = 100$, $T_0 = 400\,\mathrm{ms}$, $T_1 = 100\,\mathrm{ms}$, $T_2 = 150\,\mathrm{ms}$, $T_3 = 200\,\mathrm{ms}$, $T_0 = 10^6\,\mathrm{bps}$ and $C_1 = C_2 = C_3 = 10^7\,\mathrm{bps}$ has been chosen. The congestion windows of the competing TCPs are not limited by the senders or by the receivers. The injected TCP packets can be lost only at the bottleneck buffer, there are no random losses on the links or in other buffers. It follows that the traffic is

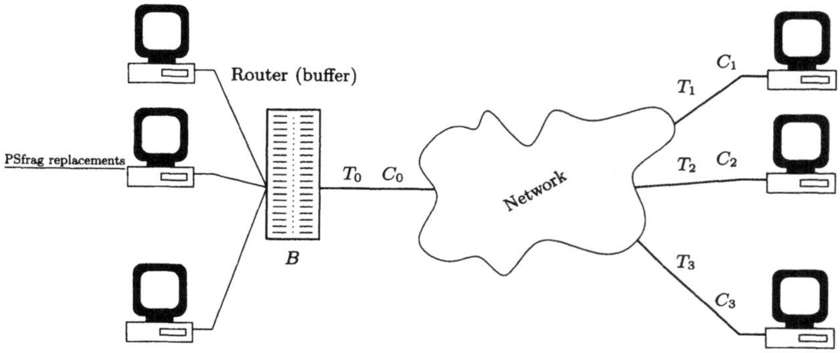

Fig. 1: Investigated network model: There are three TCP flows sharing a common buffer that can store B packets and a common line with delay D_0 and speed C_0. Then the common link splits into three different lines with different delays and speeds. In the actual simulations $B = 100$, $T_0 = 400\,\text{ms}$, $T_1 = 100\,\text{ms}$, $T_2 = 150\,\text{ms}$, $T_3 = 200\,\text{ms}$, $C_0 = 10^6$ bps and $C_1 = C_2 = C_3 = 10^7$ bps has been chosen.

controlled by strict deterministic rules, that is by the TCP Reno algorithm [9]. Numerical simulations were carried out by Network Simulator (**ns**) version 2b5 [10].

Based on ideas brought from stochastic TCP modeling a common belief is that TCP is biased against long round trip time connections and the throughputs are proportional to $\sim 1/T_{\text{RTT}}^2$. This assumption has been proven to be very good in the presence of random elements in the simulation [11]. In reality, shown in Fig. 2(a) it can be observed that the congestion window corresponding to the *largest* round trip time is significantly larger than the others. This TCP obtains *unfairly* higher throughput than the other two. The packet loss rate of the preferred TCP ($\sim 5 \cdot 10^{-5}$) is also an order of magnitude less than that of the suppressed ones ($\sim 7 \cdot 10^{-4}$). The "winner" TCP behaves seemingly periodically, while the "losers" seem to be erratic.

Obviously, there is something here which is missed completely by the stochastic model. Next, we show that this simple scenario is already chaotic and the deterministic nature of packet losses cannot be disregarded if we would like to build models that correctly predict the temporal behavior of congestion windows.

2.2 The TCP Butterfly Effect

The complete state of a TCP can be given by a number of internal variables at any moment [9]. Such variables are the congestion window, the slow start threshold, the retransmission timeout, the backoff counter, the duplicate ACK counter, and so on. However, during the optimal operation of TCP, in congestion avoidance mode, a single variable can be selected which controls almost completely the behavior of the TCP: that is the congestion window. This

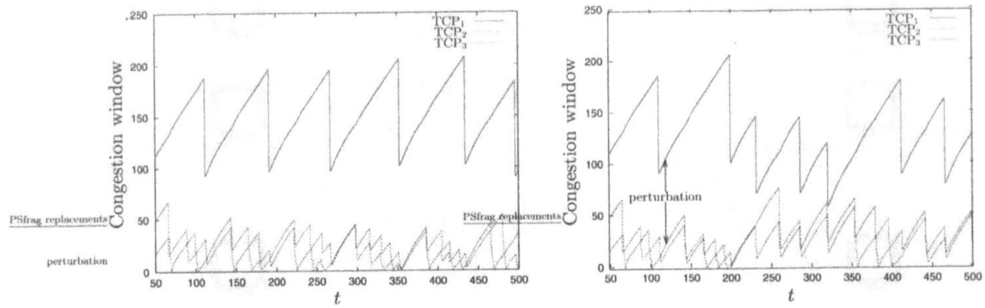

(a) (b)

Fig. 2: (a) A typical part of the congestion window time series is shown for the simulation of the scenario of Fig.1. (b) The effect of a small perturbation at $t = 120\,\mathrm{s}$ on the congestion window development.

variable has also practical importance, since it limits the maximum number of unacknowledged packets sent by a TCP into the network.

To demonstrate how *sensitive* this system can be for small perturbations the same simulation has been run for 120 s and then a perturbation of $\delta w_i(0) = 0.01$ has been added to all the congestion window values at $t = 120\,\mathrm{s}$. The result is shown in Fig. 2(b). The congestion windows remain unchanged until $t = 120\,\mathrm{s}$. Then the difference between the congestion windows of the two simulations $|\delta \mathbf{w}(\tau)|$ remains the same (~ 0.01) until the first packet drop event. Then one of the underprivileged TCPs, whose packet has been lost, halves its window. As a result, the distance between the original and the perturbed trajectories grows about an order of magnitude, since the owner of the lost packet differs in the original and in the perturbed simulations. At this point it seems that the dominant TCP is not affected at all. Finally, around $t = 240\,\mathrm{s}$ the time evolution of the dominant TCP diverges completely from the original trace due to a permutation of packets, resulting in a loss event for the dominant TCP. As we can see the rest of the simulation differs from the original one. Such sensitivity against small perturbations is called the butterfly effect in chaos theory and gives us the first clue that this system operating in congestion avoidance is actually chaotic.

3 Characterizing Chaos

Chaos is one of the most studied dynamical phenomena, and there are many ways of characterizing it. Here we introduce a few basic tools that can help us to describe the chaotic state of computer networks.

3.1 Poincaré Sections

One of the most basic tools of chaos theory in visualizing the dynamics is the Poincaré surface of section. Instead of looking at the continuous time evolution of trajectories one can select a surface in the phase space and watch only when the trajectories cross that surface. In case of TCP congestion window trajectories the evolution between two packet loss events is fairly simple. All the interesting things happen at the times of packet losses. Therefore the values of congestion windows taken at the moments of packet losses of any of the TCPs is a natural choice for a surface of section in general.

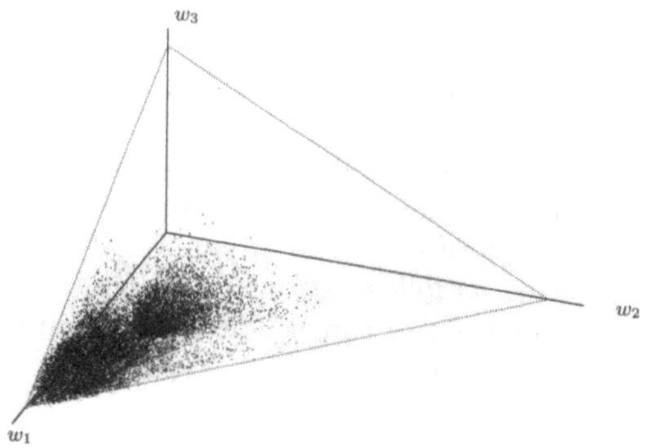

Fig. 3: Poincaré section of the phase space. Congestion window values for the three TCPs at packet loss times.

In Fig. 3 this surface of section is shown for our system. One can see that the congestion window triplets (w_1, w_2, w_3) taken at times of packet losses approximately form a two dimensional surface within the three dimensional congestion window phase space. It is easy to understand why we get such a surface: packet losses occur when the buffer is full. In the scenario of Fig. 1 the packets first fill the lines denoted by 0, 1, 2 and 3. The number of packets which can travel on these lines is given by the bandwidth delay products divided by the packet size C_0T_i/P, where the packet size in our simulations was 512 bytes. The maximum number of packets on the lines and in the buffer might be approximated by $Q = B + C_0T_0/P + \frac{1}{3}C_0(T_1 + T_2 + T_3)/P$, as all packets go through link 0, while each packet should choose either one of the three lines 1, 2 or 3. The sum of congestion windows $W = \sum_i w_i$ is approximately the number of packets in the network and packet loss occurs approximately when

$W = Q$. This equation defines the "surface of loss" inside the window phase space. This surface is also indicated in Fig. 3.

In chaotic systems the attractor is often a fractal object. Fractals are statistically self-similar geometric objects that might be characterized by suitably defined non-integer valued dimensions. For ordinary fractals this dimension is less then the Eucledian embedding dimension D of the object [12]. The fractal dimension of the points on the surface of section can be measured. To do this we can project the points onto a suitable surface. The points were projected onto the $\sum w_i = const.$ surface and the fractal dimension of this two dimensional projection was measured. A usual method for measuring the fractal dimension is when a grid of cells of size ϵ is put on the object, and the number of non-empty cells $N(\epsilon)$ is counted. The *box counting dimension* is given by

$$D_0 = -\lim_{\epsilon \to 0} \frac{\log N(\epsilon)}{\log \epsilon}. \tag{1}$$

Using the above expression we can show that the points on the surface form a non-trivial fractal with box counting dimension $D_0 = 1.69 \pm 0.02$.

Now our qualitative picture of TCP dynamics in congestion avoidance mode can be summarized as follows. Congestion windows steadily grow between packet losses. This process can usually be well approximated [13, 14] with fluid equations of the type

$$\frac{dw_i(t)}{dt} = \frac{1}{T_{\mathrm{RTT},i}(\mathbf{w})}, \tag{2}$$

where $T_{\mathrm{RTT},i}(\mathbf{w})$ is the round trip time of the ith TCP. Round trip times can also be approximated as functions of the congestion windows and then the resulting differential equations (2) can be solved self-consistently. The sum of congestion windows also grows steadily and the congestion windows cut the surface of loss at some point. Then one of the congestion windows is halved according to the TCP algorithm and the position of the point in the phase space drops below the loss surface and the process starts again.

3.2 Symbolic Description

One can see that the dynamical process between packet losses described above is relatively simple. The complicated fractal structure of the attractor and the sensitivity for perturbations should come from the fine details of the packet loss process. Each time the congestion window trajectory crosses the loss surface a packet loss event happens in one of the TCP flows. One of the symbols $S_i = \{1, 2, 3\}$ can be assigned to the ith packet loss depending on which TCP lost the packet. We can consider the sequence of the symbols $\ldots S_{i-1} S_i S_{i+1} \ldots$ generated by the time evolution of our TCPs. This symbol sequence is a coding of the real congestion window evolution. Such symbolic coding plays an important role in the theoretical description of chaotic systems. When an infinite sequence of symbols codes exactly one or zero real space trajectory the coding is called *Markov partition*. In this case the symbol sequences uniquely code the chaotic dynamics. An equivalent definition of the Markov partition is when each *periodic* symbol sequence codes exactly one or zero *periodic* orbit of the chaotic system.

In the case of TCP congestion avoidance mode we can demonstrate that the introduced symbols form a Markov partition. If an infinite periodic sequence (such as $\ldots 122312231223\ldots$) is prescribed and two different initial congestion window triplets $((w_1, w_2, w_3)$ and $(w_1', w_2', w_3'))$ are taken and they evolve according to the equations (2), they will reach the loss surface at different points. One can prove that the equations (2) are linearly stable, so they are not capable to increase the difference between two orbits. When one of the windows is halved after the trajectory crosses the loss surface, then the difference between the orbits is halved in that direction while it remains unchanged in other directions. The time evolution according to Eq. (2) and the prescribed halvings will decrease the distance between the two orbits in each period. As all the TCPs should halve their windows at least once in each cycle the distance between the two initial triplets is at least halved in each period. This way we can see that two trajectories started from different initial conditions converge exponentially to a common periodic orbit. In the end we get a unique periodic orbit corresponding to a given periodic code. So far we forced a given TCP to halve its window according to the prescribed symbol. In the end we can look at the actual packet flow generated by the window evolution. The prescribed periodic orbit can be feasible if the resulting packet flow is in accordance with the prescribed packet loss sequence or it is not feasible if the resulting packet flow generates losses in a different order than it has been assumed. This way we can decide if the calculated periodic orbit exists or not. This procedure ensures that we assign one or zero real periodic orbits to a periodic sequence and proves the existence of the Markov partition.

4 The Statistical Tool-box of Chaos

The statistical theory of chaos [15] is based on the symbol sequences introduced above. If the dynamics is regular (non-chaotic) then the dynamics is periodic or quasi periodic while the most important characteristics of chaos is that it endlessly generates topologically different new trajectories. This is reflected in the way different systems generate symbol sequences. A length n symbol sequence can continue in many ways; in average with a number of symbols to form a length $n+1$ sequence. Thus the number of length $n+1$ sequences $N(n+1)$ can be expressed as

$$N(n+1) \approx aN(n), \tag{3}$$

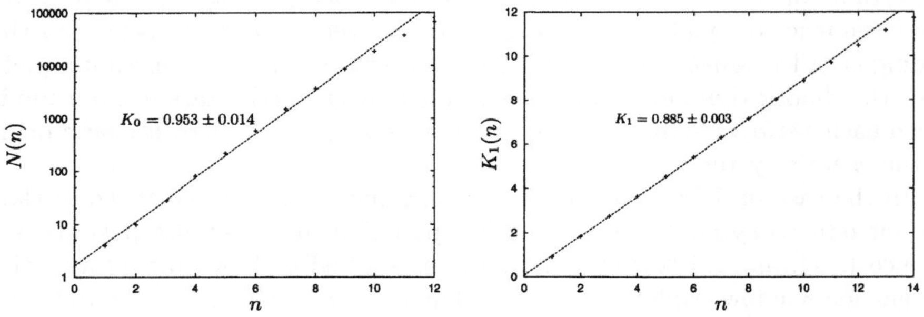

(a) (b)

Fig. 4: (a) Number of different symbols as a function of symbol length for our system on a semi-logarithmic plot. The fitted line is $N(n) = 1.669 \times 2.586^n$. (b) Kolmogorov–Sinai entropy for our system. $K_1(n) = -\sum_{i=1}^{3^n} p_i \ln p_i = 0.885 \times n + 0.078$

when the length n is large. In chaotic systems there is more than one possibility to continue a sequence in average and $a > 1$ while in regular systems $a = 1$ for long sequences. Consequently in chaotic systems the number of possible length n symbols grows exponentially

$$N(n) \sim a^n \sim e^{K_0 n}, \tag{4}$$

and the quantity $K_0 = \ln(a) > 0$ is the *topological entropy*. In non-chaotic systems the number of sequences grows sub-exponentially and the topological entropy is zero.

To prove that our TCP system really produces chaos we can measure its topological entropy. In our case we can have a maximum of 3^n different symbolic sequences of length n. Of course not all of them are realized by the dynamics since some of them are impossible. For example infinite sequences consisting of only one or two symbols are excluded since this would imply that some of the congestion windows are never halved. In Fig. 4(a) we show the number of realized symbol sequences for different lengths n up to $n = 12$. The measurement has been carried out by logging out the symbols (i.e. the index of the TCP which lost a packet) from an ns simulation of the system. We generated a sequence of 150.000 consecutive symbols and determined how many different length n sequences exist in it.

We observed that in average $a = 2.586$ symbols can follow a given symbol sequence and the topological entropy is $K_0 \approx 0.953$. This shows that our system is strongly chaotic as the number of sequences grows with a large exponent, yet it is markedly different from a stochastic system, where all combination of symbols are allowed and would result in $a = 3$ and a topological entropy of $\ln 3$.

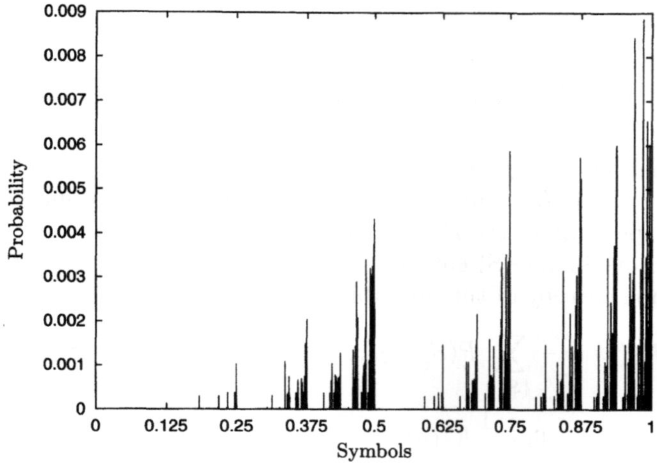

Fig. 5: Probability distribution of symbols of length 10 for our system.

The procedure described so far measures the existence of different symbol sequences only. We can characterize chaos further by calculating also the probability $P(S_1, S_2, \ldots, S_n)$ of the occurrence of the symbol sequence $S_1, S_2, \ldots, S_n$. In a system with L symbols ($S_i = \{1, 2, \ldots, L\}$) we can visualize this probability distribution by assigning the number

$$x = \sum_{i=1}^{n} (S_i - 1) L^{-i} \tag{5}$$

to each symbol sequence and plotting $P(x)$. In fact $0 \leq x < 1$ is the L-ary fractional representation of the number represented by the symbols.

In our system we carried out this analysis and the result is plotted in Fig. 5.

It can be clearly seen that the probability distribution is a fractal in the space of symbols. In fact, in all chaotic systems we should observe a multifractal distribution and the topological entropy introduced above is related to the box counting dimension of this representation. If we cut the $[0, 1[$ interval into boxes of size $\epsilon = 1/L^n$ then the number of non-empty boxes is the number of existing symbols $N(n)$. The box counting dimension is then

$$D_0 = \lim_{n \to \infty} \log N(n) / \log L^n = K_0 / \ln L.$$

In our case then the box counting dimension of the multifractal of Fig. 5 is 0.864. Note, that this box counting dimension is *not* related to the box counting dimension of the attractor discussed before.

Scaling of moments of the probability distribution give further characterization of the multifractal properties in the symbol space. We can define the

Rényi entropies [16]:

$$K_q = \lim_{n\to\infty} \frac{1}{n}\frac{1}{1-q} \ln \sum_{\{S\}_n} P^q(S_1, S_2, \ldots, S_n), \tag{6}$$

where summation $\{S\}_n$ goes over all possible symbol sequences of length n. The quantity $K_q/\ln L$ again measures the D_q generalized dimension of the multifractal spectrum of the $P(x)$ histogram. The most important entropy is the Kolmogorov–Sinai (KS) entropy $K_1 = \lim_{q\to 1} K_q$ which gives the scaling of the Shannon entropy of the probability distribution:

$$K_1(n) = -\sum_{\{S\}_n} P(S_1, S_2, \ldots, S_n) \ln P(S_1, S_2, \ldots, S_n), \tag{7}$$

$$K_1 = \lim_{n\to\infty} \frac{1}{n} K_1(n). \tag{8}$$

In Fig. 4(b) the Kolmogorov–Sinai entropy is measured for our system.

The Kolmogorov–Sinai entropy in a chaotic system is also related to the Lyapunov exponent λ of the mapping of the Poincaré section onto itself. If we consider two nearby trajectories on the Poincaré section—which is the loss surface in our case—then their initial separation in the phase space δw_0 grows each time the trajectories revisit the section. After n revisits the distance grows exponentially $\delta w_n \approx e^{\lambda n}\delta w_0$ where the average of the exponent lambda $\langle\lambda\rangle$ is the Lyapunov exponent of the Poincaré section. The KS entropy gives the Lyapunov exponent $K_1 = \langle\lambda\rangle$ in our system. The positivity of the KS entropy is another indication of chaos and the exponential sensitivity for the perturbation of trajectories.

In Fig. 6 we show the full Rényi entropy spectrum for the interval q=[0:20]. One can see that the entropies are positive for the whole range and the spectrum. The observed Rényi entropy curve is similar to those observed in typical hyperbolic chaotic systems. The large (positive) q behavior of the entropies is determined by the symbolic sequence with the slowest probability decay. Numerically we find that the asymptotic value of K_q is approximately $K_q \approx 0.49$ in our system.

Since the structure of the Markov partition for TCPs is simple even for larger networks, the topological and KS entropies are easy to measure in a simulation or can even be determined in a real network. These quantities measure the complexity of the dynamics and by evaluating them we can quantify complexity and evaluate TCP and network models in general.

5 Cellular Chaos

The results so far confirmed the hypothesis of chaotic dynamics. However, the variables of TCP in reality are discrete and not continuous. We can demonstrate this by applying such a small perturbation to congestion windows by

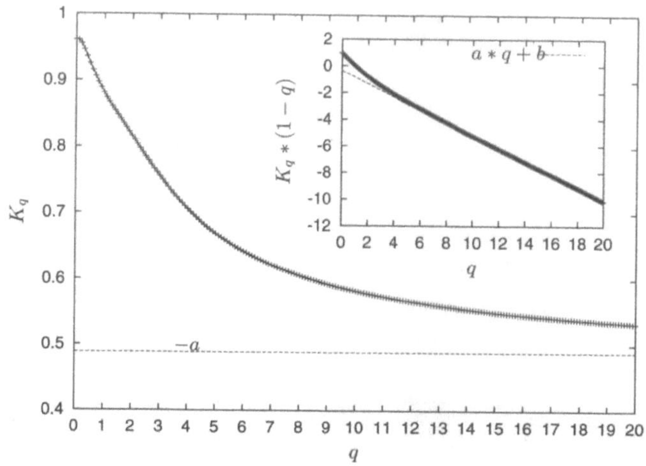

Fig. 6: Entropy spectrum for our system. Asymptotic behavior for $q \gg 1$ can be given by $K_q * (1-q) = a * q + b$, where $a = -0.49$, $b = -0.37$ (see inset).

which we do not change the loss events happening in the system. To investigate this we perturbed the trajectory of Fig. 2(a) by a small vector $\delta \mathbf{w}$ and traced the difference between the original and the perturbed trajectory. We found that there exists a set of perturbations, shown in Fig. 7, which vanishes after all windows are halved. The maximal perturbation of this type was found to be approximately $|\delta \mathbf{w}| < 0.001$. There is a well defined neighborhood around every phase point which defines the the same behavior of loss dynamics. That is, after all the congestion windows are halved, the time evolution of congestion windows becomes identical, mainly because the algorithm sets the window to its integer part. This means that the phase space is not continuous. It is divided into small "attractive cells" in which trajectories converge in finite time. If a trajectory gets into the cell of another trajectory then they follow the same trajectory later on.

This property makes TCP chaos very interesting from a theoretical point of view, since we have a globally chaotic dynamics while the fine details of the system are non-chaotic. This is not typical in natural occurrences of chaos but it is potentially important in engineered systems. The most significant aspect of the cellular structure is that all trajectories should be periodic. If a trajectory re-enters a cell visited previously then according to the attractive nature of the dynamics it will follow the same trajectory again and will repeat itself. Since the phase space can be divided into a finite number of cells a trajectory should at least repeat itself after visiting all the possible cells. In reality the repetition of a cell happens much earlier. We can make an estimate of the typical length and the distribution of the periods of trajectories

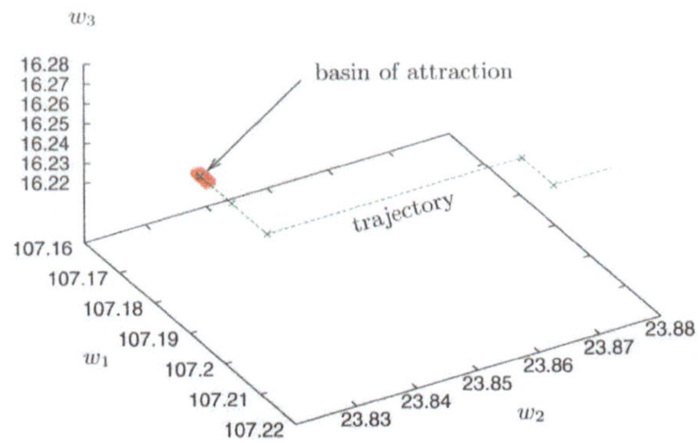

PSfrag replacements

Fig. 7: Our trial TCP trajectory (green) and the points in phase space (red) which follow the same trajectory after all three windows are halved. The basin of attraction of the starting point of the trajectory is approximately a cube of size 0.001^3.

following the theory developed by Grebogi, Ott, and Yorke [17] for chaotic systems with numerical roundoff.

These calculations assume that the dynamics is mixing, which means that the expectation value $\bar{l}$ of period length is large ($\bar{l} \gg 1$). Applying this approximation, we get the for the following relation:

$$\bar{l} = \sqrt{\frac{\pi}{8\langle p \rangle}}, \tag{9}$$

where $\langle p \rangle$ is the probability of repeating the cell in step $n+1$ that was first visited in step j, with

$$\langle p \rangle = \sum_i p_i^2, \tag{10}$$

where p_i is the probability with which the orbit visits the ith cell (i.e. the measure of the attractor in that cell).

In the next section we investigate the implications of the periodicity of the trajectories.

6 Exploring Periodic Orbits

Until now three parallel TCPs were studied in our simulation scenario. In such situation the number of cells which can be distinguished from each other in the phase space is in the order of $\sim 10^{15}$. Also, the length of a typical

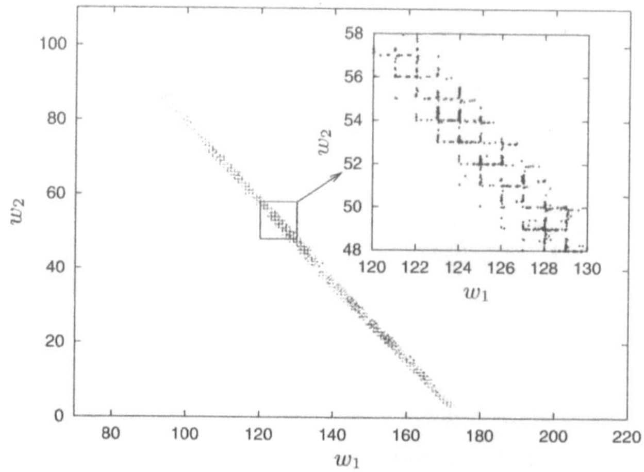

Fig. 8: Poincaré section of the phase space. Congestion window values for the two TCPs at packet loss times. A small part of the Poincaré section is zoomed to demonstrate the fine structure of the section.

periodic orbit is enormous and we do not expect to observe them in realistic simulations.

Therefore, it seems more reasonable to find periodic orbits when only two TCPs are operating in the network. For this scenario we used the same network model that we have introduced in Fig. 1 except that the TCP with the largest round-trip time was removed and the buffer size was set to $B = 50$.

Applying the tools developed in Sections 3–4 we can study the chaotic properties of this system. First the Poincaré surface of section is investigated (Fig. 8). Similarly to the three dimensional case, the sum of the congestion windows has to approximately satisfy the $\sum_i w_i = Q$ condition at packet loss times, where Q is the maximum number of packets on the lines and in the buffer. This condition defines the "line of loss" inside the two dimensional window phase space now. The detailed structure of the Poincaré section is shown in the inset in Fig. 8. It can be seen that the Poincaré section is not exactly a line but a narrow grid spreading around the ideal line.

This shape can be explained by the packet drop mechanism at the bottleneck buffer.

To measure the complexity of the TCP dynamics in this situation we apply the symbolic coding that was introduced in Section 3.2. Then, the topological (Fig. 9(a)) and the Kolmogorov–Sinai (Fig. 9(b)) entropies are estimated. For topological entropy $K_0 = 0.54 \pm 0.01$, while for the KS entropy $K_1 = 0.440 \pm 0.002$ have been obtained. Both values indicate strong mixing and the presence of chaos in the system. However, the maximum number of states is limited due to the discrete phase space, and TCP must enter into a periodic

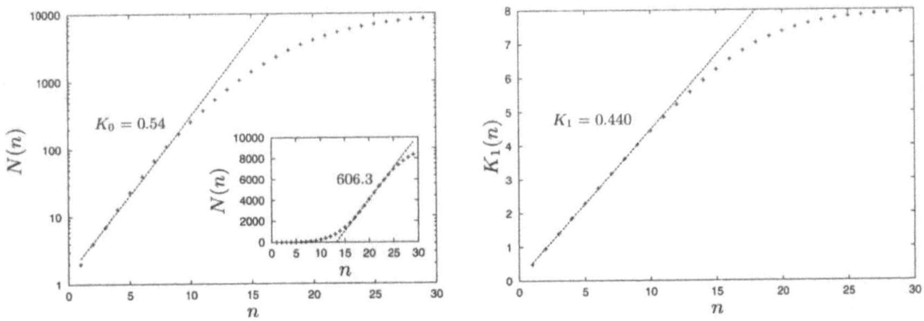

(a) (b)

Fig. 9: (a) Number of different symbols as the function of symbol length n for 2 TCP on a semi-logarithmic and on a normal plot in the inset. The fitted lines are $N(n) = 1.387 \times 1.718^n$ and $N(n) = 606.3 \times n - 8034$. (b) Kolmogorov entropy for 2 TCP scenario. $K_1(n) = -\sum_{i=1}^{2^n} p_i \ln p_i = 0.440 \times n + 0.081$

cycle after an initial transient period. If the orbit realized in the simulation is in fact periodic with some large period, then the number of existing length n sequences $N(n)$ increases only linearly with n if n is sufficiently large. In case of two TCP this linear growth is observable above $n = 15$ (see the inset in Fig. 9(a)). Also the entropy $K_1(n)$ saturates above $n = 15$ and the KS entropy K_1 goes to zero indicating that the long time behavior of the system is periodic.

In our 2-TCP scenario periodic orbits were identified. For the given simulation setup the length of the period was found to be independent of the initial conditions. The length of the period was $l = 1744$. Using this and Eqs. (10) and (9) we can estimate the order of magnitude of the number of cells in this system. The cells can be indexed by the corresponding symbol sequences. Suppose that we can index uniquely all the cells with symbol sequences of length n^*. In this case the number of cells is approximately given by $e^{K_0 n^*}$. The probability of repeating a cell is then given by

$$\langle p \rangle = \sum_{\{S\}_{n^*}} P^2(S_1, S_2, \ldots, S_{n^*}) \approx e^{-K_2 n^*}, \tag{11}$$

where K_2 is the Rényi entropy for $q = 2$. Combining (9), the estimate for the number of cells $e^{K_0 n^*}$ and (11) one obtains

$$N \approx e^{K_0 n^*} \approx \langle p \rangle^{-K_0/K_2}. \tag{12}$$

On the other hand (9) implies

$$\langle p \rangle = \frac{\pi}{8} \frac{1}{l^2}, \tag{13}$$

and finally:

$$N \approx \left(\frac{8}{\pi} \bar{l}^2 \right)^{K_0/K_2} . \tag{14}$$

Assuming that the obtained period $l = 1744$ is a good approximation of the average period and using the values $K_0 = 0.54$ and $K_2 = 0.39$ we get $N \approx 3 \cdot 10^9$. This is in accordance with expectations, since the area in the phase space visited by the congestion window trajectories is approximately a triangle with area $90 \times 90/2$ as one can see in Fig. 8, while the area of a cell is approximately 0.001^2 in accordance with the findings of Fig. 7. This implies that the visited part of the phase space contains approximately $4 \cdot 10^9$ cells.

7 Effects of Parameter Rescaling

In the previous sections we demonstrated that the dynamics computer networks connected in a simle way can be chaotic, and due to the discrete nature of the governing algorithms, the long term behavior is unavoidably periodic. So far we assumed that all the TCP devices operate in the congestion avoidance state, thus their dynamics between loss events can be well described by continous time, fluid models, like Eq. (2). In these situations the system is in a moderately congested state, when the probability of falling into backoff state is negligible.

Now we study these phenomena further by changing the properties of the network with a simple rescaling of the parameters. We return to the investigation of the 3 TCP model[4], and try to answer the question: How robust are the phenomena presented in the previous sections? In particular we introduce the scaling variable λ and use it to adjust the the link parameters C_0, T_i $(i = 0, 1, 2, 3)$ and the buffer size B:

$$\begin{aligned}
\tilde{C}_0 &= \lambda C_0, \\
\tilde{T}_i &= \lambda T_i \quad (i = 0,\, 1,\, 2,\, 3) \\
\tilde{B} &= [\lambda B]
\end{aligned}$$

where in case of the buffer size we used the integer part of the rescaled quantity. By choosing sufficiently small λ we can drive our network into a strongly congested state, where some of the TCPs stay in the backoff state for long periods of time. We seek evidences for the qualitative changes caused by tuning the scaling parameter.

First, we study the unfairness phenomenon presented in Sect. 2, and investigate it for various λ values. For example, at $\lambda = 0.03$ the system falls into a state when the three TCPs lose packets in a simple periodic manner (...123123123...). For slightly higher values various situations may arise. In

[4]Parameter values are as previously, except for $T_1 = 150\,\text{ms}$, $T_2 = 200\,\text{ms}$ and $T_3 = 250\,\text{ms}$.

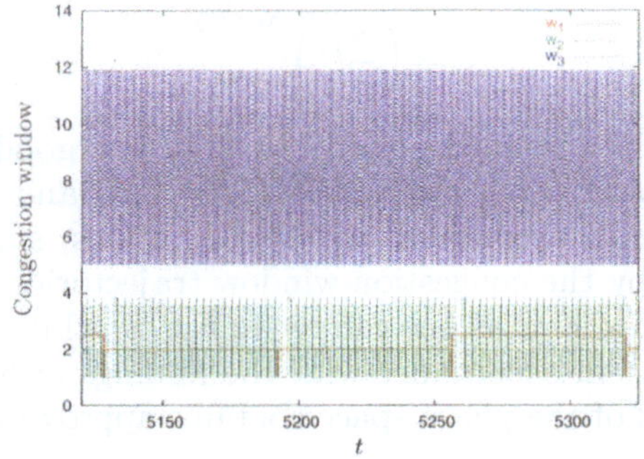

(a)

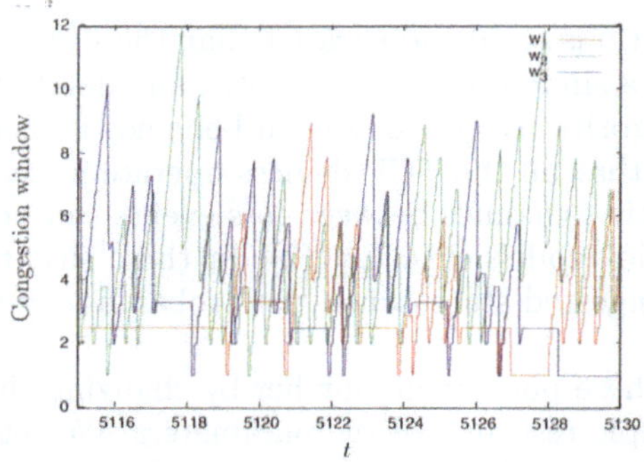

(b)

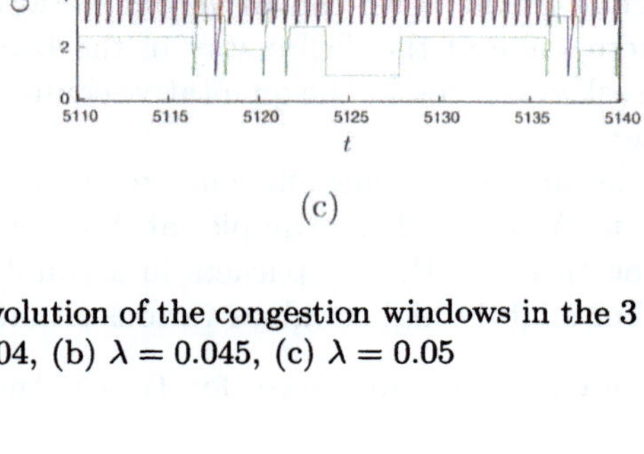

(c)

Fig. 10: The time evolution of the congestion windows in the 3 TCP case for various λ values. (a) $\lambda = 0.04$, (b) $\lambda = 0.045$, (c) $\lambda = 0.05$

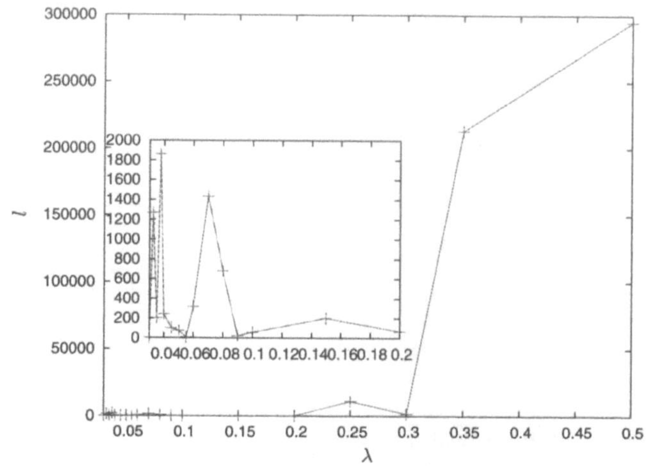

Fig. 11: The period length l of the generated symbol sequences as a function of the scaling parameter λ. Inset: magnification of the region for small λ

Fig. 10 we displayed the time evolution of the congestion window values for $\lambda = 0.04$, 0.045 and 0.05. One can observe that for $\lambda = 0.04$ one of the TCPs performs considerably better than the others, though a second one is still in congestion avoidance mode (with lower throughput), while the third remains in backoff/timeout[5] for most of the time (Fig. 10(a)). At $\lambda = 0.045$ the situation is different. Now there are two TCPs that can get into a minimal activity state (which they do alternatively), while the third one is almost always in congestion avoidance (Fig. 10(b)). Increasing our scaling parameter to $\lambda = 0.05$ we meet a new situation. Two of the TCPs share the buffer almost equally and are more or less synchronized, while the third one stays in a low activity mode, with repeated timeouts (Fig. 10(c)). All this indicates, that a slight change in the network parameters can significantly modify the resulting dynamics. As it can be seen, even the "winner" TCP can change with the change of the scaling parameter λ.

We expect that by decreasing λ the phase space shrinks, and assuming that the size of the cells does not change, the number of accessible cells decreases too. Moreover, when one ore more TCPs fall into the backoff state or waits for timeout repeatedly, then the effective dimension of the phase space is also reduced. All these suggest, that for sufficiently small values of the scaling parameter λ one should observe relatively small period lengths (compared to the original $\lambda = 1$ case). To investigate this, we measured the period length l

[5]For our purposes it is enough to consider cases when the given TCP is in a state when it only occasionally sends packets into the networks. This can happen in backoff mode, or when waiting for transmission timeout. In these cases the given TCP does not show the oscillatory behavior as in congestion avoidance.

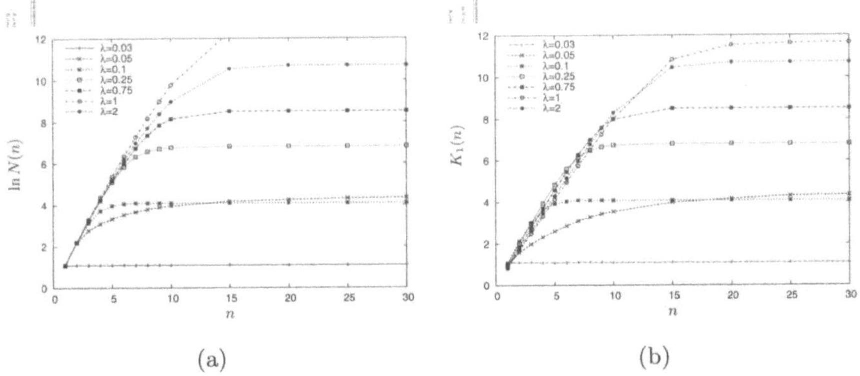

Fig. 12: Entropies of the rescaled system for different λ values. (a) Number of different symbols of length n as a function of n. (b) Kolmogorov–Sinai entropy $K_1(n) = -\sum_{i=1}^{2^n} p_i \ln p_i$

of the observed cycles in the generated symbol sequences at several λ values. The results are plotted in Fig. 11. One can see that below the value of $\lambda \approx 0.3$ the obtained period lengths are relatively small, approximately of the order 10^3. Above this value the lengths seem to diverge, their magnitude increases significantly to around $10^5 - 10^6$ in the investigated parameter range.

This phenomenon can be explained if one considers the probability of falling into one of the minimal activity (backoff/timeout) states. This probability is very small, indeed negligible for $\lambda > 0.3$. Thus for these parameter values the dynamics competing TCPs uses the whole original phase space, with all the TCPs operating in congestion avoidance mode. On the contrary, below $\lambda \approx 0.3$ the probability that one or more TCPs stops operating in congestion avoidance mode increases to a non-negligible value, which reduces the accesible phase space. In this case those TCPs that remain in congestion avoidance can still be described by the simple, continous time dynamics between losses.

Finally, we present some results concerning the Rényi entropies in the rescaled systems. The measured quantities are displayed in Fig. 12 for the topological entropy K_0, and the Kolmogorov–Sinai entropy K_1. One can clearly see, that for short symbol series the measurements follow the same curve for a wide range of the scaling parameter λ, while for low parameter values they deviate. This indicates that when the probability of backoff is low, the dynamics is governed by rules of very similar complexity. This is a direct consequence of the rescalability of the equations behind. However, when the system is driven into a qualitatvively different state, the level of complexity might change, possibly due to the effects described above.

8 Conclusions

In this paper we investigated the chaotic properties of TCPs operating in congestion avoidance mode. We showed that chaotic behavior is general even in the case of low packet loss probability. We demonstrated that the dynamics can be viewed as a smooth time evolution between packet losses and the relevant features of chaos might be described by the investigation of the Poincaré section defined by packet loss events. Chaotic dynamics can be characterized by symbol sequences and we introduced topological and Kolmogorov–Sinai entropies, whose values confirmed the hypothesis of chaos. Due to the deterministic nature of the system, in contrast to stochastic models, not all possible symbol sequences are realized, some of them are excluded by the dynamics. Accordingly the topological entropy is significantly less than its possible maximal value $\ln L$. The positive Kolmogorov–Sinai entropy indicates that the Lyapunov exponent of the system is positive and that the distribution of symbol sequence probabilities is multifractal. We also proved that due to the cellular structure of the phase space the long time behavior of the system is inherently periodic.

In the last part of this contribution, we tested our results by rescaling the system. It turned out, that the a small change in the parameters can cause relevant modification of the resulting dynamics and the fairness relations. Also, we observed the signifant decrease of the period lengths obtained from the symbolic dynamics. We attribute this to the decrease of the accessible phase space volume.

We hope that these results can help the design and testing of models of computer networks.

Acknowledgments

The authors would like to thank for the fruitful discussions with M. Boda, A. Veres, and S. Molnár. G. V. thanks the support of the Hungarian Scientific Research Fund (OTKA T032437 and T037903).

References

1. J. Yuan, Y. Ren, and X. Shan. Self-organized criticality in a computer network model. *Phys. Rev. E*, 61:1067–1071, 2000.
2. S. Valverde and R. V. Solé. Self-organized critical traffic in parallel computer networks. *Physica A*, 312:636–648, 2002.
3. A. Fekete and G. Vattay. Self-similarity in bottleneck buffers. In *Proceedings of Globecom 2001*, December 2001.
4. A. Veres and M. Boda. The chaotic nature of TCP congestion control. In *IEEE INFOCOM'2000*, March 2000.

5. K. Jiang, X. Wang, and Y. Xi. Nonlinear analysis of RED – a comparative study. *Chaos, Solitons & Fractals*, 21:1153–1162, 2004.
6. A. Veres, Zs. Kenesi, S. Molnár, and G. Vattay. On the propagation of long-range dependence in the Internet. In *ACM SIGCOMM 2000*, Stockholm, Sweden, August 2000.
7. L. Guo, M. Crovella, and I. Matta. TCP congestion control and heavy tails. Tech. Rep. BUCS-TR-2000-017, Computer Science Dep., Boston University, 2000.
8. D. R. Figueiredo, B. Liu, V. Mishra, and D. Towsley. On the autocorrelation structure of TCP traffic. Tech. Rep. 00-55, Dep. of Computer Science, University of Massachusetts, Amherst, November 2000.
9. W. Stevens. *TCP Slow Start, Congestion Avoidance, Fast Retransmit and Fast Recovery Algorithms.* RFC 2001, Jan. 1997.
10. UCB/LBNL/VINT network simulator – ns (version 2). http://www-mash.cs.berkeley.edu/ns/.
11. Y. Zhang, L. Qiu, and S. Keshav. Understanding the performance of many TCP flows. *Computer Networks*, 37:277–306, 2001.
12. T. Vicsek. *Fractal Growth Phenomena.* World Scientific, Singapore, 2nd edition, 1992.
13. T. J. Ott, J. H. B. Kemperman, and M. Mathis. The stationary behavior of ideal TCP congestion avoidance. In *Proceedings of IEEE INFOCOM'99*, New York, 1999.
14. Archan Misra and Teunis Ott. The window distribution of idealized TCP congestion avoidance with variable packet loss. In *INFOCOM'99*, March 1999.
15. D. Ruelle. *Statistical Mechanics, Thermodynamic Formalism.* Addison-Wesley, Reading MA, 1978.
16. P. Cvitanović, R. Artuso, R. Mainieri, G. Tanner, and G. Vattay. *Classical and Quantum Chaos.* Niels Bohr Institute, Copenhagen, 2001. Detailed discussion of various entropies characterizing chaotic systems can be found in.
17. C. Grebogi, E. Ott, and J. A. Yorke. Roundoff-induced periodicity and the correlation dimension of chaotic attractors. *Physical Review A*, 38(7):3688–3692, October 1988.

On Dynamics of Transport Protocols Over Wide-Area Internet Connections

Nageswara S. V. Rao[1], Jianbo Gao[2] and Leon O. Chua[3]

[1] Computer Science and Mathematics Division, Oak Ridge National Laboratory, Oak Ridge, TN 37831 raons@ornl.gov

[2] Department of Electrical and Computer Engineering, University of Florida, Gainesville, FL 32611 gao@ece.ufl.edu

[3] Department of Electrical and Computer Engineering, University of California, Berkeley, CA 94720 chua@fred.eecs.berkeley.edu

1 Introduction

The Transport Control Protocol (TCP) is widely deployed over the Internet for reliably transporting data [27], and it accounts for a significant portion of Internet traffic. While TCP has been extremely effective in transporting bulk data, it has not been as effective in remote control operations, particularly over long-haul connections. The control of fast mechanical devices using TCP over wide-area Internet connections could suffer from two problems: (a) lack of responsiveness of the device, and (b) presence of high frequency components that produce uncontrolled motions. The conventional controllers designed for electrical connections with very small delays are particularly vulnerable if simply transfered over to the Internet. The difficulty is that TCP is known to exhibit complicated dynamics over various time scales. While such behavior does not directly affect large bulk transfers, it can have serious negative effects on the controllability and stability of control loops implemented over wide-area networks.

The next generation of network applications, such as instrument grids, remotely deployed mobile robot teams and interactive simulations distributed on supercomputers, require stable control mechanisms over wide-area networks. Thus it is important to understand the dynamics of TCP at time-scales appropriate for the application at hand. TCP dynamics may also play a crucial role in other scenarios that require high throughputs over large time-scales. For example, the parallel-TCP utilizes the collective dynamics of TCP to achieve throughput rates that are significantly larger than a single TCP stream [33, 38]. Our main focus here is on the dynamics at time-scales of the order of congestion window-size updates, which may vary from a few to hundreds of milliseconds depending on connection distances and traffic levels.

TCP dynamics have been studied under various formulations, and its complicated nature been observed in many cases [40]. The work of Veres and Boda [43] illustrated, using ns-2 simulations, the chaotic dynamics of two or more competing TCP streams through a single router connected to a link. They illustrated that the congestion window dynamics of two streams generate chaos-like time series, but did not provide explanations, such as routes to chaos, for the behavior. In another direction, Ranjan and Abed [28] showed that TCP interacting with a Random Early Detection (RED) routers exhibits periodic doubling which eventually leads to chaos in the buffer occupancy rates.

Our objective is to study the dynamics of the Additive Increase and Multiplicative Decrease (AIMD) congestion control mechanism of TCP over Internet connections. TCP together with its interactions with Internet traffic constitutes a very complicated dynamical system, which does not appear to be amenable to concise and tractable models. In particular, a single differential equation that can be easily studied using standard tools from chaos theory [26] is not obvious. In this paper, we adopt a multi-faceted approach to analyzing TCP AIMD dynamics over the Internet:

(i) Using a combination of simulation and analytical modeling, we provide an explanation for its chaotic dynamics albeit for a simple scenario of single TCP stream interacting with UDP flows.
(ii) We collect message delay measurements as well as traces of TCP variables over Internet connections to illustrate the complicated dynamics, and analyze the latter using time-dependent exponent curves to show the presence of both chaotic and stochastic components.

Our analytical model is similar in spirit to that of Sparrow [39] which is composed of two unstable linear systems "glued" together; this system exhibits chaos under certain conditions. We characterize the state space of TCP using the congestion window size, end-to-end packet delay, the number of re-transmissions and acknowledgments. We model TCP by a suitable composition of two unstable regimes each of which generates bounded dynamics with a very complicated attractor. In regime one, the congestion window-size constantly grows in response to successful packet delivery, while the packet delay is relatively stable. This regime is followed by the second regime, wherein the packet delay becomes unstable due to inferred losses, in response to which the window-size is drastically reduced. TCP trajectories move back and forth between these two regimes and are cyclic in the absence of delays, buffer size limits at the routers and hosts, and packet losses. We show that the TCP dynamics of window-size updates embed a map which is qualitatively similar to the well-known tent map [2] that generates chaotic trajectories. If the network delays are negligible, the second regime is extremely short-lived resulting in the familiar saw-tooth behavior of TCP for large flows. The existence of network delays prolongs the duration of regime two, which when coupled with the background traffic and small buffers generates very complicated dynamics under certain conditions.

We adopt the following informal working definition of chaos (more formal treatments can be found for example in [44]): (a) time trajectories include non-periodic orbits; (b) trajectories are very sensitive to input conditions, namely, trajectories starting at nearby points move significantly farther apart in time, and (c) the attractor set is very complicated.

We present three types of experimental results to complement our analysis.

(a) Using ns-2 simulations we explicitly illustrate the chaotic behavior by computing the time series of window sizes and packet delays together with their Fourier spectra and the attractor sets in the Poincare plane. We consider long message streams where the source has unlimited amount of data to send, and the simulation results show chaos-like behavior.
(b) We present "indirect" Internet measurements that indicate chaos-like dynamic behavior exhibited by TCP both using wireless and wireline connections. Here we send a fixed size message at regular intervals and measure the end-to-end delay for each message. These measurements indicate very complicated end-to-end delay variations for a stream of evenly spaced messages of fixed size.
(c) We employ the time-dependent exponent curves and logarithmic displacement curves to study TCP AIMD congestion window-size traces collected over Internet connections. We show that these dynamics have two dominant parts, a stochastic component in response to network traffic and a deterministic chaotic component due to the non-linearity of protocol. These dynamics can be largely characterized as anomalous diffusions with a large exponent.

In Section 2, we describe the state-space of TCP, and its dynamics in dealing with a bottleneck link and a router with small buffer. In Section 3, we describe results based on ns-2 simulations, and describe indirect measurements collected over Internet in Section 4. In Section 5, we describe the traces and their analysis of TCP variables for Internet connections. Earlier versions of partial results from Sections 2-4 and 5 appear in [30] and [8], respectively.

2 Dynamics of TCP

In this section we first describe a simplified model of TCP, followed by a description of its state space variables. Then we show the decomposition of TCP dynamics into two distinct unstable regimes, which gives rise to chaotic dynamics.

2.1 Simplified TCP Model

TCP provides a connection-based reliable transport mechanism from a source to a destination [27]. The sender interacts with the destination to ensure a reliable delivery of packets; each received packet is specifically acknowledged by

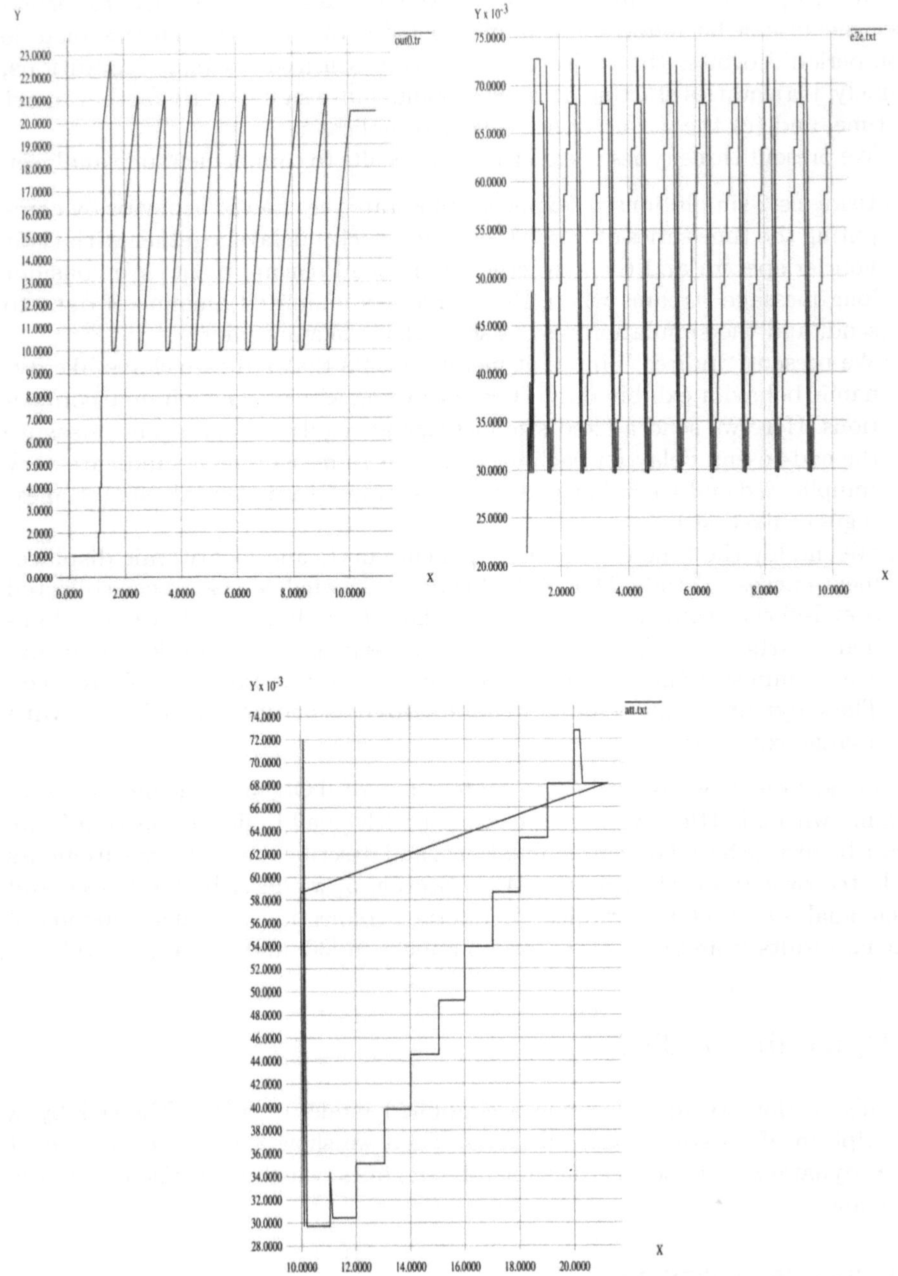

Fig. 1: Top left and right plots show $w(t)$ and $e(t)$ vs t, respectively, and bottom plot shows $w(t) - e(t)$ attractor.

the destination. At any given time, there are no more than a certain number of unacknowledged packets, given by the window size, at the source. These include the packets that are in flight, dropped at a router or destination, lost on communications links, or whose acknowledgments have been delayed or lost. The individual packets and their acknowledgments can potentially travel different paths and delayed by different amounts (indefinitely if they have been dropped). The packets are put into the proper order at the destination. The receiver maintains a certain window, called the receive window or buffer, of packets which are not in the correct order. Packets arriving at the destination when the receive buffer is full are simply dropped. Source maintains the flow window which corresponds to the receive buffer. The packet flow rate is restricted by the flow window which is typically fixed at the time of initiation of the connection.

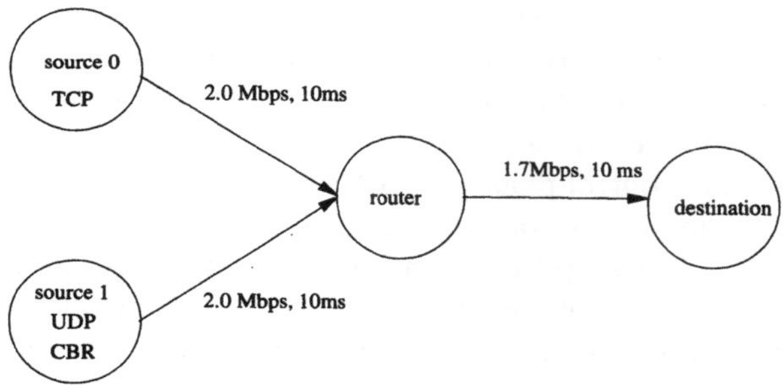

Fig. 2: Simulation setup.

The source attempts to adjust its sending rate in response to the traffic condition on the connection, as per a process called the congestion control. There are a large number of variations of the basic TCP [7] in the manner the congestion control is implemented. We restrict our focus to a simplified model of a basic AIMD version that was originally described in [18]. TCP maintains the congestion window $w(t)$ which limits the number of unacknowledged packets at the source (in addition to the flow window constraint). Since flow window is fixed in the basic version, the dynamics are due to the congestion window; note however, if methods such as dynamic right-sizing combined with parallel streams [31] are applied, the dynamics are not solely dependent on congestion window.

TCP responds to the network traffic by adjusting $w(t)$ which in turn controls its flow rate. TCP dynamics consist of two phases as shown in the top left plot of Figure 1 corresponding to the setup in Figure 2: (a) initial slow-start phase, and (b) the subsequent congestion control phase. During the slow start phase the window size grows fast until a loss is inferred or a set threshold W_t is reached. In congestion control phase, $w(t)$ is incremented with each acknowledgment until a loss occurs, and then it is reduced by half. The value of $w(t)$ has a significant effect on the packet end-to-end delay (defined formally in the next section). If $w(t)$ is small most messages will reach the destination and hence packet delay is fairly stable. At the other extreme, if $w(t)$ is large, losses and retransmissions occur which increase the end-to-end packet delay.

Consider that the delays are extremely small such that source knows immediately The congestion window size $w(t)$ is recomputed based on the response to the packets sent. After every acknowledgment of a new packet, $w(t)$ is recomputed follows:

if $w < W_t$, then $w \leftarrow w + 1$ during slow start
else $w \leftarrow w + 1/w$ during congestion control

After a loss is inferred at the source, the window size is reduced by half such that $w \leftarrow w/2$. The above method is referred to as the AIMD method for congestion control. We make the assumption that the loss is inferred at the source almost instantaneously and the acknowledgments are not lost. This is a simplified description of the congestion control but captures the essential components of TCP needed here.

In the congestion control phase, there are two distinct regimes denoted by $\mathcal{R}_1$ and $\mathcal{R}_2$ as typified in the top left plot of Figure 1. In $\mathcal{R}_1$ TCP starts with a low value of $w(t)$ and keep incrementing as along as the packets are being acknowledged. When packet loss is inferred due to explicit notification or time-out, regime $\mathcal{R}_2$ is entered, wherein w is drastically reduced, particularly so in case of multiple losses.

2.2 State-Space Characterization of TCP

In addition to $w(t)$, we characterize the TCP dynamics using three other variables. Let the time instant t_1, called the *epoch*, correspond to the time a packet transmission started by TCP for first time for this packet. For epoch t_1, let $e(t_1)$ denote the *end-to-end delay* of the packet under the assumption that the acknowledgments are not lost. We extrapolate $e(t)$ for t in between two consecutive epochs t_1 ans t_2 by $e(t) = e(t_1)$. Let $r(t)$ and $a(t)$ denote the number of retransmissions and acknowledgments, respectively since the start; note that they both are non-decreasing functions of t. Usually, the sending rate of TCP is specified by $w(t)$ per round trip time; for simplicity of presentation we assume that $w(t)$ has been scaled for unit time. Thus the number of packets sent from the source during time interval $[T_1, T_2]$ is given by $\int_{T_1}^{T_2} w(t)dt$.

We consider that TCP is responding to a simple network scenario as in Figure 3 with a single bottleneck link with maximum packet delivery rate of w_b. The packet transmission through this link is controlled by a drop-tail router with buffer size B_τ. There is an underlying traffic that arrives at the router which competes for the link bandwidth and buffer space. Let $B(t)$ denote the number of elements in the buffer at time t with $B_\tau - B(t)$ denoting

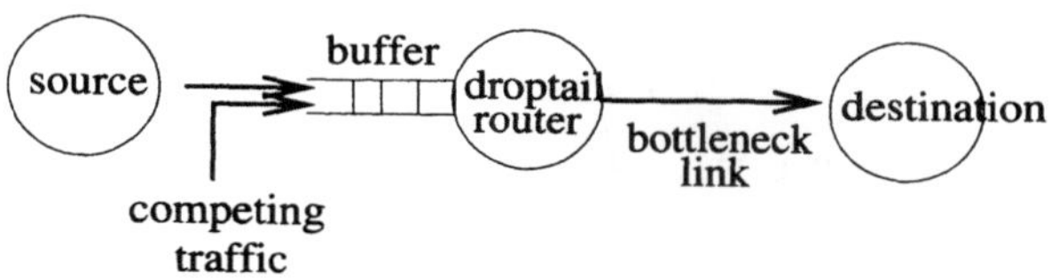

Fig. 3: TCP interacting with a drop-tail router.

the free buffer space. We assume that there is a processing delay of a packet at the router that increases slowly at a rate μ with the sending rate at the source.

Consider that there is no competing traffic at the router and source rate is fixed (by some non-TCP mechanism) such that $w(t) = w_f \leq w_b$. Then no packets will be dropped and dynamics of $w(t)$ are constant at w_b. In general w_b is not known and varies with the competing traffic. Roughly speaking, TCP attempts to estimate w_b, and to keep $w(t)$ lower than the estimated value to avoid packet losses and retransmissions.

Consider that the delays are extremely small such that source knows immediately after a packet is dropped and also that $B_\tau = 0$. In such case, dynamics of $w(t)$ for TCP are periodic: in regime $\mathcal{R}_1$, $w(t)$ starts around $w_b/2$ and increases with the acknowledgments until it exceeds w_b; then it infers a single packet loss and enters regime $\mathcal{R}_2$, where $w(t)$ is reduced by half. Here regime $\mathcal{R}_2$ is very short lived since it is the result of a single instantaneous loss. Now consider that the delays are non-zero and it takes T_1 units of time before the loss is inferred since the source rate exceeds the available bandwidth. Consider that $w(t_1) = w_b$ such that at time t_1 the sending rates becomes equal to the bottleneck bandwidth. TCP sends

$$n_{[t_1,t_1+T_1]} = \int_{t_1}^{t_1+T_1} w(t)dt$$

packets in the interim period $[t_1, t_1 + T_1]$ of which

$$n_d = \int_{t_1}^{t_1+T_1} (w(t) - w_b)dt = [n_{[t_1,t_1+T_1]} - w_b T_1]_+$$

will be dropped, where $[x]_+$ is x if $x > 0$ and is 0 otherwise. If buffer size is non-zero, then no packets will be dropped until $t = t_1 + t_{B_\tau}$ such that

$$B_\tau = \int_{t_1}^{t_1+t_{B_\tau}} [w(t) - w_b]_+ dt$$

if the entire buffer B_τ is available. Notice that the availability of buffer delays the time the packet is dropped by t_{B_τ} in this case. Thus the number of packets dropped will be

$$n_{B_\tau} = \int_{t_1+t_{B_{tau}}}^{t_1+T_1} [w(t) - w_b]_+ dt \tag{2.1}$$

for $t_{B_\tau} < T_1$. If the entire buffer is not available, the computation of the number of dropped packets is more involved. Let t_r denote the time the packet that left the source at time t_1 reaches the buffer. Then let $t_b \geq t_1$ be the earliest time such that the packet sent from source at t_b arrives at full buffer at time t_f such that $B(t_f) = B_\tau$, i.e. earliest time that the packet leaving the source faces a full buffer. Let $w_S(t)$ denote the rate measured at the source with which the packets will eventually pass through the bottleneck link. Note that if there is no competing traffic we have the simple relation

$$w_S(t) = \begin{cases} w(t) & \text{if } w(t) \leq w_b \\ w_b & \text{if } w(t) > w_b \end{cases}$$

Then the number of dropped packets is given by

$$n_b = \int_{t_b}^{t_b+T_1} (w(t) - w_S(t))dt.$$

Again note that the available buffer space delays the packet loss by time $t_b < T_{B_\tau}$, and since $w_S(t) \leq w_b$ we have $n_b \geq n_{B_\tau}$. Then after all the n_b dropped packets are accounted for, the resultant window-size is set to $w(t_1)/2^{n_b}$ for the regime $\mathcal{R}_2$.

2.3 Dynamic Regimes of TCP

Consider that the source is sending an infinitely long message. In case of zero delays, TCP dynamics in congestion control mode constitute a stable cycle with period $T_{\mathcal{R}_1} + T_{\mathcal{R}_2}$, where $T_{\mathcal{R}_i}$ is the duration of regime $\mathcal{R}_i$. This stable cyclic behavior is well-known in TCP literature as shown in the top left plot of Figure 1 which corresponds to the ns-2 simulation with no background UDP traffic; despite its ubiquity in literature, we observed such nice periodic behavior only under very special conditions in simulation. The corresponding packet end-to-end delay is shown in the top right plot of Figure 1 where in $e(t)$ increases slowly during $T_{\mathcal{R}_1}$ but jumps in large steps briefly due to retransmissions; nevertheless, it is still cyclic. In the bottom plot of Figure 1, we show the periodic sampling of $e(t)$ and $w(t)$ corresponding to the Poincare map in $w(t) - e(t)$ plane. After an initial transient period, the trajectories of $w(t)$ and $e(t)$ settle into periodic motions, and the corresponding attractor in $w(t) - e(t)$ plane is represented by a closed curve.

We can conceptualize the dynamic behavior of TCP, in term of the alternating regimes $\mathcal{R}_1$ and $\mathcal{R}_2$, which are of varying duration in general. For t corresponding to $\mathcal{R}_1$, the dynamics can be approximated locally as follows

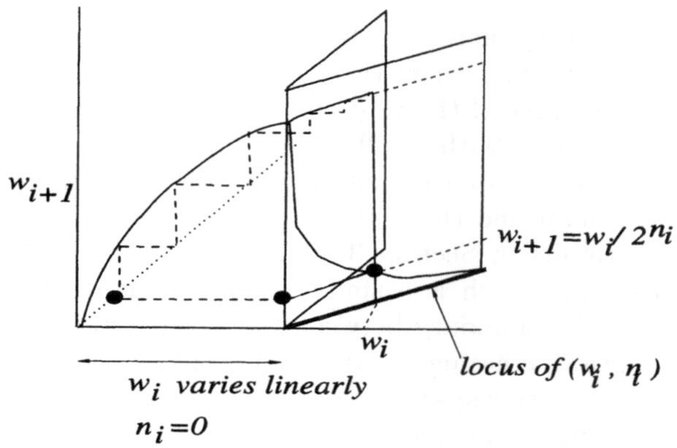

Fig. 4: Examples of map M for zero delays and zero buffer size for various B_τ.

$$\begin{bmatrix} \frac{dw}{dt} \\ \frac{de}{dt} \\ \frac{da}{dt} \end{bmatrix} = \begin{bmatrix} 1/w\frac{da}{dt} \\ \mu\frac{da}{dt} \\ \frac{da}{dt} \end{bmatrix}$$

Note that $\frac{dr}{dt}(t) = 0$ during this period, and the acknowledgments arrive at a non-zero rate. Computation of Eigenvalues of the Jacobian matrix shows that at least one of them will have positive real part; hence the system is unstable. One can intuitively draw such conclusion since the arrival of acknowledgments constantly increases $w(t)$ at a local rate of $1/w$ and also increases $e(t)$ at a slow rate of μ.

Now consider the regime $\mathcal{R}_2$ wherein the source infers a packet loss due to the buffer overflow. The packets sent during $(t_1, t_1 + T_2)$ will result in overflow where $w(t_1) = w_b$, for t_1 in $\mathcal{R}_1$; the exact number dropped packets depends on the occupancy of the buffer. The behavior can be approximated locally by

$$\begin{bmatrix} \frac{dw}{dt} \\ \frac{de}{dt} \\ \frac{dr}{dt} \end{bmatrix} = \begin{bmatrix} \frac{w}{2^{-\frac{dr}{dt}}} \\ \eta\frac{dr}{dt} \\ \frac{dr}{dt} \end{bmatrix}$$

for some η that depends on the delays. In this regions $\frac{da}{dt}(t) = 0$ and $\frac{dr}{dt} > 0$. Intuitively, this is an unstable regime since the packets are dropped which increases $e(t)$ and drastically reduces $w(t)$. This can also be verified by explicitly computing the Eigenvalues of the Jacobian matrix.

The TCP dynamics are due to the "gluing" together of the regimes $\mathcal{R}_1$ and $\mathcal{R}_2$. The above equations in region $\mathcal{R}_1$ can be derived by ignoring the control terms in the fluid models of [22, 16] which are known in the literature. The dynamics in region $\mathcal{R}_2$ are a direct result of using $e(t)$ as a state variable, which plays a very critical role in our analysis; we are unaware of earlier works using this variable and highlighting its role. Note that both regimes are unstable in that there is no stationary point in either. Thus the trajectories in one regime will enter the other and vice versa thereby generating bounded trajectories. The period behavior is one of the way such trajectories can manifest but the background traffic coupled with small buffer sizes result in more complicated trajectories. Such behavior appears not only in the time series of $w(t)$ and $e(t)$ but also in the attractor and the Fourier spectra of $e(t)$ as will be discussed in the simulation results in Section 3. Note that as the background traffic is varied the attractor, which is a simple curve in Figure 1 becomes much more complicated. The boundary between the two regimes is defined by the transition between the increasing and decreasing values for $w(t)$. We identify a particular subset of the state space where $w(t)$ reaches a value that just causes buffer overflow in regime $\mathcal{R}_1$. Under no delays and buffers this condition is met when $w(t) = w_b$, and in general is given by $n_d = 0$ in Eq (2.1), which is not easily visualized. As $w(t)$ is increased beyond this value in $\mathcal{R}_1$, TCP will infer loses and transit to regime $\mathcal{R}_2$. The resultant $w(t)$ value in $\mathcal{R}_2$ depends on n_d given by $w \leftarrow w/2^{n_d}$, which in turn depends on $B(t)$ during the appropriate period.

To understand the time evolution of $w(t)$ we define *w-update* map $M : [1, W_{\max}] \mapsto [1, W_{\max}]$ such that $M(w_i) = w_{i+1}$ gives the earliest changed value of $w(t)$ since the time the condition $w(t) = w_i$ is met. Let $w_i \in \mathcal{R}_j$ for $j = 1, 2$ denote that when $w(t) = w_i$, TCP is in regime $\mathcal{R}_j$. Then this map is specified as follows:

$$M(w_i) = \begin{cases} w_i + 1/w_i & \text{if } w_i \in \mathcal{R}_1, w_{i+1} \in \mathcal{R}_1 \\ w_i/2^{n_i} & \text{if } w_i \in \mathcal{R}_1, w_{i+1} \in \mathcal{R}_2 \\ w_i/2^{n_i} & \text{if } w_i \in \mathcal{R}_2, w_{i+1} \in \mathcal{R}_2 \\ w_i + 1/w_i & \text{if } w_i \in \mathcal{R}_2, w_{i+1} \in \mathcal{R}_1 \end{cases}$$

where n_i is the number of packet losses inferred during the period. In $\mathcal{R}_1$ we have $n_i = 0$ hence it can be represented as a simple map. In $\mathcal{R}_2$ it is more complicated and can be visualized as in Figure 4. For $w(t) > w_b$, there could be packet losses and but $w(t)$ attains a value higher than w_b. This process can be imagined in terms of a set L along which n_i increases together with $w(t)$. At any point on this set L, corresponding to (w_i, n_i) the map M is specified by $M(w_i) = w_i/2^{n_i}$. The plot of M can be imagined along the set $L \times [1, N_{\max}]$ so that L forms the x-axis and $M(w_{i+1})$ is along the y-axis. Empirical computation of M is shown in Figure 5 as a function of w_i alone based on the simulation results in Section 3; note however that this map reflects only the limited number of simulated trajectories.

Periodic trajectories can be easily generated as illustrated in Figure 4 in which the trajectory stays in $n_i = 0$ plane for part of the time and on the set L for the rest of the time in a cycle. This map shares some of the basic properties of the tent map [2] in that it has a monotonically increasing part in $n_i = 0$ plane and a decreasing part in the half plane $n_i > 0$. The map M is more complicated due to the presence of n_i which makes it two-dimensional in the half plane $n_i > 0$. It is instructive to visualize the map M in terms of a

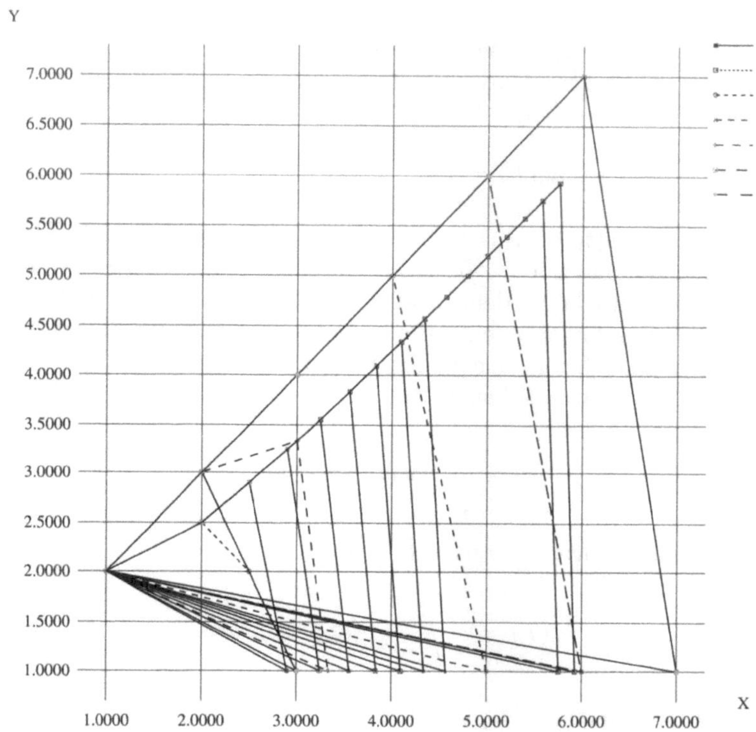

Fig. 5: Empirical map M plotted along w_i-axis.

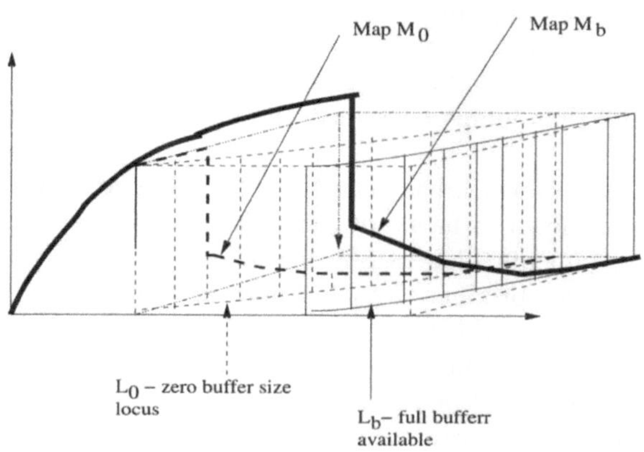

Fig. 6: Illustration of map M in terms of family of two-dimensional maps $M_b(t)$, $0 \leq b \leq B$.

family of maps such that $M_b : [0, W_{\max}] \times [0, N_{\max}] \to [0, W_{\max}] \times [0, N_{\max}]$, is the two-dimensional map obtained in case of no background traffic and buffer size b, $0 \leq b \leq B$, i.e. entire buffer is available for this TCP stream as shown in Figure 6. Due to non-zero delay, $w(t)$ exceeds w_b at time t_1 for a time period T which will result in the over flow of

$$n_0(T) = \int_{t_1}^{t_1+T} (w(t) - w_b)dt$$

packets. Then $M_0(w(t_1 + T)) = w_i/2^{n_0}$ whose value depends on T. This map can be imagined along a locus L_0 of all points given by $\{(w(t_1 + t), n_0(t)) : t \in [0, T]\}$. The map M_0 is defined at any point on L_0 as $M_0(w(t_1 + t)) = w(t_1+t)/2^{n_0(t)}$. We similarly define the locus L_b when the buffer size is b which is entirely available. Let t_b denote the period during which $w(t_1)$ increases before first packet is dropped such that

$$b = \int_{t_1}^{t_1+t_b} (w(t) - w_d)dt.$$

Then the number of overflown packets is given by

$$n_b(T) = \int_{t_1+t_b}^{t_1+t_b+T} (w(t) - w_b)_+ dt.$$

Then the M_b is defined as $M_b(w(t_1 + t_b + T)) = 2^{n_b(T)}$ whose value depends on T. Then we define the original map M by using M_b with the appropriately available buffer size b. The Poincare iterates $\{w_i\}$ under M depend on the available buffer space when $w(t)$ is in the regions $\mathcal{R}_2$ and is given by $M_{B_\tau - B(t)}$. Any trajectory of $w(t)$ stays in in the plane $n_i = 0$ while incurring loses and transits to a locus that lies in between L_0 and L_{B_τ} for a duration determined by $B(t)$ and the delay to the buffer. Then it reduces $w(t)$ in response to the resultant loses and thus jumps back onto the plane $n_i = 0$, and this process repeats.

Unlike the tent map which has period 3 orbits, M starts at a much higher periodicity, can be shown to have any periodicity as well as aperiodic orbits using standard arguments such as Sharkovskii's Theorem [2]. The detailed study of the dynamics produced by this map are under investigation, but its general nature indicates that it generates behavior qualitatively similar to that produced by the tent map. Maps that are qualitatively similar to M have also been presented in [39] to show the chaotic behavior. The tent-map has a constant derivative of magnitude 2, and the map M has a slope of approximately $1/w$ in $\mathcal{R}_1$ but of much higher magnitude in region $\mathcal{R}_2$. Informally speaking, the larger slope in $\mathcal{R}_2$ is responsible for the exponential separation of the nearby trajectories and results in a positive "aggregate" Lyapunov exponent under the conditions during which the trajectories stay in $\mathcal{R}_2$ for a significant portion of the time.

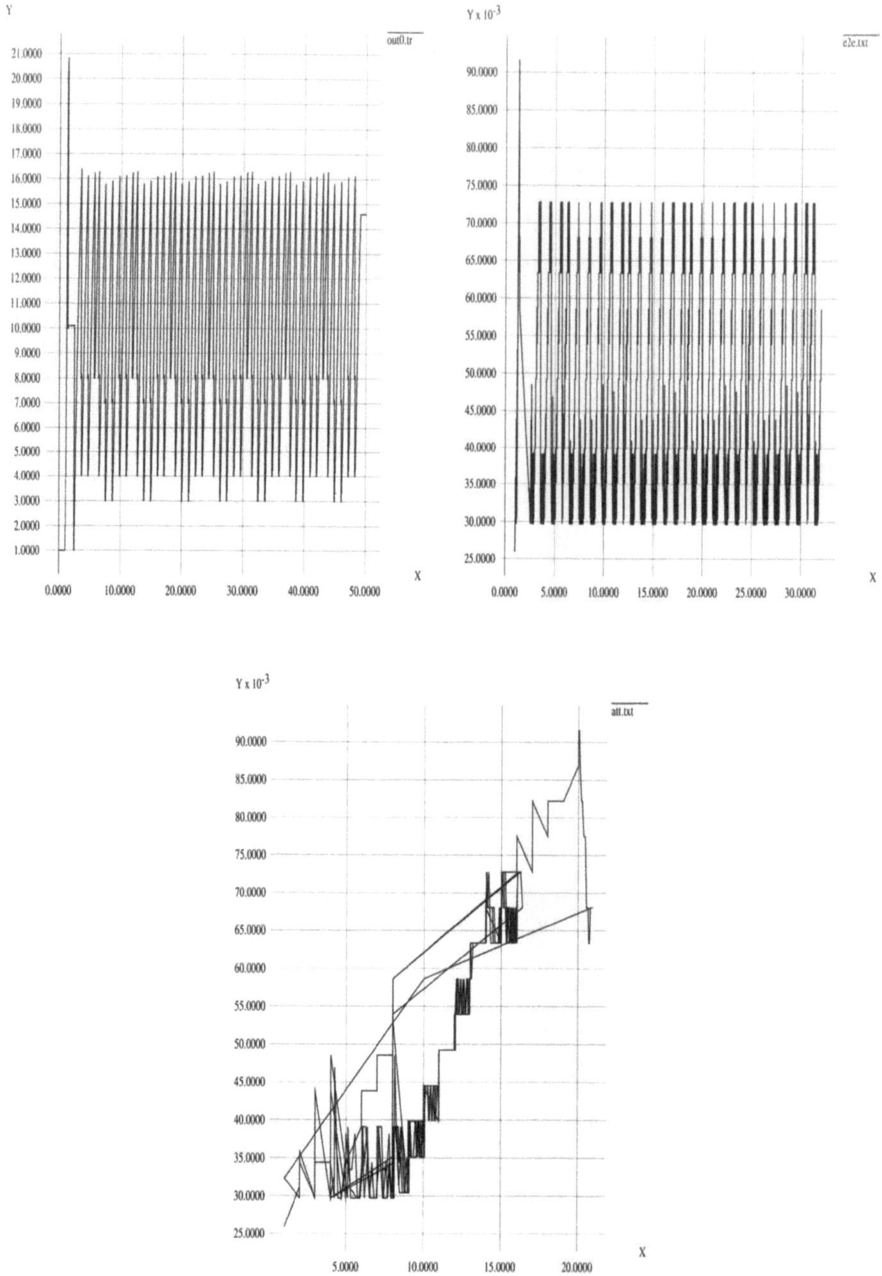

Fig. 7: Long TCP stream with UDP rate of 0.5Mbps. Top left plot is $w(t)$, top right plot is $e(t)$ and bottom plot is $w(t) - e(t)$ attractor.

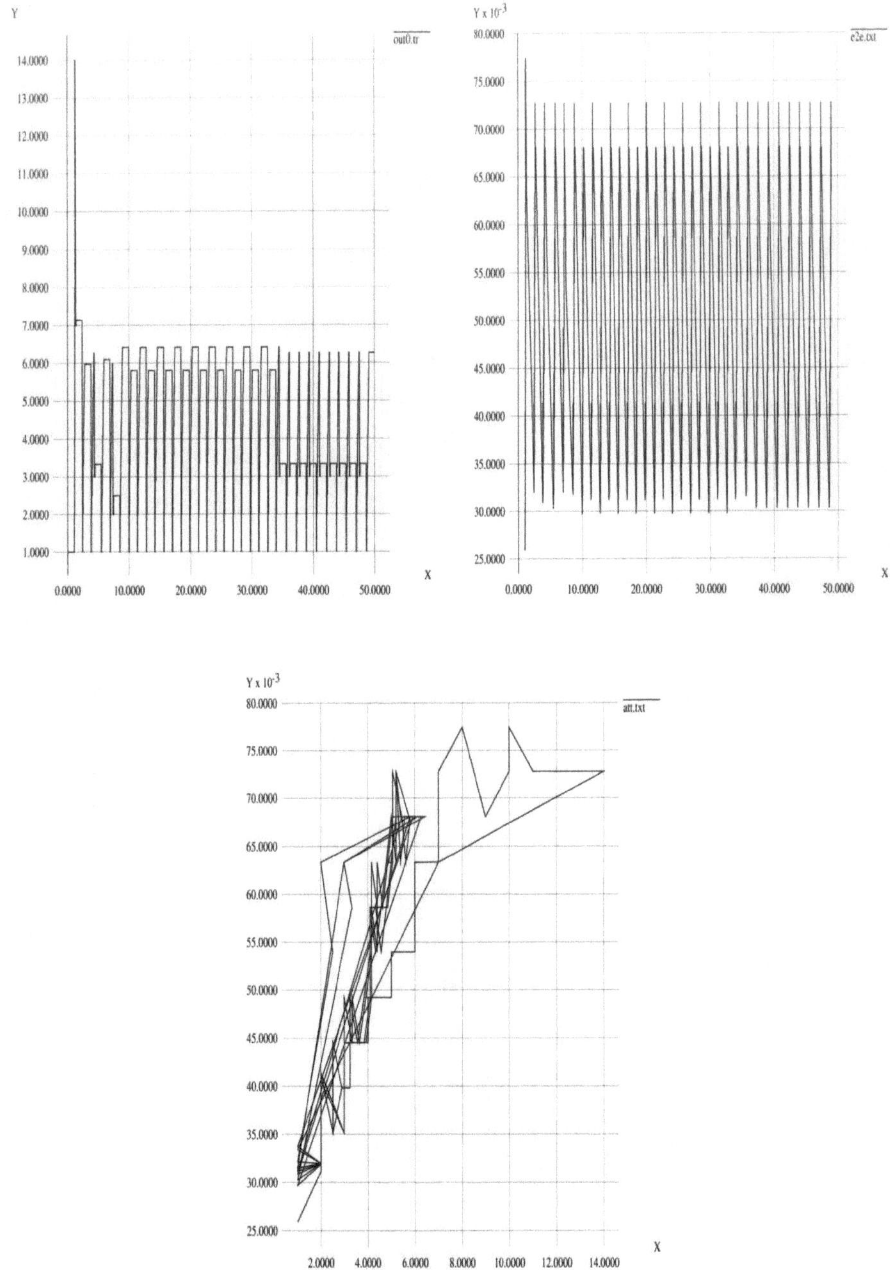

Fig. 8: Long TCP stream with UDP rate of 1.5Mbps. Top left plot is $w(t)$, top right plot is $e(t)$ and bottom plot is $w(t) - e(t)$ attractor.

3 Simulations of TCP competing with UDP

In this section, we describe simulation measurements to illustrate the complicated dynamics of TCP AIMD method. We plot various TCP parameters under controlled scenarios that are not realizable under Internet environments. Figure 2 shows the network setup used in ns-2 simulations to perform our experiments. Source nodes 0 and 1 generate TCP and UDP traffic, respectively, that goes to a drop-tail router over 2.0 Mbps links and then to a bottleneck link of 1.7 Mbps before reaching the destination. Source node 1 generates UDP traffic at a constant bit rate (CBR) and serves as the background traffic that the TCP connection must adapt to. To simulate different levels of congestion, the traffic-generation rate is varied from 0 to 1.5 Mbps in steps of 0.5 Mbps. This set-up provides an easy way to increase the traffic on the bottleneck link so that different TCP behaviors can be observed.

The source node is attached to an FTP client that acts as an infinite source of TCP traffic. The results with no UDP traffic are shown in Figure 1; as discussed before both $w(t)$ and $e(t)$ are periodic with the attractor described by a simple curve. For other UDP rates, both $w(t)$ and $e(t)$ did not show repeated patterns during the observation period, and the attractor became significantly more complicated. When the UDP rate is 0.5Mbps, the dynamics are more complicated as shown in Figure 7. While the trajectories remain bounded in the $w(t)-e(t)$ plane, the shapes of the attractors at various UDP rates appeared quite varied and did not follow any particular evolutionary pattern. At all UDP loads both $w(t)$ and $e(t)$ had significant periodic components but were not actually period – the digression from the periodic nature increased as the UDP rate is increased. We only showed some typical plots in Figures 7 and 8 but the general nature of the $w(t)$ and $e(t)$ remained more or less the same. Since the observation time is limited, it is quite possible that these plots are small parts of periodic trajectories. But the radical difference in the attractor shape together with the map M discussed in the last section provides a strong indication that TCP can generates aperiodic orbits for $w(t)$ and $e(t)$.

The effects of TCP dynamics are more pronounced in transfers of small messages such as control signals. To study such effects, we consider a simple on-off ftp source simulating the square-wave function for the message transfers. A typical case is shown in Figure 9 where the duration of each ftp session is 0.15 sec with 0.15sec duration between the consecutive sessions (more cases can be found in [30]). The UDP CBR rate is 1.6Mbps. The top left plot shows $w(t)$ which is aperiodic during the observation interval. The top right plot shows $e(t)$ which is also aperiodic but shows much more variation compared to the similar plots in case of long streams; the bottom left plot shows its magnitude plot of the Fourier coefficients, which shows significant non-periodic parts. The bottom right plot shows the attractor which is also more complicated than in the long stream case.

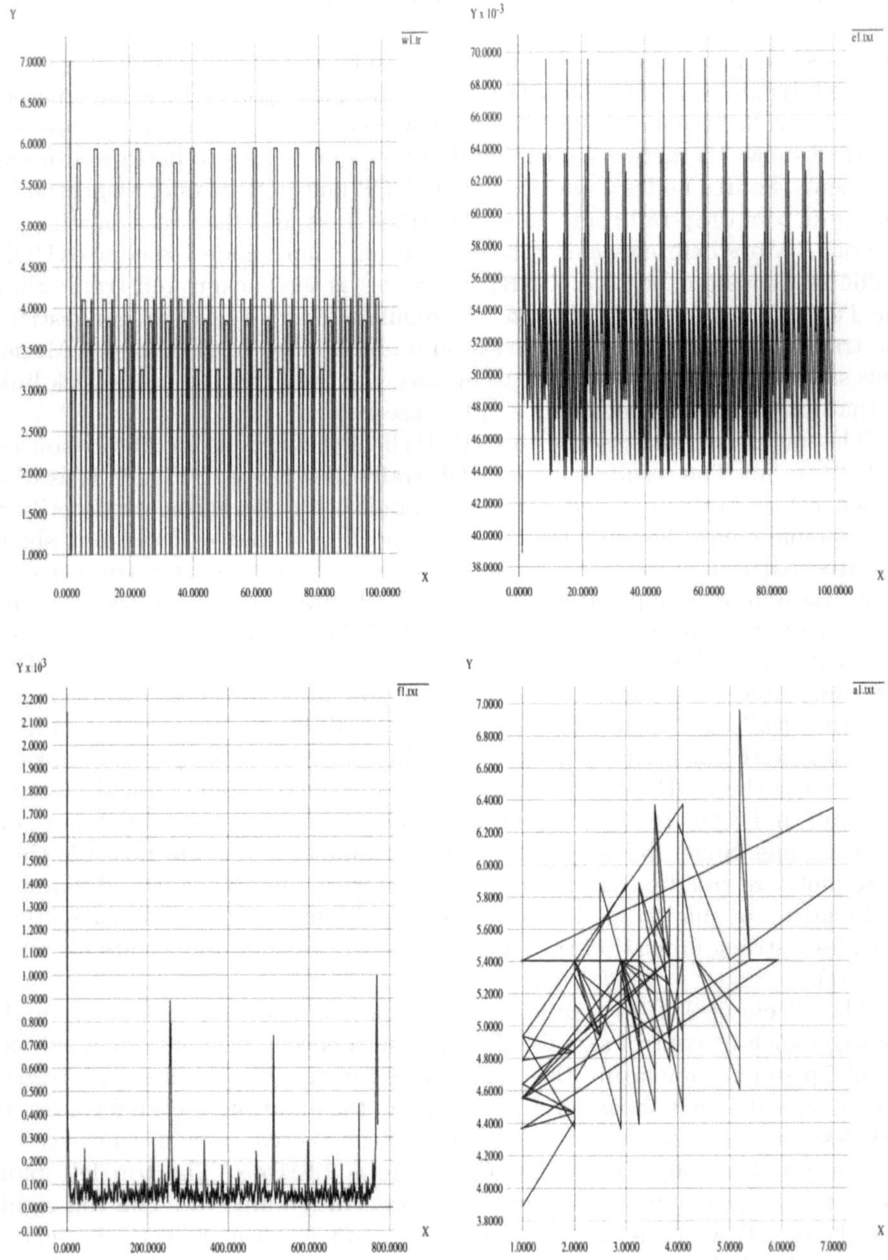

Fig. 9: Bursty TCP stream with 0.15 sec square wave with UDP rate of 1.6 Mbps. Top left plot is $w(t)$, top right plot is $e(t)$, bottom left plot is Fourier spectrum of $e(t)$, and bottom right plot is $w(t) - e(t)$ attractor.

4 Message Delay Measurements

In this section, we describe Internet measurements corresponding to the end-to-end delays of fixed size messages. While these measurements do not directly correspond to TCP variables (unlike in next section) they do highlight the complicated dynamics of message delays that are of importance to the applications.

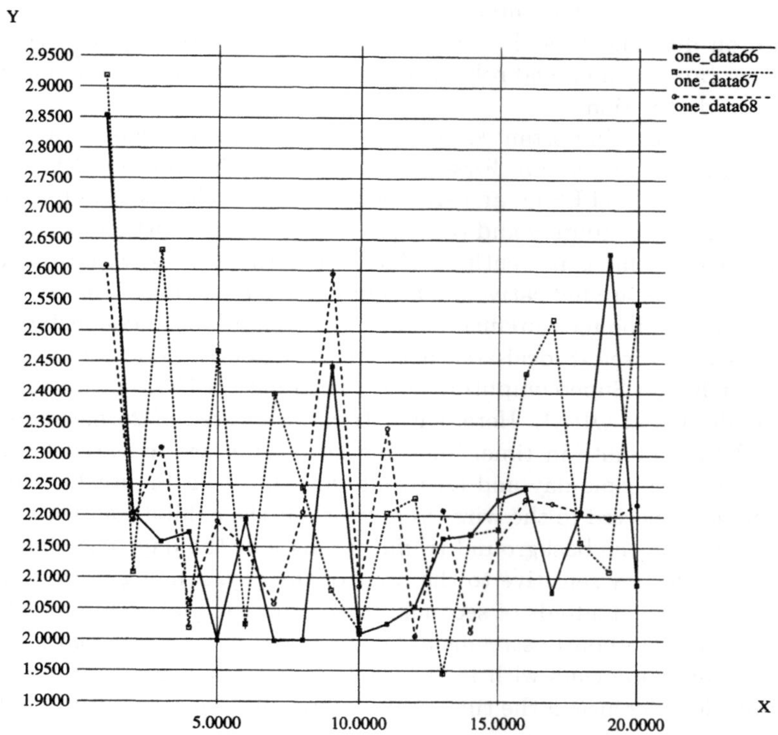

Fig. 10: Wireline network measurements.

We collected Internet measurements between Oak Ridge National Laboratory (ORNL) and University of Oklahoma (OU) by sending streams each consisting of messages of fixed size 10K bytes at the regular intervals of 10 seconds. Each message is received at the destination and sent back to the source. The round-trip time is computed at the source and divided by two to estimate one way end-to-end delay for the message; this is only an estimate since the forward and return paths of a TCP stream could be different, and even in each way different packets may travel different physical paths. Due to lack of access to the TCP internal variables, we only consider the end-to-end delays of the messages rather than the packet-level delays discussed in the last and next sections. But these two quantities are closely related: the end-to-end delay for the message is the difference between the time last packet of the message is received at the destination and the time first packet left the source. Thus the dynamics of the one will be reflected in the other. Figure 10 corresponds to three streams each with 20 messages. The individual streams are separated by an hour. Notice that there is a considerable variation in the end-to-end delay even for the messages send only a second apart. Such variation is present in measurements collected within a day as well as those collected on different days. These dynamic variations are very important because they can play a crucial role in the stability of control loops between the source and destination.

There are some important differences in these measurements and simulation results of last section. First, the Internet paths from ORNL and OU consists of the parts of ESnet and Abiline networks, each of which in turn consists of a number of routers and perhaps more than one bottleneck link. Also the competing traffic is not entirely UDP. Nevertheless, the variations in the end-to-end delays are indicative of the qualitative behavior of the dynamics.

We now describe measurements collected from a laptop connected to the Internet over a wireless local-area network. The source node was located at the exhibit hall of Supercomputing 2001 conference in Denver and the destination node was at ORNL. Here, each stream consisted of 25 messages each separated by a minute in time. Such measurements are repeated hourly 13 times during the same day and all results are shown in Figure 11. Here the source node connected to the wireless network at the exhibit hall which had a large variation in the background traffic throughout the day. The time scales of variation are very sensitive to the hour of the day, ranging from hundreds of milliseconds to hundreds of seconds. In the top left plot of Figure 11 measurements of all 13 hourly streams are shown, and in the other plots some of the hourly measurements with highest delays are removed. The three orders of magnitude difference in the time-scales of various plots is mainly due to the competing traffic, which was low in the morning and at night but was much higher during the day. It is very interesting to note the end-to-end delays are very dynamic at every scale. Thus for control application in such scenarios, both the time-scales as well as the dynamic variations in the end-to-end message delays are very important. Simply knowing the traffic levels (and hence the hourly time scales) is not sufficient to predict the end-to-end delays, since they showed quite a variability at every time scale.

5 Analysis of Internet TCP Traces

Typical TCP dynamics over the Internet connections are a result of its nonlinear dynamics interacting with the Internet traffic, which is stochastic and

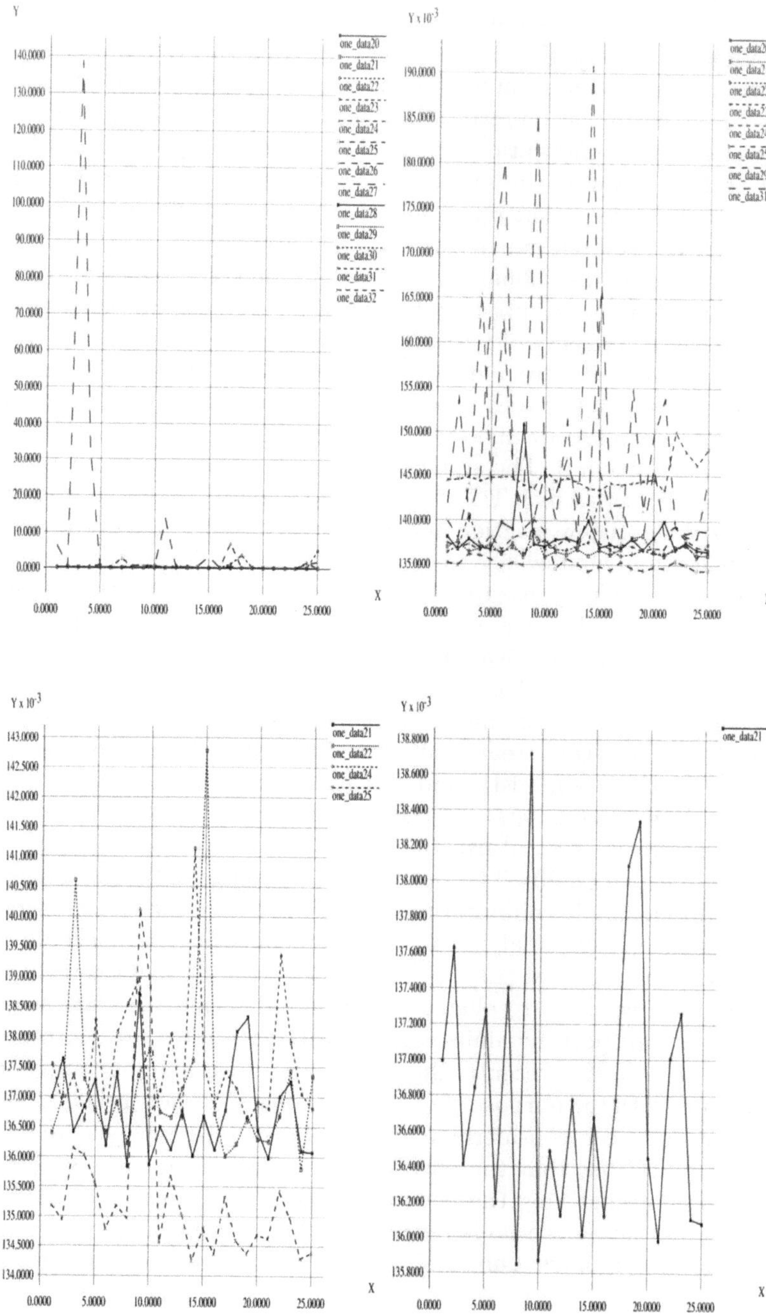

Fig. 11: Wireless network measurements.

often self-similar. This leads one to naturally expect the transport dynamics to be inevitably complicated as indicated by a number of preliminary studies [5, 43]. While these works demonstrate that the transport dynamics can indeed be chaotic under certain circumstances, their scope has been often limited since they rely on simulation results [43], secondary measurements of previous section or network configurations atypical of Internet [28]. In terms of actual Internet traffic, however, the question of the chaotic and stochastic nature of transport dynamics is left open: it is unclear if such dynamics are indeed commonplace or even relevant to operational networks.

There are two major difficulties in understanding the transport dynamics over Internet connections. First, it has been technologically hard to collect high quality measurements of network transport variables on "live" Internet connections. Second, it is very difficult to analytically handle the interaction between the deterministic and stochastic parts of transport dynamics. The deterministic dynamics are due to TCP AIMD congestion control method as described in Section 2. The stochastic component is due to the often self-similar traffic [25] with which TCP competes for bandwidth and router buffers. Roughly speaking, the interaction between the two is in terms of the additions to TCP congestion window-size in response to acknowledgments, and multiplicative decreases in response to inferred losses (ignoring the initial slow-start phase). Any protocol, however simple its dynamics are, is generally expected to exhibit apparently complicated dynamics due to its interaction with the stochastic network traffic. Recall that TCP exhibits chaotic or chaos-like behavior even when the competing traffic is much simpler such as a single competing TCP [43] or UDP stream [30]. The difficulty is to understand the dynamics when both deterministic and stochastic phenomena are in effect, and equally importantly to understand their impact on actual Internet streams.

In this section we first utilize the recently developed net100 [23] instruments to collect high quality TCP $w(t)$ traces for actual Internet connections. We then utilize a simple computational model to study the effects of random packet losses on $w(t)$. We analyze the measurements using the concepts of time dependent exponent curves [13, 15, 14] and logarithmic displacement curves [9]. Our major purpose is to elucidate how the deterministic and stochastic components of transport dynamics interact with each other on the Internet. It is of interest to note that recently there have been several important works on TCP dynamics with the purpose of improving congestion control [6, 21, 17, 19]. In fact, there has been considerable effort in developing new versions and/or alternatives to TCP so that the network dynamics can be more stable. Conventional ways of analyzing the network dynamics are unable to readily determine whether newer methods result in stable transport dynamics or help in designing such methods. Our analysis shows two important features in this direction:

(a) Randomness is an integral part of the transport dynamics and must be explicitly handled. In particular, the step sizes utilized for adjusting the congestion parameters must be suitably varied (e.g. using Robbins-Monro conditions) to ensure the eventual convergence [32]. This is not the case for TCP which utilizes fixed step sizes.

(b) The chaotic dynamics of the transport protocols do have a significant impact on practical transfers, and the protocol design must be cognizant of its effects. In particular, it might be worthwhile to investigate protocols that do not contain dominant chaotic regimes, particularly for remote control and steering applications.

The traditional transport protocols are not designed to explicitly address the above two issues, but justifiably so since their original purpose is data transport rather than finer control of dynamics.

We have collected a number of $w(t)$ traces using single and two competing TCP streams. This data was collected on two different connections from ORNL to Georgia Institute of Technology (GaTech) and to Louisiana State University (LSU). The first connection has high-bandwidth (OC192 at 10Gbps) with relatively low backbone traffic and a round-trip time of about 10 milliseconds. Four traces were collected for this connection, two with a single TCP stream and the others with two competing TCP streams. The second connection has much lower bandwidth (10 Mbps) with higher levels of traffic and a round trip-time of about 26 milliseconds. The sampling time is approximately 1 millisecond, with an error of 10s of microseconds (due to a "busy waiting" measurement loop). Eight traces were collected for the second connection. The results based on these eight traces are qualitatively very similar to the ORNL-GaTech traces, and we only discuss the results for the latter. To appreciate better what these data look like, we have plotted in Figure 12(a-d) segments of these four datasets. Power spectral analysis of these data does not show any dominant peaks, and hence, the dynamics are not simply oscillatory. Since our data was measured on the Internet with "live" background traffic, it is apparently more complicated and realistic than ns-2 traces [43, 30].

5.1 Time-Dependent Exponent Curves

Next let us analyze data, $w(i)$, $i = 1, \cdots, n$, using the concepts of time dependent exponent $\Lambda(k)$ curves [13, 15, 14]. For this purpose, we first employ the embedding theorem [24, 41, 35, 34] and construct vectors of the form: $V_i = [w(i), w(i+L), ..., w(i+(m-1)L)]$, where m is the embedding dimension and L the delay time. The embedding theorem [41, 35] states that when the embedding dimension m is larger than twice the box counting dimension of the attractor, then the dynamics of the original system can be studied from a single scalar time series. We note that the embedding dimension used in [43] is only two, which has to be considered not large enough (this may call for a closer examination of their conclusions).

The $\Lambda(k)$ curves are defined by [13, 15, 14]:

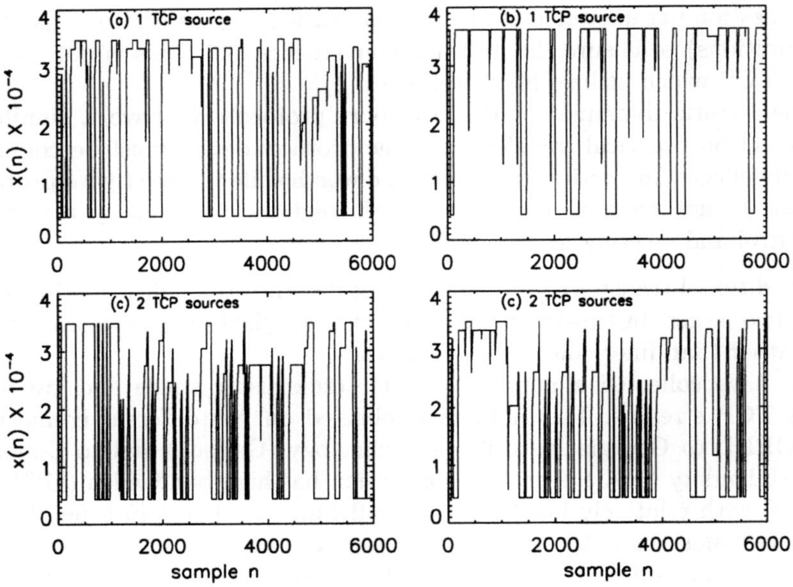

Fig. 12: Time series of $w(t)$ (in Bytes) for ORNL-GaTech connection. Before plotting on the figure, the actual value of $w(n)$ has been divided by 10^4.

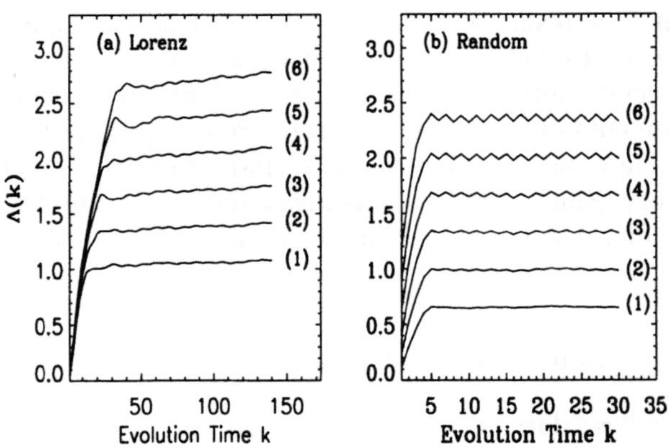

Fig. 13: $\Lambda(k)$ curves for (a) the chaotic Lorenz attractor with $m = 4$, $L = 3$ and sampling time 0.03, and (b) a random process with $m = 6$, $L = 1$. Curves numbered $1, \cdots, 6$ correspond to shells with sizes $(2^{-(i+1)/2}, 2^{-i/2})$, $i = 5, 6, \cdots, 10$.

$$\Lambda(k) = \left\langle \ln \left(\frac{\|V_{i+k} - V_{j+k}\|}{\|V_i - V_j\|} \right) \right\rangle \tag{1}$$

where $V_i, i = 1, 2, \cdots$ are vectors constructed from a scalar time series using the embedding theorem. The computation is carried out for a sequence of shells, $r \leq \|V_i - V_j\| \leq r + \Delta r$, where r and Δr are prescribed small distances. The angle brackets denote the ensemble average of all possible (V_i, V_j) pairs, and k is called the evolution time. For true low-dimensional chaotic systems, the curves $\Lambda(k)$ for different shells form a common envelope, and the slope of the envelope is an estimate of the largest positive Lyapunov exponent. An example is given in Figure 13(a) for the well-known chaotic Lorenz system. We note the common envelope at the lower left corner of Figure 13(a). The existence of a common envelope guarantees that a robust positive Lyapunov exponent will be obtained (by different methods) no matter which shell is used in the computation. For non-chaotic systems, the common envelope is absent. As an example, Figure 13(b) shows the $\Lambda(k)$ curves for a set of uniformly distributed random variables. Here, we note there is no common envelope in the lower left corner of Figure 13(b). At this point it is appropriate to comment that it is often assumed that a numerically estimated positive value for the Lyapunov exponent and non-integral value for the fractal dimension is a sufficient indicator of chaos in a time series. This assumption stirred up a passionate wave of research into many areas in natural and social sciences as well as engineering, resulting in a scene that *chaos is everywhere*! Since $\Lambda(k)/(k\delta t)$, where δt is the sampling time, is more or less equivalent to the largest positive Lyapunov exponent which can be obtained using conventional ways, we thus see that under this common assumption, random processes or noise can be easily interpreted as chaos. Hence, Gao and Zheng's method [13, 15, 14] represents one of the more stringent tests for deterministic chaos.

Before we move on, we point out a few interesting features of the $\Lambda(k)$ curves: (i) For noise, only for k up to the embedding window size $(m-1)L$ will $\Lambda(k)$ increase. This can be easily seen from Figure 13(b). Thus, whenever $\Lambda(k)$ increases for a much larger range of k, it is an indication of non-trivial deterministic structure in the data. (ii) For periodic signals, $\Lambda(k)$ is essentially zero for any k. (iii) For quasi-periodic signals, $\Lambda(k)$ is periodic with an amplitude typically smaller than 0.1, hence, for practical purposes, $\Lambda(k)$ can be considered very close to 0. (iv) When a chaotic signal is corrupted by noise, then the $\Lambda(k)$ curves break themselves away from the common envelope. The stronger the noise is, the more the $\Lambda(k)$ curves break away till the envelope is not defined at all. This feature has actually been used to estimate the amount of both measurement and dynamic noise [9].

Now we are ready to compute and understand the $\Lambda(k)$ curves for $w(t)$ traces. The set of $\Lambda(k)$ curves, corresponding to Figure 12, are plotted in Figure 14. In the computations, 3×10^4 points are used, and $m = 10$, $L = 1$. The eight curves, from the bottom to top, correspond to shells of sizes $(2^{-(i+1)/2}, 2^{-i/2})$, $i = 8, 9, \cdots, 15$. We make the following observations:

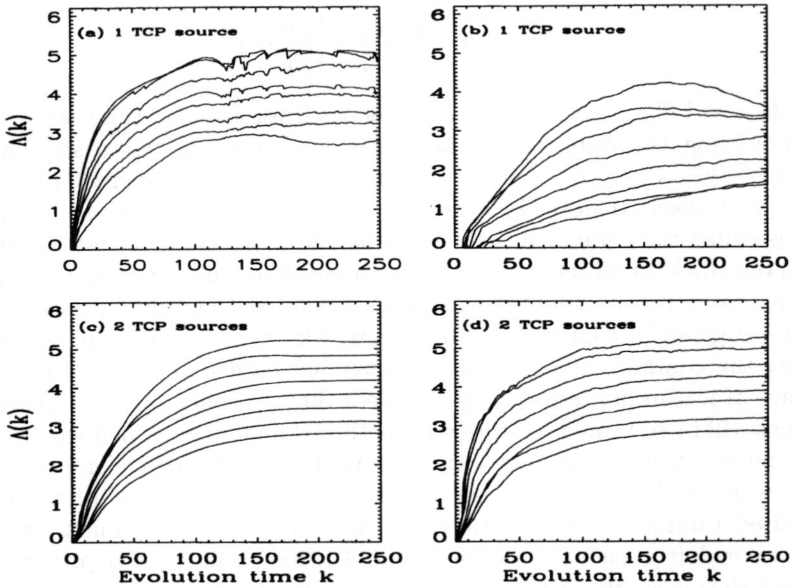

Fig. 14: $\Lambda(k)$ curves for $w(t)$ data corresponding to Figure 12. In the computations, 3×10^4 points are used, and $m = 10$, $L = 1$. Curves from the bottom to top correspond to shells with sizes $(2^{-(i+1)/2}, 2^{-i/2})$, $i = 8, 9, \cdots, 15$.

(i) The dynamics are complicated and cannot be described as either periodic or quasi-periodic motions, since $\Lambda(k)$ is much larger than 0.

(ii) The dynamics cannot be characterized as pure deterministic chaos, since in no case can we observe a well-defined linear envelope. Thus the random component of the dynamics due to competing network traffic is evident and can not simply be ignored.

(iii) The data is not simply noisy, since otherwise we should have observed that $\Lambda(k)$ is almost flat when $k > (m-1)L$. Thus, the deterministic component of dynamics which is due to the transport protocol plays an integral role and must be carefully studied. The features (ii) and (iii) indicate that the Internet transport dynamic contains both chaotic and stochastic components.

(iv) There are considerable differences between the data with only 1 TCP source and with 2 competing TCP sources. In the latter case, the $\Lambda(k)$ curves sharply rise when k just exceeds the embedding window size, $(m-1)L$. On the other hand, $\Lambda(k)$ for Figure 14(b) with only one TCP source increases much slower when k just exceeds $(m-1)L$. Also important is that the $\Lambda(k)$ curves in Figure 14(c,d) are much smoother than those in Figure 14(a,b). Hence, we can say that the deterministic component of the dynamics is more visible when there are more than 1 competing TCP sources (along the lines of [43]).

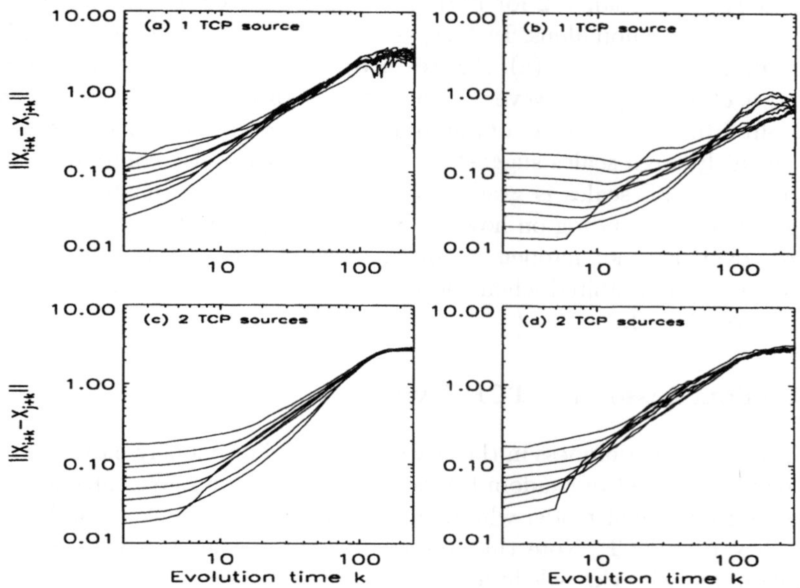

Fig. 15: A log-log plot of the displacement curves for Figure 12.

5.2 Anomalous Diffusions

Since the increasing part of the $\Lambda(k)$ curves are not very linear, let us next examine if in those regions of k, the displacement $||V_{i+k}-V_{j+k}||$ curves actually increase with k in a power-law manner,

$$||V_{i+k} - V_{j+k}|| \sim k^{\alpha},$$

where α is called the diffusional exponent. Note that $||V_{i+k} - V_{j+k}||$ is the numerator of Eq. (1). For Brownian motions, $\alpha = 1/2$ [12]. This leads to a classification of diffusional processes [10]:

(i) $\alpha = 1/2$: Normal Diffusion.
(ii) $\alpha > 1/2$: Anomalous Diffusion. This type of diffusion plays a key role in the study of noise-induced chaos. In fact, in that context, it may be termed "pre-noise-induced chaos".
(iii) $\alpha < 1/2$: Sub-Diffusion. When the diffusional process takes place near a limit cycle, this behavior is often due to a strong convergent flow to the limit cycle [11].

Now it is a simple matter to plot the evolution of $||V_{i+k} - V_{j+k}||$ with k in a log-log scale. See Figure 15(a-d). Very interestingly, we have now indeed observed a bunch of better defined linear lines. Furthermore, the diffusional exponent is larger than 1/2 for Figure 15(a,c,d). Hence, we can say that the diffusion is often anomalous. In fact, α is about 1.2 for both Figure 15(c,d) and about 1.0 for Figure 15(a). Figure 15(b) is an exception: for not too small shells, α is close to 1/2. However, when the scale or the shell size is very small, α is about 1.0. This crossover of anomalous diffusion to normal diffusion from small to fairly large scales suggests that the deterministic component of the dynamics at small scales is gradually masked by the stochastic component at large scales. Depending on how noisy the background traffic is, sometimes this crossover may not happen, as in the case of Figure 15(a). In fact, such crossover is often inhibited when there are competing TCP sources, as we see from Figure 15(c,d).

5.3 Random Losses and TCP Dynamics

The presence of randomness in the traces described previously motivates us to understand the effect of random losses on TCP dynamics. We consider a very simple computational model where each packet is concluded to be lost with probability p by TCP. Note that p captures the packet loss from source to destination as well acknowledgment loss from destination back to source. By using simple AIMD equations, we computed $w(t)$ traces for different values of p, and also computed the corresponding time-dependent exponent plots. For low loss rates, the plots resemble periodic trajectories as shown in the top plots of Figure 16. For high losses, the plots resemble random trajectories as shown in the bottom plots of Figure 16. In both these extreme cases, the Λ curves do not resemble those of a chaotic system such as one shown for Lorenz system in Figure 13. The chaos-like behavior is displayed in the medium loss range as shown in Figures 17 and 18 for cases $p = 0.1$ and $p = 0.2$. The expanded initial portions of Λ curves shown in Figure 18 show a clearer common envelope similar to that of the Lorenz system. Since packet losses on the Internet connections do have a random component and losses are often at moderate levels in typical connections, it is not unreasonable to expect TCP traces to exhibit chaotic components illustrated here. We however note that data in Figures 16 - 18 is based on a simple computational model, whereas that in Figure 12 corresponds to actual Internet measurements. In addition, it is assumed in the former the changes in $w(t)$ occur at exact integer intervals, whereas in the latter $w(t)$ is sampled at regular millisecond intervals but the changes could occur within the intervals. Despite the differences in detail, both the traces illustrate the close interplay between the TCP AIMD dynamics and the underlying randomness in the network connection.

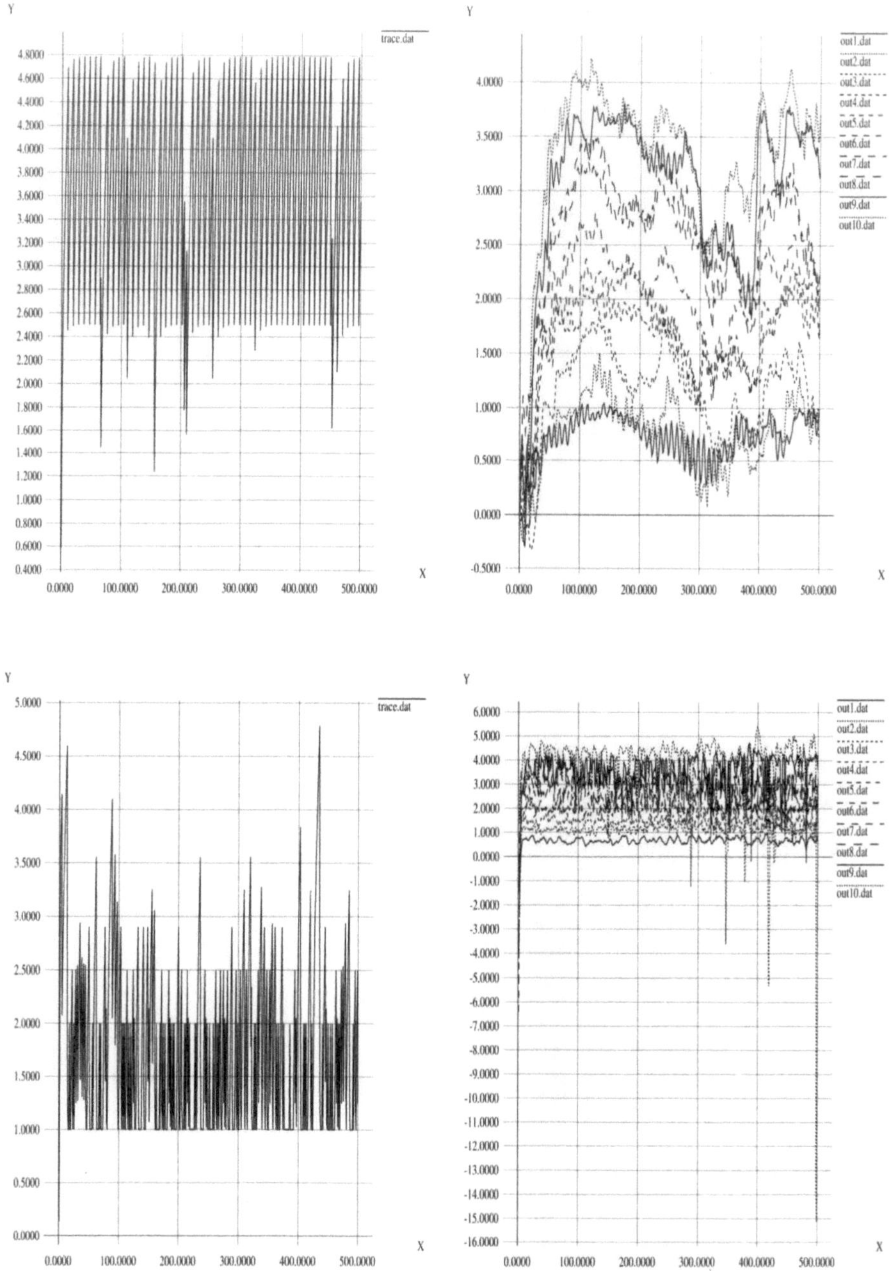

Fig. 16: Top plots correspond to $w(.)$ on the left and $\Lambda(.)$ on the right for $p = 0.03$. Bottom plots correspond to $w(.)$ on the left and $\Lambda(.)$ on the right for $p = 0.5$.

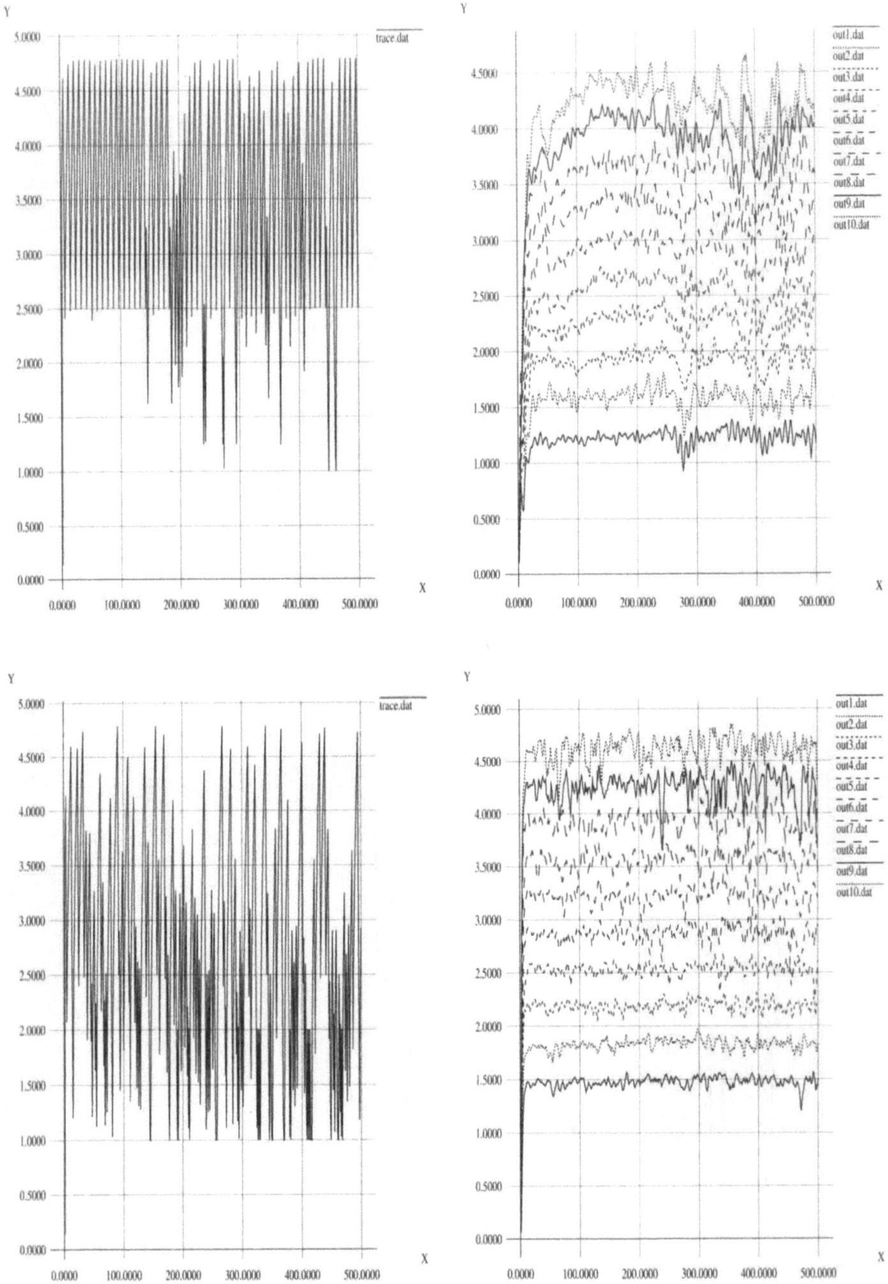

Fig. 17: Top plots correspond to $w(.)$ on the left and $\Lambda(.)$ on the right for $p = 0.1$. Bottom plots correspond to $w(.)$ on the left and $\Lambda(.)$ on the right for $p = 0.2$.

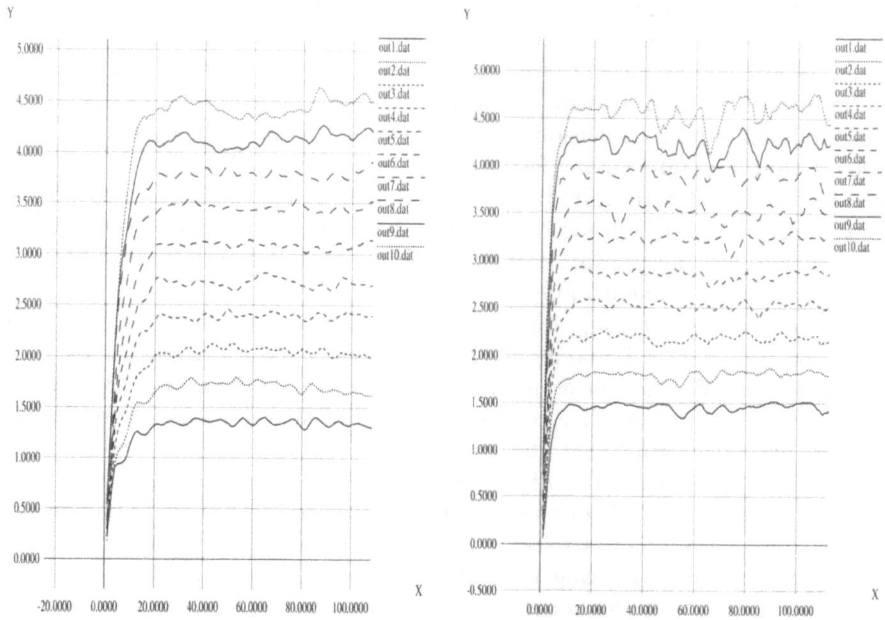

Fig. 18: Expanded view of initial parts of $\Lambda(.)$ for $p = 0.1$ on the left and $p = 0.2$ on the right.

6 Conclusions

We presented analytical and experimental results that indicate the chaotic behavior of TCP AIMD at certain time scales. We presented a state-space description of TCP using the window size and end-to-end packet delays. By composing two almost linear but unstable models, we presented a model for TCP dynamics. This model shows that TCP generates bounded but highly complicated dynamics when it interacts with the routers with small buffers. We discussed ways in which chaotic trajectories can be generated by TCP, by identifying a tent-like map embedded in the dynamics of TCP. We then collected TCP traces over Internet connections and analyzed them using exponential delay curves. We consider this work to be a small step towards understanding the complicated nature of TCP dynamics over wide area networks. While such dynamics do not significantly impact large data transfer applications, they could be very important in closed-loop control applications implemented over wide-area networks.

There are several open questions in the area of understanding TCP dynamics. It would be interesting to see the differences in the dynamics generated by various versions of TCP. Our initial ns simulations of Vegas and Reno versions indicate a behavior qualitatively similar to the results presented here with Tahoe version, but the precise parameter values corresponding to various behaviors are quite different. Also we assumed here that $w(t)$ can take and can be initialized with arbitrary real numbers, and it would be interesting to see the effects of finite precision implementation on the TCP dynamics. Note however that a simple linear feedback system can exhibit chaotic behavior when implemented with finite precision [42], and hence it is non trivial to predict the effects of finite precision on TCP dynamics. It would be particularly interesting to see the relevance of such effects in ns-2 simulations, which are finite precision implementations.

Another important issue for further investigation is the practical implications of the chaotic behavior of TCP on real network applications. It is clear that dynamics are most important for control applications. By analyzing a number of high quality congestion window-size data measured on the Internet, we have found that the transport dynamics are best described by the diffusional processes. The diffusions are often anomalous, especially when there are competing TCP sources. It is interesting to further examine how one might suppress the stochasticity of the network by executing more controls on the network when making measurements, such as using a large number of competing TCP sources together with a constant UDP flow. Our analysis motivates that both chaotic and stochastic aspects be paid a close attention to in designing Internet protocols that are required to provide the desired and tractable dynamics. In some cases, multiple path methods can be used to increase the effective bandwidth [29] which can then be traded-off with the jitter by using simple end-filtering. While some chaotic dynamics can be accounted for using such methods, the area of providing stable control loops over wide area networks has received very little attention. A stochastic approximation method was presented in [32] to achieve provably stable throughput over Internet connections; this method does not employ AIMD but uses the classical Robbins-Monro method to adjust the congestion window sizes. It would be interesting to see if there are methods that drive TCP dynamics away from the chaotic regions or control it if chaos become inevitable.

In terms of analysis, a detailed and explicit analysis of the proposed map M will provide more insights into the TCP dynamics. Also, it would be interesting to see if the TCP dynamics can be shown to be amenable to standard methods [26] such as snapback repellers [36, 20], Shilnikov Theorem [37] or delayed feedback effects [4, 3]. It would be of future interest to apply methods such as computation of various Lyapunov exponents on the time series [1] data of window sizes and end-to-end delays, particularly for Internet measurements.

Acknowledgments

This research is sponsored by the High Performance Networking Program of the Office of Science, U.S. Department of Energy, under Contract No. DE-

AC05-00OR22725 with UT-Battelle, LLC, the Defense Advanced Projects Research Agency under MIPR No. K153, and by National Science Foundation under Awards No. ANI-0229969 and ANI-0335185.

References

1. H. D. I. Abarbanel. *Analysis of Observed Chaotic Data.* Springer-Verlag New York, Inc., 1996.
2. K. T. Alligood, T. D. Sauer, and J. A. York. *Chaos: An Introduction to Dynamical Systems.* Springer-Verlag New York, Inc., 1997.
3. U. an der Heiden and H. O. Walther. Existence of chaos in control systems with delayed feedback. *Journal of Differential Equations*, 47:273–295, 1983.
4. O. Diekmann, S. A. van Gils, S. M. Verduyn Lunel, and H. O. Walther. *Delay Equations: Functional-, Complex- and Nonlinear Analysis.* Springer-Verlag New York, Inc., 1995.
5. A. Erramilli and L.J. Forys. Traffic synchronization effects in teletraffic systems. In *Proc. ITC-13*, 1991.
6. V. Firoiu and M. Borden. A study of active queue management for congestion control. In *Proc. of Infocom*, 2000.
7. S. Floyd. A report on some recent developments in TCP congestion control. *IEEE Communcations Magazine*, April 2001.
8. J. Gao and N. S. V. Rao. TCP AIMD dynamics over Internet connections. 2004. under review.
9. J.B. Gao. Recognizing randomness in a time series. *Physica D*, 106:49, 1997.
10. J.B. Gao, C.C. Chen, S.K. Hwang, and J.M. Liu. Noise-induced chaos. *Int. J. Mod. Phys. B*, 13:3283–3305, 1999.
11. J.B. Gao, S.K. Hwang, and J.M. Liu. Effects of intrinsic spontaneous-emission noise on the nonlinear dynamics of an optically injected semiconductor laser. *Phys. Rev. A*, 59:1582–1585, 1999.
12. J.B. Gao and W.W. Tung. Pathological tremors as diffusional processes. *Biological Cybernetics*, 86:263–270, 2002.
13. J.B. Gao and Z.M. Zheng. Local exponential divergence plot and optimal embedding of a chaotic time series. *Phys. Lett.*, 181:153–158, 1993.
14. J.B. Gao and Z.M. Zheng. Direct dynamical test for deterministic chaos. *Europhys. Lett.*, 25:485, 1994.
15. J.B. Gao and Z.M. Zheng. Direct dynamical test for deterministic chaos and optimal embedding of a chaotic time series. *Phys. Rev. E*, 49(3807), 1994.
16. C. V. Hallot, V. Misra, and D. Towsley. Analysis and design of controllers for AQM routers supporting TCP flows. *IEEE Trans. on Automatic Control*, 2002.
17. C. Hollot, V. Misra, D. Towsley, and W. Gong. A control theoretic analysis of red. In *Proc. of Inforcom*, 2001.
18. V. Jacobson. Congestion avoidance and control. In *Proc. of SIGCOMM*, pages 314–329. 1988.
19. P. Kuusela, P. Lassilaand, and J. Virtamo. Stability of tcp red congestion control, 2000. http://www.tct.hut.fi/tutkimus/com2/publ/.
20. F. R. Marotto. Snap-back repeller imply chaos in r^n. *Journal of Mathematical Analysis and Applications*, 63:199–223, 1978.

21. V. Misra, W. Gong, and D. Towsley. A study of active queue for congestion control. 2000.
22. V. Misra, W. B. Gong, and D. Towsley. Fluid-based analysis of a network of AQM routers suporting TCP flows with an application to RED. In *Proc. of SIGCOMM.* 2000.
23. Net100 project. http://www.net100.org.
24. N.H. Packard, J.P. Crutchfield, J.D. Farmer, and R.S. Shaw. Geometry from a time series. *Phys. Rev. Lett.*, 45:712, 1980.
25. K. Park and W. Willinger. *Self-Similar Network Traffic and Performance Evaluation.* Wiley, New York, 2000.
26. T. S. Parker and L. O. Chua. Chaos: A tutorial for engineers. *Proceedings of the IEEE*, 75(8):982–1008, 1987.
27. L. L. Peterson and B. S. Davie. *Computer Networks.* Morgan Kaufman Pub., 2000. Second Edition.
28. P. Rajan and E. H. Abed. Nonlinear instabiliities in TCP-RED. Technical Report TR 2001-37, Institute for Systems Research, 2001. www.isr.umd.edu.
29. N. S. V. Rao. Overlay networks of in-situ instruments for probabilistic guarantees on message delays in wide-area networks. *IEEE Journal on Selected Areas in Communications*, 22(1), 2004.
30. N. S. V. Rao and L. O. Chua. On dynamics of network transport protocols. In *Proc. of Workshop on Signal Processing, Communications, Chaos and Systems*, pages 29–43, 2002.
31. N. S. V. Rao and W. Feng. Performance trade-offs of tcp adaptation methods. In *International Conference on Networking*, pages 499–510, 2002.
32. N. S. V. Rao, Q. Wu, and S. S. Iyengar. On throughput stabilization of network transport. *IEEE Communications Letters*, 8(1):66–68, 2004.
33. D. Salamoni and S. Luitz. High-performance throughput tuning/measurements. In *PPDG Collaboration Meeting.* 2000.
34. T. Sauer. Reconstruction of dynamical-systems from interspike intervals. *Phys. Rev. Lett.*, 72:3811–3814, 1994.
35. T. Sauer, J.A. Yorke, and M. Casdagli. Embedology. *J. Stat. Phys.*, 65:579–616, 1991.
36. K. Shiraiwa and M. Kurata. A generalization of a theorem of Marotto. *Nagoya Mathematics Journal*, 82:83–97, 191.
37. C. P. Silva. Shilnikov theorem – A tutorial. *IEEE Transactions on Circuits and Systems -I: Fundamental Theory and Applications*, 40(10):675–682, 1993.
38. H. Sivakumar, S. Bailey, and R. L. Gorssman. Psockets: The case for applicaiton-level network striping for data intensive applications using high speed wide area networks. In *Proc. of SC2000.* 2000.
39. C. Sparrow. Chaos in a three-dimensional single loop feedback system with a piecewise linear feedback function. *Journal of Mathematical Analysis and Applications*, 83:275–291, 1981.
40. Y. Srikant. The mathematics of internet congestion control. 2004.
41. F. Takens. Detecting strange attractors in turbulence. In D.A. Rand and L.S. Young, editors, *Dynamical Systems and Turbulence, Lecture Notes in Mathematics*, volume 898, page 366. Springer-Verlag, Berlin, 1981.
42. T. Ushio and C. S. Hsu. Simple example of digital control systems with chaotic rouding errors. *International Journal of Control*, 45(1):17–31, 1987.
43. A. Veres and M. Boda. The chaotic nature of tcp congestion control. In *Proc. INFOCOM 2000.* 2000.

44. S. Wiggins. *Introduction to Applied Nonlinear Dynamical Systems and Chaos.* Springer-Verlag New York, Inc., 1990.

Dynamical Properties of Externally Driven TCP traffic

Gábor Simon[1,2], Péter Pollner[1], Péter Hága[1,2] and István Csabai[1,2]

[1] Department of Physics of Complex Systems, Eötvös University, Budapest 1117, Hungary {gaba,pollner,haga,csabai}@complex.elte.hu
[2] Collegium Budapest, Institute for Advanced Study, Budapest 1014, Hungary

1 Introduction

One of the most influential achievements of the last sanctuary was the emergence of a decentralized, highly heterogenous, global communication network, the so called Internet. The current Internet supports various services of web browsing, transferring of files, interactive shell and multimedia applications, that became a common activity for many people. The spreading of e-business and the increasing number of home offices points to a tendency that the community of future will get increasingly reliant on the firm functioning of the Internet, and demands quality of service. These facts make the study of the physical Internet and the phenomena associated to the various layers of its traffic, a very important topic.

This contribution investigates the dynamics of packet flows obeying the transmission control protocol (TCP), which is a reliable, end-to-end data transfer scheme, that carries most of the traffic of todays Internet. The reliability and the success of TCP stems from the feedback mechanism provided by acknowledgement packets. This feedback is used to explore the actual network conditions and to adopt to their changes. By the term "externally driven" in the title we mean the interaction of TCP with "blind", connectionless flows, such as a stream of packets that obey the user datagram protocol (UDP), or the Internet control message protocol (ICMP). This problem is relevant at least for two practical situations. First for the case of multimedia applications, that use UDP, and share common network resources with background TCP traffic (TV and radio channels over the Internet). Second for the case of active probing measurements that are the main tool for revealing various characteristics of Internet paths and usually employ UDP or ICMP packets. Here we will concentrate on the later case, because although active probing is widely used in practice its impact on the measured TCP background traffic is not well explored and understood.

The contribution is divided into three main parts. We start the first part with a brief review of the Reno version of TCP, present simulation methodology and the details of the investigated network model. Then in the second part we study the effect of periodically sent UDP packets on a single persistent TCP connection, and on an aggregate flow consisting of two persistent TCP connections. Results are shown for different packet rates and sizes. Finally in the last part we introduce the problem of measuring a system whose characteristics are changing as a result of the measuring process, and present a comprehensive study of the influence of active probing on realistic time-varying TCP traffic.

1.1 A Short Review of the TCP-Reno Algorithm

The data in a TCP connection is sent in constant sized packets, whose size depends on the type of the particular application (typically 512, 536 or 1500 Bytes). To ensure the reliability of the connection TCP maintains sequence number and timer information and a feedback mechanism through acknowledgement (ACK) packets. Upon receipt of a data packet with a given sequence number the client sends back an ACK with the sequence number of the next expected data packet. Loss of packets is detected either by timer expire based on round trip time estimation, or if the number of ACKs with the same next expected sequence number exceeds a threshold (typically 3), which signals out of order packets. Details of the timer implementations can be found in [2, 3, 1, 15].

An important variable of the algorithm is the congestion window (W), which tells the number of unacknowledged packets injected into the network. The value of this parameter is continuously adjusted by the algorithm between 1 and the saturation window W_s, to utilize the available bandwidth, and also to respond quickly to the changes in the network conditions. In real applications W_s is set in order to avoid situations where the source host sends data faster than the destination host could process it.

The available bandwidth is probed at the initiation of the connection in the so-called *slow start* mode. In this regime the server sends two packets back-to-back for every received ACK if its sequence number is found in order. This behavior results in an exponential increase of injected packets that quickly reaches the available capacity in the network and ends with a packet loss.

After this the algorithm switches into *congestion avoidance* mode, which is intended to be the main working regime of a TCP connection. In this mode if W is a non-integer, a new packet is injected in the network every time an ACK is received, while for integer W values two packets are sent back-to-back in a response to a received ACK. This scheme results in an approximately linear growth of the congestion window variable and of the number of transmitted packets.

In case of timer expire the algorithm switches to *exponential back-off*. This mode is used to keep silent the connection in the case of high network congestion, until some free capacity develops again. The actual TCP-Reno algorithm defined in technical standards [18] and its practical implementations include lots of fine details that we shall not reproduce here, we only present a simplified algorithm of the congestion window evolution in the different working regimes of TCP that we hope is still adequate for the understanding of its mechanism.

1. *Initiation:* Let W_t be a threshold window, set $W = 1$ and $W_t = W_s$.
2. *Slow start:* After every acknowledgement of a new packet if $W < W_t$, set $W = W + 1$. If $W = W_s$, then W remains constant until a packet loss occurs.
3. *Congestion avoidance:* After every acknowledgement of a new packet if $W \geq W_t$ and $W_t < W_s$, set $W = W + 1/[W]$, where $[...]$ denotes the integer value of its argument. If $W = W_s$, then W remains constant until a packet loss occurs.
4. *Duplicate acknowledgements:* If the number of acknowledgements with the same sequence number exceeds 3, retransmit the next expected packet, set $W_t = W/2$, then set $W = W_t$. After the retransmission is acknowledged the algorithm continues in congestion avoidance mode.
5. *Exponential back-off:* In case of timer expire retransmit the timed-out packet, set $W_t = W/2$, then set $W = 1$ and double the timeout value. After this the algorithm goes to slow start.

1.2 Method of Investigation and Topology Description

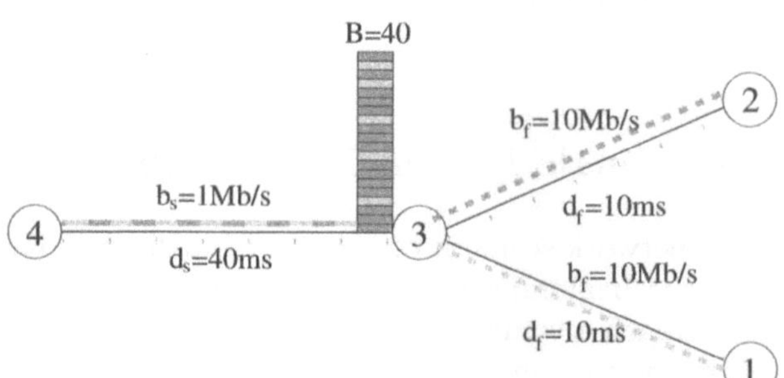

Fig. 1: The investigated network scenario. The edge nodes 1, 2, and 4 are computers sending and receiving data, while node 3 is a router. The nodes are connected by slow and fast full-duplex links characterized by bandwidths b_s, b_f and propagation delays d_s, d_f while the router is characterized by the number of packets it can store B.

Wherever it was possible we attempted to provide analytical considerations, however most of our findings are results from simulations. When investigating a complicated dynamical system like TCP it is essential to consider all the subtle details of the protocol, thus the only acceptable abstraction seemed to be to make the simulations on packet-level. We use the latest release of the network simulator ns-2.27 [15]. Although bugs in the implementation are reported (and are corrected constantly) in the past, still ns represents the most widely used and accepted public network simulator. Throughout the text we tried to give all the details of simulations, so that results could be reproducible and comparable to future investigations.

The structure of the Internet is best approximated by a hierarchical network of internal routers that forward incoming packets towards their destination, whereas clients and servers are situated at the periferia. In the absence of background cross traffic at the routers of the path the performance of a connection is mostly affected by the characteristics of the bottleneck link (the link with the smallest bandwidth) and by the end-to-end propagation delay. We approximate the path of a connection with the simplified model of Fig. 1 described below. The nodes are connected by full-duplex links characterized by bandwidth b and propagation delay d given in the figure. TCP traffic consisting of 540 Byte data packets flows from node 2 to node 4, whereas 40 Byte ACKs flow in the reverse direction. The periodic UDP stream originates from node 1 and terminates in node 4. Both the TCP and the UDP packets queue together in the first-in first-out (FIFO) buffer of the bottleneck router (node3). The bottleneck router uses drop-tail policy, which is the most frequent type in the current Internet, and can store a maximum of $B = 40$ packets without a loss. In the simulations we set the saturation window variable W_s of TCP to a high value that can never be reached in our model network, thus the saturated behavior characterized by the constant maximal congestion window is not studied here.

2 The Case of a Single Driven TCP Connection

Consider the simple network scenario of Fig. 1 with a single TCP connection that is initiated at the beginning of the simulation and that has unlimited data to send. We call such a connection a persistent TCP. Figure 2 shows the time-evolution of the logged congestion window variable, the number of TCP packets in the buffer and the packet drop events. After the initial slow-start the connection settles into the periodic regime of congestion avoidance mode. The period associated to the connection can be defined by the time elapsed between two packet drop events. Because of the simplicity of the studied network scenario and the absence of cross-traffic it is possible to accurately describe the behavior of a single TCP in the congestion avoidance phase analytically. We calculate the period (τ) of a single TCP using the following approach. The evolution of W between the minimal value and the maximal value without a packet loss (W_{max}) is calculated from a fluid-model, where we treat W as a continuous variable. Since it takes a full round-trip time (RTT), and a receipt of 3 duplicate ACKS for the sender to be informed of the packet loss, thus meanwhile packets are still injected into the network and W is inflated over

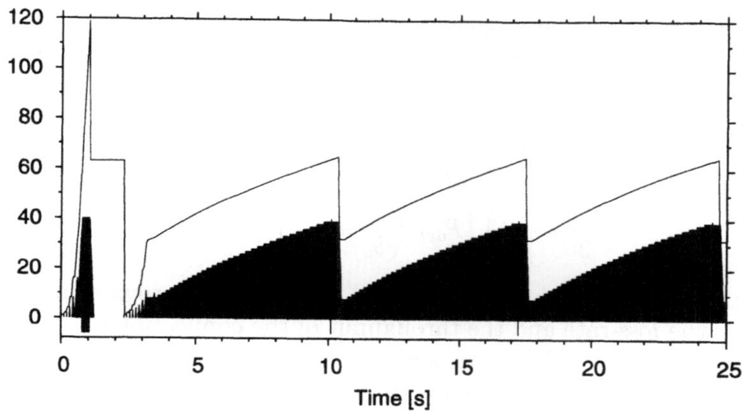

Fig. 2: The performance of a single TCP over the simple scenario of Fig. 1. The continuous line indicates the logged congestion window variable, the filled pattern shows the number of TCP packets in the bottleneck buffer, and the negative impulses signal packet-drop events.

W_{max}. When the third duplicate ACK is received TCP retransmits instantly the lost packet, then halves the window $W = W_{max}/2$ and waits until the ACK of the retransmitted packet arrives. This again takes a full RTT, until TCP can continue in congestion avoidance. Thus the period of the TCP to a good approximation is

$$\tau = \int_{\frac{1}{2}W_{max}}^{W_{max}} RTT(W)dW + 2RTT(W_{max}) + 3\delta t, \tag{1}$$

where $\delta t = P/b_s$ is the time it takes for a TCP packet of size P to be shifted out of the bottleneck buffer. In the above formula the additional terms after the integral take into account the time elapsed until the detection of packet loss and during fast recovery. The expressions for W_{max} and $RTT(W)$ are determined by the characteristics of the network path. In our case the maximal congestion window reads

$$W_{max} = B + N_{link}, \qquad N_{link} = \frac{b_s}{P}2(d_f + d_s) + 1. \tag{2}$$

Evaluating this expression with the particular numbers from Fig. 1, we see that $N_{link} < B$, which means that in the congestion avoidance mode there is always queuing in the bottleneck buffer. This implies that the rate at which TCP packets arrive at the receiver will be determined by δt. Since TCP packets arrive every δt seconds, thus the return of ACKs and the release of new data packets also goes with this rate. Because the condition $N_{link} < B$ holds, thus the round-trip time as a function of the congestion window is given by

$$RTT(W) = P\left(\frac{W - N_{link}}{b_s} + \frac{1}{b_f}\right) + P_{ack}\left(\frac{1}{b_s} + \frac{1}{b_f}\right) + 2(d_f + d_s), \tag{3}$$

which is the sum of the propagation and transmission delays and queuing at the bottleneck buffer. Putting together (1), and (3) the final formula for the period of a single TCP in our simple network is

$$\tau = \frac{3P}{b_s}\left(\frac{W_{max}^2}{8}+1\right) + \frac{P}{b_s}\left(2B - \frac{W_{max}^2 - BW_{max}}{2}\right) + \left(\frac{W_{max}}{2}+2\right)\left[P_{ack}\left(\frac{1}{b_s}+\frac{1}{b_f}\right) + 2(d_f+d_s) + \frac{P}{b_f}\right]. \tag{4}$$

Once the period of the TCP is known one can also provide simple expressions for the loss-rate and the throughput of the connection in the congestion avoidance mode. The loss-rate is simply one packet in each period $L = 1/\tau$, while the throughput to a good approximation is

$$T = \frac{P}{\tau}\left(\int_{\frac{1}{2}W_{max}}^{W_{max}} WdW + \frac{3}{2}W_{max} - 1\right) = \frac{P}{\tau}\left(\frac{3}{8}W_{max}^2 + \frac{3}{2}W_{max} - 1\right). \tag{5}$$

In the following table we compare the calculations for the period, loss-rate and throughput of the TCP to the values found from simulations. The calculated quantities are indicated by a "c" index. As is apparent from the table one can

Table 1: The performance of the analytic model vs. simulation results.

τ^c [s]	L^c [1/s]	T^c [Packet/s]	τ[s]	L [1/s]	T [Packet/s]
7.10	0.141	229.7	7.19	0.144	229.1

develop fairly accurate analytic models for a single TCP connection. However what happens if a connectionless packet stream is also present in the network, and TCP must share the limited resources? A TCP connection interacts with packets from external sources exclusively in the buffers along the path. We can distinguish two types of interaction.

1. Delaying of a TCP packet due to queueing behind packets from external sources. These extra delays are equal to $q_p \times P_p/b_i$, where q_p is the number, and P_p is the size in bits of external packets in the queue before the given TCP packet, and b_i is the bandwidth of the link downstream of the i-th buffer. This effect introduces phase-shifts in the TCP dynamics.
2. Due to the presence of packets from external sources in the queues, the buffers may fill up at a different TCP sequence number, than they would in the absence of these sources. As a result not only the dynamics is phase-shifted but also the loss events may be redistributed to TCP packets with different congestion window variable.

These effects are well illustrated in the traces of Fig. 3, that are simulation results for the congestion window variable in the presence of 50 Byte UDP

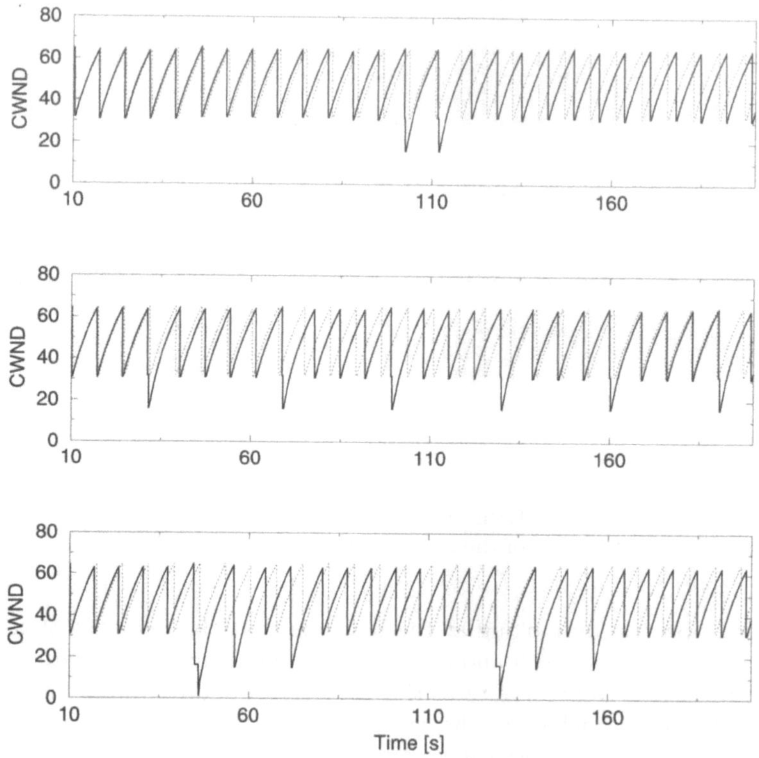

Fig. 3: From top to bottom the figures show traces of the congestion window of the TCP connection (continuous curves) interacting with UDP packets sent at rates of Δt_p =1 s, 0.2 s, and 0.1 s respectively. For comparison purposes also shown in every figure is the trace of the congestion window in the absence of external packets (dashed curves).

packets sent regularly with different inter-departure times Δt_p. After we see that external packets may alter greatly the window dynamics of a TCP, the next step would be to perform systematic analyses that would reveal the dependence on the rate and the size of the external packets. Since we can not show traces for all the different values of these quantities, thus we introduce a simplified description of the dynamics by extracting the times elapsed between two TCP packet loss events (Δt_l). For a simple periodic trace like for instance the TCP in the absence of the external packets this procedure yields a single number, while more complex traces are characterised by many different inter-loss periods. This technique has much in common with that of using the Poincaré section to simplify the description of the time-evolution of a dynamical system. The results of such an investigation are shown in Fig. 4.

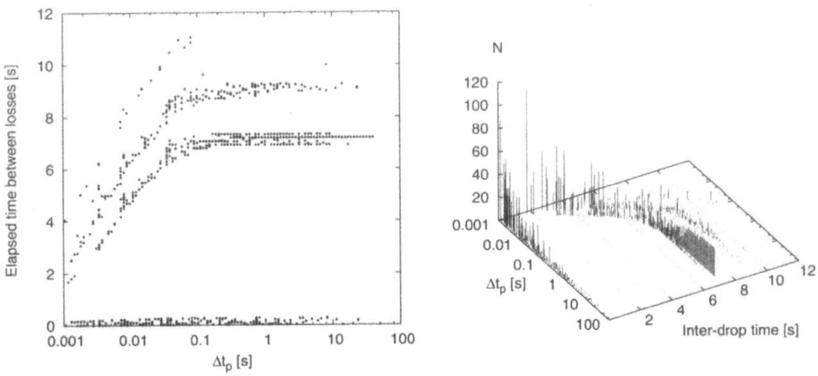

(a) (b)

Fig. 4: The time elapsed between two loss events (a), and the histogram of their occurences (b) as a function of the interdeparture time of the 50 Byte UDP packets.

A striking observation looking at the figure is that the response of the TCP does not result in a smooth and continuous dependence on the rate. One can see that until relatively large rates the Δt_l corresponding to the unperturbed period τ remaines unchanged, however at the same time the traces also develop new discrete inter-loss times. As the rate of UDP packets increases the statistical weight of the different Δt_l values in a trace increases, while the weight of $\Delta t_l = \tau$ decreases. Note that the occurence of different values of inter-loss times in a trace does not neccessarily mean that the time evolution is non-periodic, it only implies that the periodicity with the original period τ is destroyed. Indeed, having looked at many traces we observe that the different Δt_l values are arranged in a periodic pattern, which results in a subperiodic time evolution with $\tau(\Delta t_p) > \tau$. For instance the picture in the middle of Fig. 3 shows a beginning of a trace with periodicity of $\approx 4\tau$. Note that a subperiodicity that involves many different Δt_l values and a period that is greater than the duration of the TCP connection will appear as completely chaotic for an observer.

Apart from the rate dependence of the inter-loss time, we also investigated how the change of the dynamics on small time-scales of a few round-trip times influences the averaged quantities of throughput and TCP lossrate. The rate dependence of these quantities for the cases of 50 Byte and 1000 Byte UDP packets are shown in Fig. 5. As can be seen the global trend is that the lossrate increases, while the throughput decreases with the rate, however these curves are quite fluctuating, and are likely not even continuous functions of the rate. As it is expected a stream with higher packet size but the same rate causes higher TCP lossrate, and accordingly higher throughput degradation. Given

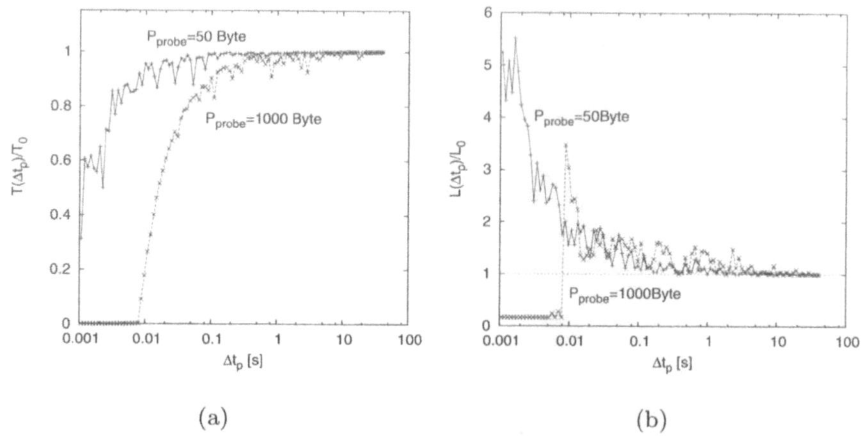

Fig. 5: The throughput (a), and the lossrate (b) of the connection normalised to the unperturbed values as a function of the inverse probe rate, for two different probe packet sizes.

the complexity of the results it seems to us unlikely that a feasable analytic model can be developed that would accurately reproduce the observed features of the window evolution. However the reproduction of the global trends of the rate and packet size dependence of T and L from a statistical model is more promising, that we will attempt to develop elsewhere.

3 The Case of Multiple Driven TCP Connections

In the previous section we investigated the response of a single persistent TCP connection to a stream of regularly sent UDP packets and found that the external packets induce subperiodic window evolution. Here we are interested in the effect of the external packets on the dynamics of aggregate traffic composed of many parallel TCP connections. The types of interaction of a TCP listed in Sect. 2 still hold, however in the case of an aggregate traffic the notion of external packets also covers TCP packets from the other connections. Another difference compared to the single TCP case is, that here the TCP packet loss events are not only redistributed among different sequence numbers of the same connection, but also between the different connections. Since at a given time the TCPs are characterised by different congestion windows, thus the redistribution of the packet drop events efficiently modifies the traffic pattern. We will only show results for the case where the aggregate is composed of two TCPs, because the qualitative behavior that we shall describe below is found to hold for higher number of parallel TCPs as well.

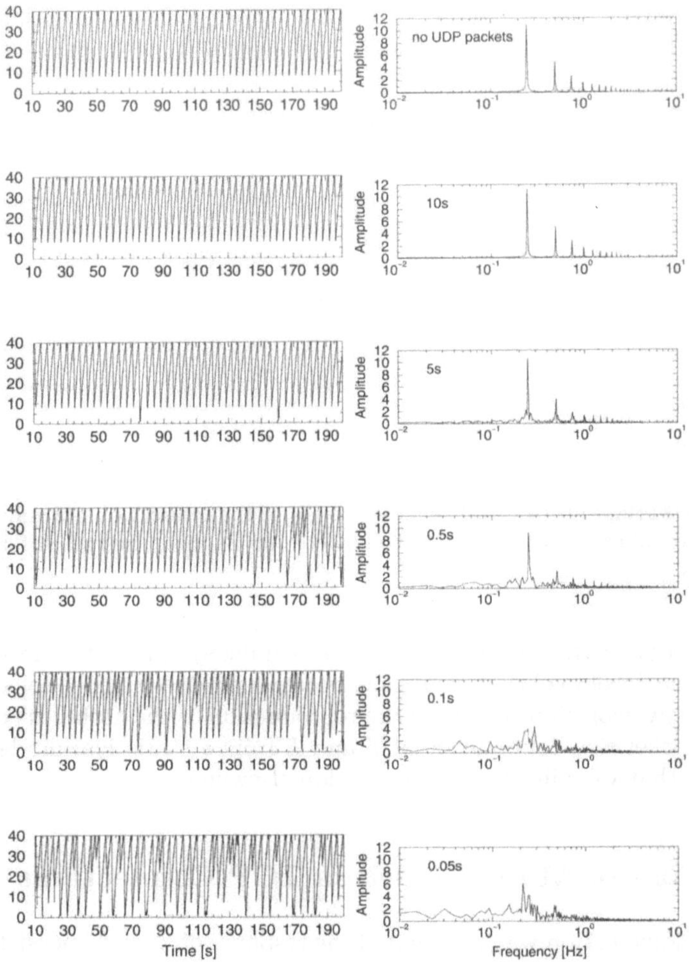

Fig. 6: The behavior of aggregate TCP traffic composed of two parallel TCP connections sharing the same network resources in the presence of 50 Byte UDP stream with inter-departure times, indicated in the corner of the figures. On the left we show the number of TCP packets in the buffer as a function of time, while in the right column we give the Fourier spectra of the time-series.

The top trace in Fig. 6 shows the time-series of the number of TCP packets in the bottleneck buffer without external UDP packets. In this case the interaction of TCP connections results in synchronisation, and in a periodic aggreagate traffic. The periodicity is also apparent on the Fourier spectrum of the trace where we observe a sharp peak and its higher harmonics with diminishing amplitude. We also made simulations at much higher number of parallel TCPs and in all cases found synchronization. Thus if the external packets are also TCP packets from connections between the same end-points, then their interaction results in a qualitatively different behavior than in the case of connectionless flows. On the other hand if our TCP besides of another TCP also interacts with a connectionless stream of packets, then we observe a similar alteration of the aggregate dynamics as for the single TCP case.

One can see in Fig. 6 that with the increase of the rate of UDP packets a continuos component appears in the Fourier spectrum that gets more and more pronounced, and at high rates dominates the spectrum. The continuous component in the spectrum is a marker of chaos, however investigations of longer traces reveal that the dynamics is again periodic, only the associated period is much longer than that of the unperturbed aggreagte TCP traffic. Note that so far every investigated scenario was complitely deterministic, and the resulting complicated behavior was due to the complexity of the system. Simulations with different number of parallel TCPs also revealed that the aggregate traffic gets more sensitive to UDP packets as the number of connections is increased. This implies that the same rate of UDP packets, at which synchronization persist at a low number of TCP connections, destroys the synchronization in the case of a higher number of TCPs.

The dynamic behavior of chaos at small time-scales and periodicity at large scales is similar to the "cellular chaos" described in the contribution of Vattay *et al* [4]. The difference is that they investigated a system of interacting TCPs with different source-destination pairs and slightly different minimal round-trip times, wheras in our case chaos is induced solely by the interaction with connectionless packet streams.

4 The Impact of Active Probing Measuerements on TCP Traffic

Active probing is a class of measuring methods that are used to infer various characteristics of network paths. The common in these methods is that they involve ICMP or UDP probe packets that are injected into the network, while a reciever side analyses the returned responses. This general framework admits the determination of the topology of a network [16], the link-bandwidths on a path [7, 5], and the statistics of packet size, packet-loss and delay along a route [6, 17]. These last mentioned quantities are especially important for quality of service considerations, because together they determine the rate at which applications can send data on that route.

In general an active probing stream may contain N probe packets of different sizes $P_p(i)$, $i \in \{1..N\}$ sent with different inter-departure times $\Delta t_p(i) = t^d(i+1) - t^d(i)$, $i \in \{1..N-1\}$. The actual choice of $P_p(i)$ and $\Delta t_p(i)$ determines the architecture of the probe stream that may varie according to the particular quantity under investigation (e.g. bandwidth, distribution or spectrum of end-to-end delays). In most of the applications it is customary to send constant sized packets regularly ($P_p(i) \equiv P_p$ and $\Delta t_p(i) \equiv \Delta t_p$), however other architectures are also common. For instance to measure bottleneck bandwidth one can send packet-pairs with an inter-pair time chosen randomly from an exponential distribution, while the packets in a pair are sent back-to-back [8]. Another example is the packet tailgating technique of [9], where pairs consisting of a packet with the highest possible size immediately followed by a packet with the smallest possible size are sent to measure the bandwidth of each link on a path.

Here we only investigate the simplest scenario of regularly sent uniform sized packets, that can be used for instance to measure time-series and spectrum of end-to-end delay. By measuring delay between two end nodes one can also estimate the amount of TCP traffic queueing in the routers of the path through the following simple delay model

$$D(i) = \sum_{j=1}^{L} \left(d_j + \frac{P_p}{b_j} \right) + \sum_{j=1}^{L} \left(P_p \frac{q_p(i)}{b_j} + P \frac{Q(i)}{b_j} \right), \tag{6}$$

where $D(i)$ is the end-to-end delay experienced by the i-th probe, d_j and b_j are the propagation delay and the bandwidth of the j-th link on the path, P_p and P are the sizes of probe packets and background TCP packets, and finally $q_p(i)$ and $Q(i)$ stand for the actual number of probe packets and TCP packets in the queue of the j-th buffer at the instant when the probe packet arrives to it. The first sum in (6) contains the constant part of the delay (propagation delays and transmission times), while the second is the fluctuating part.

Since the background traffic fluctuates one can assume that instances can be found when there is no queuing in the buffers, and in this cases (6) yields the constant part of the delay. Since queuing delays are always positive or zero, thus by minimum filtering, one can obtain the constant part, and by its subtraction also the component due to queuing.

4.1 The Dilemma

In general for a measurement process the desire is to gain the maximum information about a system without considerably changing its state. In the previous sections we saw that a periodic stream of connectionless packets can alter greatly the dynamics of persistent TCP traffic. The question might arise, can TCP traffic be measured by probes, without drasticly changing its dynamics?

In our case this dilemma translates to the problem of adjusting two parameters, the packet size P_p and the inter-departure time Δt_p (inverse probe rate) of the probe packets, while keeping the influence of the probe stream minimal. From (6) it is clear that one should use the smallest possible probe packet size, which is constrained by the fixed sizes of protocol headers. On the other hand the choice of the probe rate is a harder question, because small rates may be inadequate to achieve good resolution, while high rates introduce big perturbation of the original TCP flow.

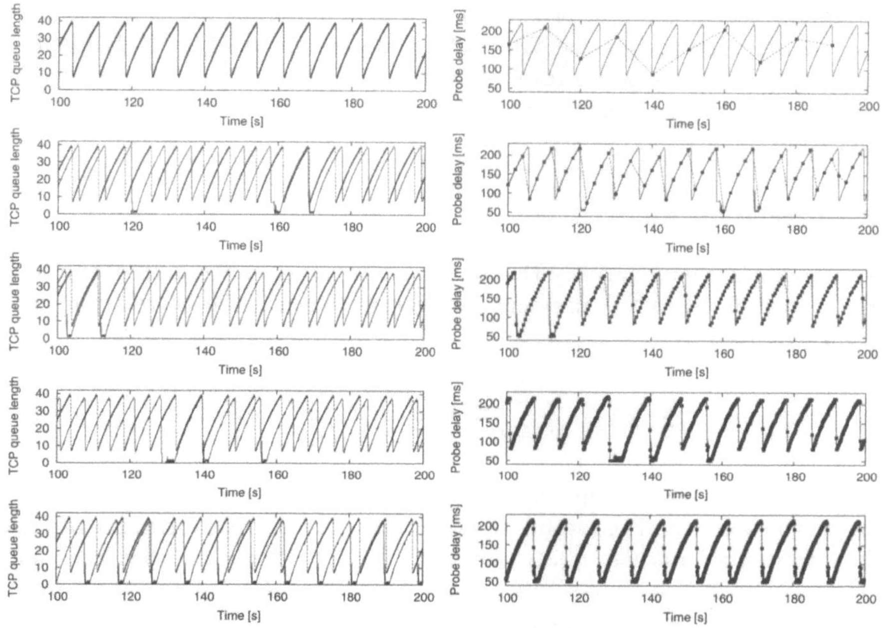

Fig. 7: On the left results are shown for the unperturbed TCP queue-length time series of the bottleneck buffer $Q^u(k)$ (dashed curves), together with the actual TCP queue-length time series $Q(l)$ perturbed by the probe stream (continuous curves). The right side compares the measured end-to-end delay time serie $D_m(i)$ (symbols connected by dashed line) with the end-to-end delay $D(l)$ calculated from (6). From top to bottom the traces correspond to inter-departure times of 10 s, 2 s, 0.5 s, 0.1 s, and 0.05 s respectively. The probe packet size was 50 Bytes.

4.2 Measuring a Single Persistent TCP

To illustrate the dilemma we present simulation results made over the topology of Fig. 1. A single persistent TCP connection was simulated for 200 seconds, with and without a probe stream composed of UDP packets sent at various rates. The compared characteristics are the time series of the unperturbed TCP queue length in the bottleneck buffer $Q_u(k)$ and of the actual TCP queue length in the presence of the measurement probe stream with a given rate $Q(l)$. Both of these time-series were generated on an event basis. Every time a new TCP packet entered the bottleneck buffer, the number of TCP packets in the queue were logged together with the timestamp of the event. For the same probe rates we also compared the time series of the measured end-to-end delay $D_m(i)$ and the end-to-end delay calculated from the model of (6) $D(l)$. Note that among these time-series only $D(l)$ and $Q(l)$ has the same number of elements that is also indicated by the common index.

In Fig 7 one can see that in parallel with the gain of information from measured probe delays, the influence of the probe stream rapidly increases , and is non-negligible even for small rates. It is apparent that (6) is a useful delay model, however for the satisfactory measurement of TCP dynamics it is necessary to use a probe rate, at which the state of the TCP connection is so much perturbed that it does not resemble anymore that of the original connection.

4.3 Correlation Analysis

To characterize quantitatively the local similarities or differences of the time-series investigated in the previous subsection, we use the correlation index $C(X,Y)$ defined by

$$C(X,Y) = \frac{1}{N}\sum_{i=1}^{N}\frac{(X_i - E(X))(Y_i - E(Y))}{D(X)D(Y)}, \tag{7}$$

where X and Y are two time-series, and E and D denotes respectively their expectation values and the standard deviations. The correlation index can take values from the interval $[-1,1]$, where $C(X,Y) = 1$ means that the shape of the two compared time-series are the same, and $C(X,Y) = 0$ means no correlation. In order to use the correlation index, X and Y must contain equal number of elements. To achieve this in our case we used linear interpolation of the time series with less number of elements at points from the time series with more elements. Using the correlation index we can quantify the qualitative trends illustrated in Fig. 7 and present results of the probe rate dependence with high resolution. Note that the value of the correlation index is invariant to the scaling of the time series along the ordinate axis and to the addition of a constant, thus one can directly compare the time series of the measured end-to-end delay and of the actual TCP queue length.

In Fig. 8 results are shown for the probe rate and packet size dependence of the $C(Q_u, Q)$ and $C(D_m, Q)$ correlation indices. For both of the probe packet sizes the qualitative picture is similar. In the case when only a few probes are sent during the time of the simulation the perturbation of the probes is negligible $C(Q_u, Q) \approx 1$, however the information that we gain about the underlieing TCP dynamics from the few probes is insufficient $C(D_m, Q) \approx 0$. On the other hand at high probe rates the measurement process gets almost perfect, but at the same time the perturbation caused by the probes alters so much the dynamics of the TCP connection that we cannot say anything about the state of the TCP in the absence of the measurement. At very high rates, one can observe a decrease of $C(D_m, Q)$ from near 1. This feature is due to the fact that for high enough rates the contribution of the probe packets to the queuing delay gets non-negligible compared to the component due to the TCP traffic. Another notable detail is that the dependence of the correlation indices, especially of $C(Q_u, Q)$ is rather fluctuating, thus very little changes in the rate induce high variability in the outcome of the success of the measuring process.

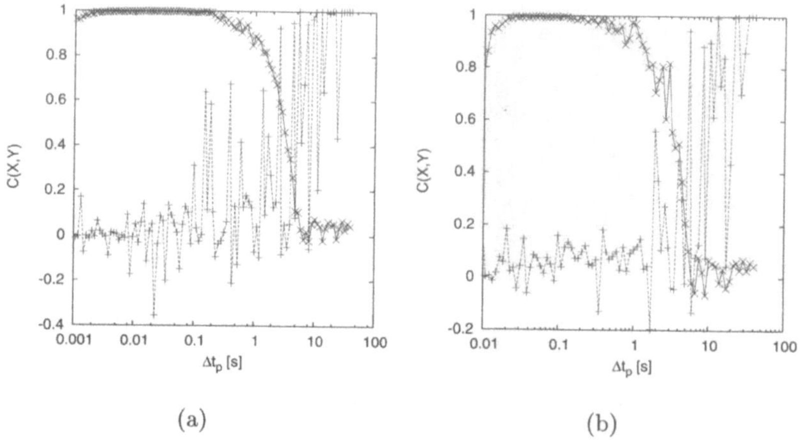

Fig. 8: The (+) symbols connected by a dashed line indicate the correlation of the unperturbed and the perturbed TCP queue length time series, while the (X) symbols connected by a continuous line show the correlation between the measured end-to-end delay and the actual TCP queue length time series. The results are shown for 100 different probe rates and for probe packet sizes of 50 Bytes (a) and 1000 Bytes (b).

4.4 Measuring Time-dependent Aggregate TCP Traffic

In the previous subsection we investigated the problem of measuring the traffic of a single persistent TCP connection. That scenario, although interesting from theoretical point of view, is too abstract to draw conclusions for the case of realistic Internet traffic. In reality the TCP traffic on a path is an aggregate composed of many connections that interact with each other and are characterized by different ON and OFF periods. To account for this we model the background traffic by a superposition of many individual TCP sessions. The initiation times of the sessions were chosen to be a random variable distributed uniformly during the time of the simulation. It is observed in real measurements that Internet traffic is often characterized by self-similarity or long range dependence [10, 11]. To be able to investigate such a scenario we choose the durations (ON periods) of the connections from a Pareto distribution with an average of 100 s and a shape parameter of 1.1.

Note that the strategy of choosing predefined durations of the connections may not always reflect the reality. In the Internet a TCP connection may end for several reasons. First it can end because the file that is being downloaded is transfered completely. In this case the durations are set by the file sizes and available bandwidth at the time of downloading. Another possibility is that users decide to end a connection before the downloading could be complete. This can happen for psychological reasons, for instance in the case of

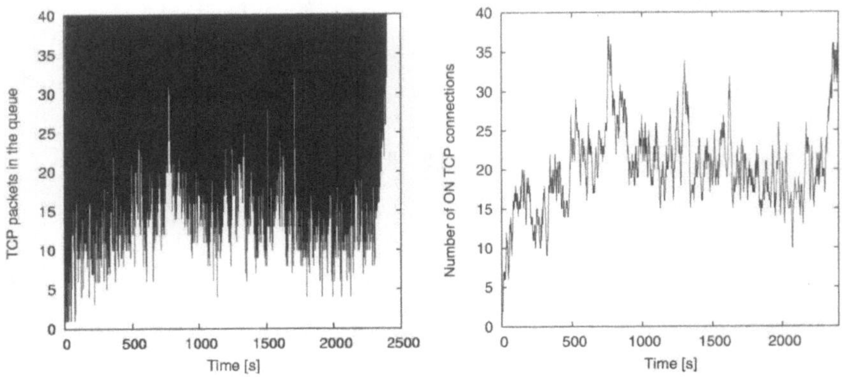

(a) (b)

Fig. 9: The modeling of aggregate TCP traffic. In (a) we show the trace of an unperturbed TCP queue length time series of 1200 TCP connections initiated uniformly between 0 and 2400 s, while the durations of the connections were chosen from a Pareto distribution with an average of 100 s and a shape parameter of 1.1. The number of ON connections at a given time for this trace are shown in (b).

congestion, when TCP connections can get into long back-off states. Since in this study we are interested mostly in trends and qualitative aspects of the influence of active probing, for the sake of simplicity we do not consider these fine details of aggregate traffic modeling.

To investigate the success and the influence of measuring the fluctuating aggregate TCP traffic with probe packets we performed similar correlation analysis as in the case of the single persistent TCP. Figure 7(a) shows the results for the correlation indices of $C(Q_u, Q)$ and $C(D_m, Q)$, where we adopt the same notations as in 4.3, except that the queue length time series here correspond to the aggregate TCP traffic. Simulations were performed for 10 different seeding of the random generators to see the general trends of the results, whereas for a given seeding we used the same set of random numbers to generate the unperturbed and the perturbed background TCP traffic. Additional details of the simulations are given in the caption of the figure.

One can compare the results of Fig. 10(a) with the case of the single persistent TCP in Fig. 8. It is apparent that the probe rate dependence of the correlation between the measured end-to-end delay and the actual TCP queue length time series is quite similar to the case of the single TCP scenario. On the other hand the behavior of $C(Q_u, Q)$ is markedly different in these cases, since here we never observe high correlation index near 1. This means that in the case when many TCP connections interact with each other at any given time the system is more sensitive to the perturbations, and the effect

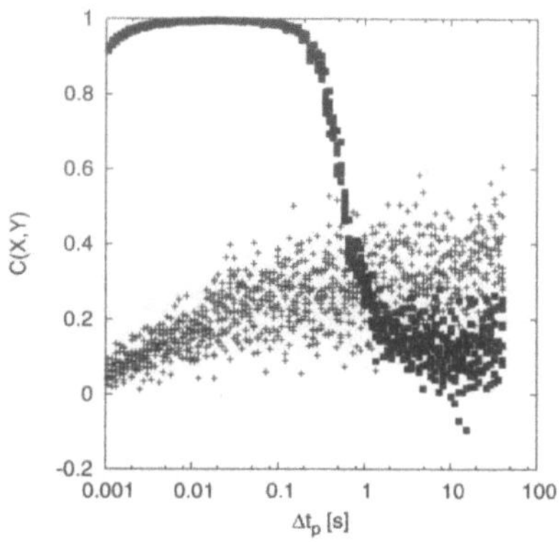

(a)

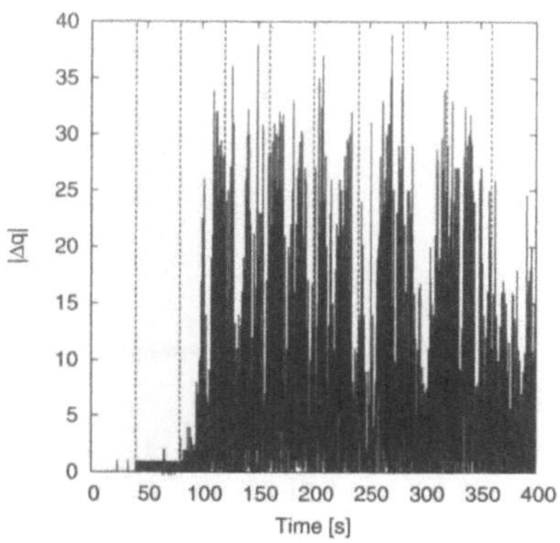

(b)

Fig. 10: Correlation indices $C(Q_u, Q)$ (pluses) and $C(D_m, Q)$ (filled boxes) as a function of the probe rate for 10 different random number seeding (a). Here the simulations were made for 400 s with 200 TCP connections, and the parameters of the Pareto distribution giving the ON periods were 100 s (average) and 1.1 (shape). The probe stream was composed of 50 Byte UDP packets. In (b) we show the growth of the perturbation with time at a single low probe rate. The vertical lines indicate the instants when the UDP probes were sent.

of a few probe packets are already sufficient to alter greatly the trace of the background traffic. This *butterfly-effect* is also illustrated in Fig. 10(b), where we show the absolute value of the difference between the TCP queue length time series without probes and with probes sent every 40 s. One can see that after the second probe the error between the two time-series already grows to the size of the system limited by the buffer size.

One can expect that similar results will be observed in other cases too, as long as there is queueing in the buffers and at least a few TCP connections interact with each other all the time. Smaller sensitivity and accordingly higher values of $C(Q_u, Q)$ can be expected if the periods of non-zero queue length come to an end before the perturbation from probe packets could grow considerably. On each event when the queue empties the system looses the memory of the past and the errors developed before these events can not propagate into the next period with nonzero queue length.

4.5 The Impact on the Statistical Properties

In the previous subsection we saw that even a few probes could change drastically the local structure of the aggregate TCP trace. This implies that it seems impossible to measure a high resolution trace of TCP background traffic using active probing without considerably changing its state. However in most of

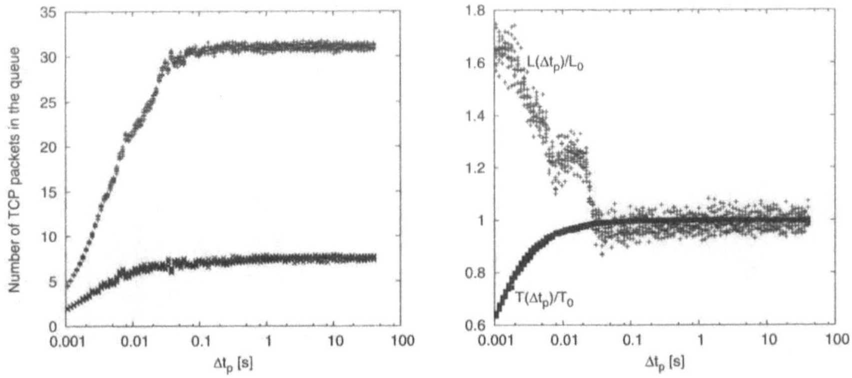

(a) (b)

Fig. 11: In (a) the (+) symbols stand for the mean, and the (X) symbols for the standard deviations of the TCP queue length time series as a function of the inverse probe rate. In (b) the filled squares indicate the overall throughput, while the (+) symbols show the overall loss-rate of the aggregate TCP traffic normalized by the value found in the absence of the probes. The results in each case are shown for 10 different random number seedings.

the cases we are not interested in the dynamics on small scales, but instead in statistical properties of the measured traffic, such as the distribution, mean or standard deviation of the TCP queue length, the overall throughput or loss-rate.

In the followings we present results for the dependence of these characteristics on the probe rate. Figure 11 shows the mean and standard deviation of the TCP queue length time series, and the normalized throughput together with the loss-rate as the rate of the probes is varied. An immediate difference compared to the results of the correlation analysis is that in a broad range of the probe rates these statistical properties are unchanged by the influence of the probes. One can observe a threshold around $\Delta t_p \approx 0.024$s, where sudden changes occur in the average TCP queue length and in the normalized loss-rate, whereas the change in the standard deviation of TCP queue length and in the normalized throughput is much smoother.

We also analyzed the change of the distribution of the TCP queue length and of the measured probe delay time series. Figure 12 shows the results for a single random number seeding. As could be expected from Fig. 11(a) the distribution of the TCP queue length is not altered very much even by the perturbation of sending the probes every 0.1s. However for higher probe rates the shape of the distribution shows a qualitatively different picture. One can also observe that at high probe rate the shape of the measured probe

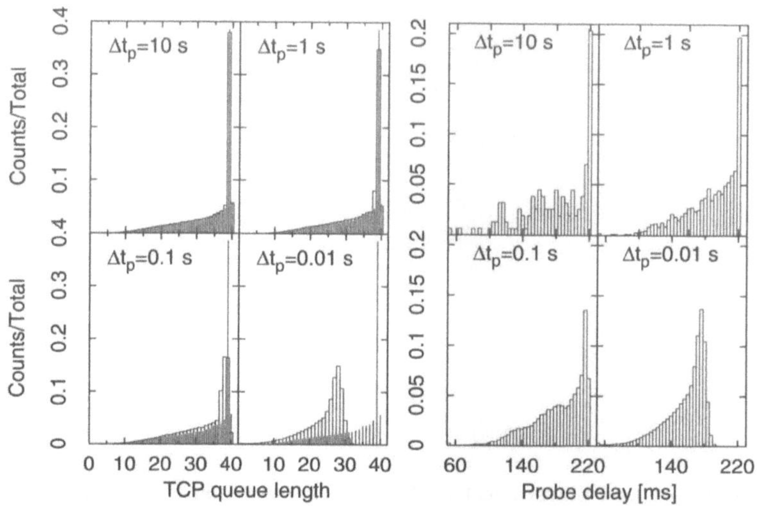

Fig. 12: The left group of figures compares the distributions of the unperturbed TCP queue length (impulses) with that of the TCP queue length perturbed by a probe stream (boxes). The right group of figures show the distribution of the measured end-to-end delay time series. The probe rates are indicated in the corner of each figure.

delay distribution accurately resembles that of the actual TCP queue length distribution, whereas for low rates the number of samples is insufficient to obtain a good statistics.

4.6 Implications for Spectral Measurements and Hurst-parameter Estimation

Active measurements with regularly sent uniform sized probes apart from measuring the time-series of different characteristics of Internet paths, also admit the determination of their power-spectrum. Fitting the low frequency limit of the power spectrum $S(f)$ is one of the few methods to estimate the degree of self-similarity of a time-series (see for instance [12]). A self-similar process is characterized by the Hurst exponent $0.5 \leq H < 1$, and its value is related to the asymptotic shape of the power spectrum at low frequencies, as $S(f) \sim f^{\beta}$, where $\beta = 1-2H$. The value $H = 1$ corresponds to the interesting limiting case of $1/f$ noise [13, 14]. In previous subsections we saw that the influence of probes changes the local structure of the background TCP traffic, but until relatively high rates leaves unchanged the global properties, such as the average TCP queue length in the bottleneck buffer. Based on these observation the question may arise, how the measuring process influences the

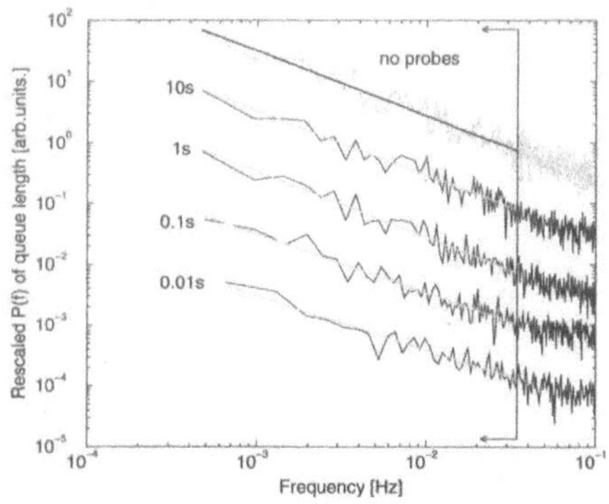

(a)

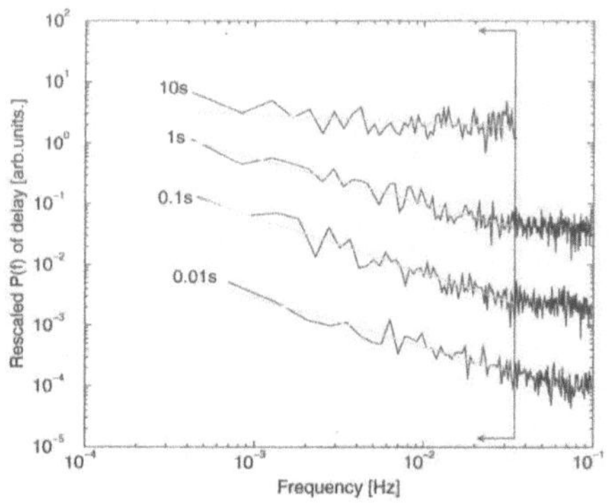

(b)

Fig. 13: The averaged power spectra of the TCP queue length (a) and of the end-to-end delay (b) time-series at different probe rates indicated in the figures. The continuous curves are the power-law fit to the power spectra, fitted in the interval of frequencies to the left from the vertical lines

shape of the power spectrum, and the Hurst exponent of a self-similar time-series.

To answer this question we performed 2400s long simulations in the network model of subsection 1.2. The background TCP traffic consisted of 1200 connections initiated uniformly during the simulation time, with Pareto-distributed ON-periods (average: 100 s, and shape: 1.1). We studied the effect of probes sent at rates Δt_p =10 s, 1 s, 0.1 s, and 0.01 s. The simulations were repeated for 10 different random number seedings and the final spectra were averaged over these cases. Figure 13 shows the shape of the power spectra of TCP queue length with and without probes and the power spectra of the measured probe delays. The results of the exponents of the power-law fit together with the calculated Hurst exponents are given in table 2.

Based on the numbers in the table one can draw the following conclusions. As a result of the increasing probe rate the shape of TCP queue length power spectra flattens, while the shape of the measured delay power spectra steepen. In order for the measured Hurst exponent be appropriate for the TCP queue length, one needs to use high rate of probes, however at such rates the perturbation is already too high, so that the measured Hurst exponent will underestimate the self-similarity of the original (unperturbed) TCP queue length. This result suggests that the measuring process can influence the background TCP traffic in a way that lowers its degree of self-similarity.

Table 2: The β exponents of the low frequency limit power law fit of the power spectra, and the Hurst exponents calculated from $H = -(\beta - 1)/2$ at different probe rates Δt_p. The notations D and Q stand for the measured delay and TCP queue length time series respectively. The first row of the table gives the result for the unperturbed case.

Δt_p[s]	$\beta(D)$	$\beta(Q)$	$H(D)$	$H(Q)$
-	-	-1.06±0.04	-	-
10	-0.13±0.04	-1.06±0.05	0.57±0.02	-
1	-0.74±0.04	-1.08±0.04	0.87±0.02	-
0.1	-0.93±0.04	-0.96±0.05	0.97±0.02	0.98±0.03
0.01	-0.80±0.05	-0.89±0.05	0.9±0.03	0.95±0.03

5 Summary

In this contribution we investigated through packet level simulations the interaction of TCP traffic with a connectionless stream of regularly sent uniform sized packets. Results indicate that external packets induce sub-periodic congestion window evolution for a single TCP. In contrast to the unperturbed behavior, where the period is the time elapsed between two TCP packet drops, in the sub-periodic case multiple drops occur during the period, with different inter-loss times. Numerical results are presented for the rate-dependence of the throughput and loss-rate of the connection. These curves show highly fluctuating behavior on small scales superimposed on a monotonic trend on the large scales.

We also studied the interaction of the connectionless packet stream with an aggregate traffic composed of parallel TCP connections. Interestingly if only the TCP connections were present in the system the interaction between them always resulted in synchronization and a periodic aggregate traffic. This periodicity is gradually destroyed in the presence of connectionless packets with increasing rate, yielding an aggregate traffic that is chaotic on practical time-scales but periodic on the longer run.

In the remaining part of this chapter we investigated the problem of measuring TCP traffic by active probing which involves packets from connectionless streams. It is found that in order to resolve the dynamics of a single TCP from measured probe delays one is forced to use a high rate of probes at which the TCP connection is so much altered that it does not resemble anymore the original trace that one wanted to measure. In other words the influence of the measuring process on the measurable system is non-negligible. Investigations of time-varying aggregate TCP traffic reveal that this system is more sensitive to the perturbation of probes than a single connection, and even a couple of packets change completely the local structure of the time evolution. However despite the locally changed dynamics for a broad range of probe rates the aggregate traffic yields similar results for averaged quantities, like the distribution of the TCP queue length, overall throughput and loss-rate. Active probing is also used to measure the Hurst parameter of self-similar traffic. Comparing the spectrum of the original TCP traffic with those in the presence of the probes we revealed that the measuring process underestimates the degree of self-similarity of the original TCP traffic.

References

1. G. R. Wright and W. R. Stevens: *TCP/IP Illustrated, Volume 2, The Implementation* (Addison Wesley, 1995)
2. V. Jacobson: Congestion avoidance and control. Computer Communication Review, vol 18, no 4, pp 314-329 (1988)
3. P. Karn and C. Partridge: Improving Round-Trip Time Estimates in Reliable Transport Protocols. ACM Trans. on Computer Systems, vol 9, no 4 pp 364-373 (1991)
4. G. Vattay, K. Diriczi, A. Fekete, L. Kocarev, M. Maródi, and J. Stéger. Statistical Properties of Chaos in Communication Networks. In: Kocarev L, Vattay G (eds) Complex Dynamics in Communication Networks, Springer...
5. A. B. Downey Using pathchar to Estimate Internet Link Characteristics. In proceedings of ACM SIGCOMM, 1999.
6. J.-C. Bolot. Characterizing End-to-End Packet Delay and Loss in the Internet. Journal of High-Speed Networks, vol. 2, no. 3, pp.305-323, Dec. 1993.
7. R. L. Carter and M. E. Crovella. Measureing Bottleneck Link Speed in Packet-Switched Networks, Performance Evaluation, vol.27&28, pp. 297-318, 1996.
8. A. Pásztor, and D. Veitch. A precision infrastructure for active probing, in *PAM2001, Workshop on Passive and Active Networking*, Amsterdam, The Netherlands, 2001, pp. 33-44.
9. K. Lai and M. Baker. Measuring Link Bandwidths Using a Deterministic Model of Packet Delay, in Proceedings of ACM SIGCOMM 2000, Stockholm, Sweden, 2000.
10. W. E. Leland, M. S. Taqqu, W. Willinger, and D. V. Willson, On the Self-Similar Nature of Ethernet Traffic (Extended Version), IEEE/ACM Transactions on Networking, 2(1), pp.1-15, 1994.
11. M. Crovella and A. Bestavros. Self-Similarity in World Wide Web Traffic: Evidence and Possible Causes, IEEE/ACM Transactions on Networking, pp.835-846, December 1997.
12. A. Veres, Zs. Kenesei, S. Molnár, and G. Vattay. On the Propagation of Long-Range Dependence in the Internet, in Proceedings of ACM SIGCOMM 2000, Stockholm, Sweden, Aug.-Sep. 2000.
13. I. Csabai, 1/f Noise in Computer Network Traffic, J. Phys. A: Math. Gen.,27 (1994) 417.
14. M. Takayasu, H. Takayasu, and T. Sato, Critical behaviours and 1/f
15. "Ucb/lbnl/vint network simulator - ns (version 2.27)," http://www-mash.cs.berkeley.edu/ns/.
16. Cooperative Association for Internet Data Analysis (CAIDA) web page http://www.caida.org/Tools/
17. V. Paxson, Measurements and Analysis of End-to-End Internet Dynamics. PhD thesis, University of California, Berkeley, April 1997.

18. [RFC793], [RFC1122], [RFC1323], [RFC2018] and [RFC2581]. http://www.ietf.org/rfc.html.

Data Traffic, Topology and Congestion

David K. Arrowsmith[1], R. J. Mondrag[2], and M. Woolf[1]

[1] School of Mathematical Sciences, Queen Mary, University of London, London E1 4NS, UK d.k.arrowsmith@qmul.ac.uk, m.woolf@qmul.ac.uk
[2] Department of Electronic Engineering, Queen Mary, University of London, London E1 4NS, UK r.j.mondragon@elec.qmul.ac.uk

1 Introduction

We consider the interaction between the topology of a network and the traffic carried along its channels. The binding elements between the topology and the traffic dynamics are the routing mechanisms. In a packet-based network, like the Internet, the transmission of information is carried out in discrete packets. The path that a packet follows when travelling the network is determined by the routing algorithm. Usually, from the topological properties of the network and statistical properties of the traffic, the routing algorithm tries to minimise the packet delivery time and maximise the throughput; this implies that packet flow affects the behaviour of the routers which in return regulate the flow. The dynamics of a packet network can also be regulated by controlling the packet production at the various packet sources and varying the server capacities. An example is *Transmission Control Protocol* or *TCP*, where the source of traffic adjusts its rate of packet transmission as a function of the round trip delay time.

Previously, the statistical properties of packet traffic, which are dependent on the control and routing algorithms, were described by a model where the traffic input was *Poisson-like* where the auto-correlation decays exponentially fast. Traffic with this decay property, of which the Poisson type is a particular example, is referred to as having *Short Range Dependence* (*SRD*). From studies carried out in the early 1990's [17] it is known that *Poisson-like* models do not capture all the statistical properties of packet traffic. Packet traffic exhibits spurts of activity over a large number of time scales. These bursts last from milliseconds to days and they look similar independently of the time scale. This phenomenon is known as *self-similar* traffic. One characteristic of this self-similar traffic is that it has *Long Range Dependence* (*LRD*), i.e. the traffic is strongly correlated at all time scales of engineering interest. This observation was a surprise as, previously, the properties of packet traffic were described as *SRD* processes. We begin in section 2 by briefly discussing some of the properties of *LRD* packet traffic.

Even though some researchers [26, 40] have suggested that the burstiness in packet traffic is connected to the behaviour of individual users within the network, the modelling of packet traffic is based on its measured characteristics more than on the underlying mechanisms responsible for the self-similarity. The bursty traffic is often described by stochastic methods based on Gaussian self-similar processes, for example fractional Brownian motion and fractional Gaussian noise [35, 31, 46, 25] or, on the superposition of *on/off* sources with *heavy tailed on* or *off* periods [48] and chaotic maps [14, 15, 39]. All these models describe successfully the burstiness of the traffic but their approach is very different. In section 3 we introduce a non-linear chaotic map as a model to generate packet traffic with varying statistical properties. We also discuss several equivalent deterministic models for packet production models which have calculational advantages. The packet production dynamics is also extended to employ the *TCP* window dynamics [13].

The packet delivery time is the time that elapses between the creation of a packet at its source s, to the arrival at its destination d. This time is known as the *end-to-end* time, packet *lifetime* or *latency*. A packet travels through the network visiting different nodes. If one of the nodes is busy, the packet is stored in the queue at that node. Eventually, as the node serves its queue, the packet is forwarded toward its destination. Usually longer routes and/or congested queues mean longer delivery times. The routing algorithm tries to reduce the packet delivery time by selecting short, lightly utilised routes. In such a network, the traffic characteristics are not drastically changed as the packet transverses the network. The delivery time for a packet from its source to its destination is finite. As the load increases, the delivery time will typically increase accordingly. There is a critical load where the delivery time diverges, or at least increases dramatically. At this point the network is congested.

In a regular-symmetric network, it is possible to predict the traffic load where congestion occurs (a dynamical characteristic) by only considering the average of the shortest path lengths from all sources to all destinations (a topological characteristic) [21, 19, 41, 51]. In section 4 the relevant topological characteristics when studying congestion are introduced and the network dynamics is introduced in section 4.4.

In section 5 we make the connection between the traffic characteristics, the topology of the network and congestion by looking first at *Poisson-like* traffic and then at the differences when the input traffic is *LRD*. In section 6 we briefly introduce some mechanisms to control *LRD* traffic, by limiting the size of the queues [5], by reducing the rate of packet production [45] and by using *TCP*-like control [52].

2 Long-Range Dependence

In 1994, *LRD* was shown to be a feature of Internet packet traffic by Leland *et al.*, [29]. The *LRD* behaviour manifests itself along a communication channel as *bursty* activity in the packet rate (no. of packets/unit time) which persists on all relevant time scales. The bursty traffic makes it much more difficult to implement effective traffic congestion protocols (e.g. *TCP*).

The statistical nature of *LRD* traffic is formally defined in [8]. A key requirement is that the autocorrelation of packet traces, $\gamma(k)$, where the lag is k, satisfies a power law decay of the form $\gamma(k) \sim Ck^{-\beta}$, where $\beta \in (0,1)$ and C is constant. Equivalently, $\gamma(k) \sim Ck^{-2+2H}$, where $H = 1-\beta/2 \in (\frac{1}{2}, 1)$ is the *Hurst* parameter. By comparison, *Poisson-like* traffic has an exponential rate of decay $\gamma(k) \sim C'\alpha^{-k}$ with $\alpha > 1$ and C' a constant. The Hurst parameter distinguishes between *LRD* traffic for $H \approx 1$ and the *onset* of *SRD* traffic for $H \approx 1/2$, when the autocorrelation decay changes to exponential.

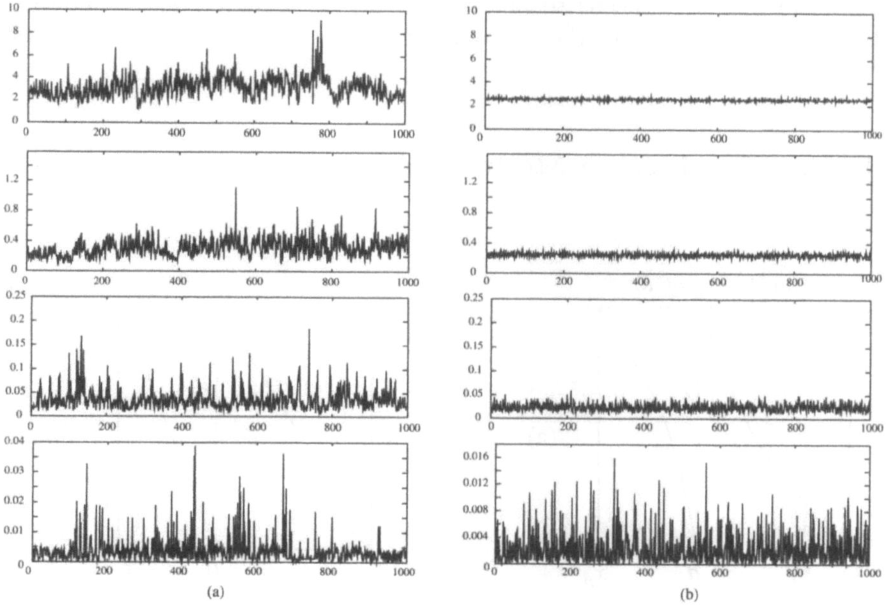

Fig. 1: The batch averages of packets/unit time for (a) a real *LRD* traffic trace (Bellcore data from http://ita.ee.lbl.gov/html/contrib/BC.html), and (b) a Poisson based trace for the same load. Each for sizes $N = 1, 10, 100$ and $N = 1000$. A relatively large variance is retained in case (a).

The essential contrast between *SRD*, Poisson-like traffic, arising typically from traditional voice traffic, and the bursty nature of Internet *LRD* traffic is seen in Fig. 1. The effect of scaling is shown for (a) long-range dependent and (b) short-range dependent traffic for a time series of a random variable $X_n, n = 0, 1, 2, \ldots$ The data is averaged in batch sizes of $N = 1$, 10, 100 and $N = 1000$.

The standard deviation in the *Poisson* traffic varies as the square root of the batch size, or magnification, and we see a 'smoothing' of the traffic as N increases in Fig. 1(b). Thus the mean is an increasingly effective indicator of the instantaneous load, i.e expected packet rate, in the traffic. By comparison, for *LRD* traffic, we see that the variation around the mean remains relatively high for large N in Fig. 1(a). Even when averaged over longer time intervals by several orders of magnitude, we can still obtain packet rates which are close to 0 and 1.

One of the consequences of *LRD* traffic is that it increases queue lengths and latency dramatically. The length of a queue fed with *LRD* traffic sources decays as a power law, compared with an exponential decay if it is fed with *Poisson* traffic sources. The effects of *LRD* cannot be 'removed' by a control mechanism and *LRD* needs to be allowed for, both in computer models of network behaviour and in the routing algorithms used to control data flow through networks.

3 Packet Production Models

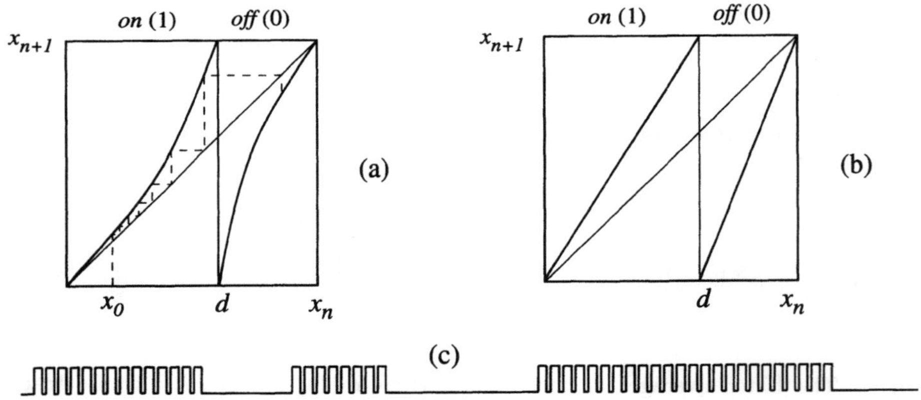

Fig. 2: (a) The graph of the map f (Eq. 1) consists of two segments and each has a tangency with the line $y = x$. The iteration of the map f with initial condition x_0 forms a 'web' generating the iterative sequence, or orbit, x_n, where $x_n = f(x_{n-1})$, $n = 1, 2, \ldots$. Note that the tangencies at $x = 0, 1$ give a 'slow' change in the values of the sequence x_n, and therefore the output y_n provides long sequences of consecutive '0's or '1's. (b) Graph of f for $m_1 = m_2 = 1$, this is known as the Bernoulli shift map and it generates *Poisson* traffic. If the map parameters are chosen to satisfy $m_1 = m_2 \in \{1, \frac{3}{2}\}$, then the map generates *SRD* traffic.
(c) The traffic on a packet network is represented by a binary sequence of zeros and ones (Eq. 2).

Previous simulations of *SRD* packet traffic generation at each host have used *SRD* Poisson (or Markovian) distributions. In this case a packet is created at a host only if a random number on the unit interval, $\mathbb{I} = \{x | x \in [0,1]\}$, is below a discriminator value λ. Hence, for a uniform random distribution the average rate at which packets are produced at a host is λ.

An alternative to this is to use chaotic maps to model the *LRD* nature of real packet traffic. We used the family of maps $f = f_{(m_1.m_2.d)} : \mathbb{I} \to \mathbb{I}$ defined in the unit interval $\mathbb{I}$ by $x_{n+1} = f(x_n)$ where

$$x_{n+1} = \begin{cases} x_n + (1-d)\left(\dfrac{x_n}{d}\right)^{m_1}, & x_n \in [0,\, d], \\ \\ x_n + d\left(\dfrac{1-x_n}{1-d}\right)^{m_2}, & x_n \in (d,\, 1], \end{cases} \tag{1}$$

described in previous papers (see Erramilli *et al*, [14]), and related maps in [47]. Here $d \in (0,1)$ and the parameters $m_1,\, m_2 \in (\frac{3}{2},\, 2)$ induce *intermittency* at each of the points $x = 0$ and $x = 1$, by producing tangencies to the diagonal in the graph of f. The orbital 'escape time' in neighbourhoods of 0 and 1 become power law dependent. If this map is iterated a large number of times, the values of x_n will form a non-uniform continuous distribution on the interval $\mathbb{I}$. The parameter d is used as a discriminator, as λ is for the Poisson case. If x_n falls between 0 and d, a packet is generated; and if x_n falls between d and 1, no packet is generated. Thus we have a discrete output map associated with the function in Eq. (1) which is

$$y_n = \begin{cases} 1: & x_n \in [0,\, d] \;-\; \text{packet generated}, \\ 0: & x_n \in (d,\, 1] \;-\; \text{no packet generated}. \end{cases} \tag{2}$$

The above model, which represents the traffic as a binary sequence is also known as a *packet train* model [27] (see Fig. 2(c)). The intermittency behaviour of the map f induces so-called *memory* in the digital output y_n giving the long range correlation effects required for the packet traffic. This feature is shown by the slow decay of variance with respect to n, the size of batched output, see Fig. 1(a) and [29]. The power-decay of the variance arises from the small orbital increments of the intermittency map which in turn provides memory in the digital output. An example of this phenomenon is illustrated in Fig. 2(a) where a sequence of the iterated values x near the origin have small increments. This effect is stronger for orbits passing even closer to $x = 0$. The time of escape (i.e. into the region $x > d$) of an orbit from a neighbourhood of the origin has a power-law dependence on its initial position, [47].

The nonlinear nature of f means that in this case the load $\lambda = \lambda(m_1,\, m_2,\, d)$ (i.e. the average value of the output y per iteration) is not equal to d, but is given by $\lambda = \int_0^d \nu(x)\, dx$, where ν is the natural invariant density distribution of the map f on the interval $[0, d]$. The distribution ν has no closed form and is often obtained numerically via the Perron-Frobenius operator [32]. Thus the various statistical properties of traffic generated in this way are determined by the map's parameters m_1, m_2 and d including the auto-correlation behaviour as we shall see in section 3.3.

3.1 Closed Form Map

There are two other important models of the above intermittency type which have useful mathematical characteristics not available in the standard model described above. The first extension was introduced by Pruthi [38], and the closed form model appears, at first sight, to be more intractable than the original model in Eq. (1). Essentially, the function in (1) is replaced by

$$x_{n+1} = \begin{cases} \dfrac{x_n}{(1 - c_1 x_n^{m_1 - 1})^{\frac{1}{m_1 - 1}}}, & x_n \in [0, d], \quad (a) \\[2ex] 1 - \dfrac{1 - x_n}{(1 - c_2 (1 - x_n)^{m_2 - 1})^{\frac{1}{m_2 - 1}}}, & x_n \in (d, 1], \quad (b) \end{cases} \tag{3}$$

where

$$c_1 = \frac{1 - d^{m_1 - 1}}{d^{m_1 - 1}}, \qquad c_2 = \frac{1 - (1 - d)^{m_2 - 1}}{(1 - d)^{m_2 - 1}}. \tag{4}$$

A Taylor expansion of the functions in Eq. 3 (a) and (b), around the points $x = 0$ and $x = 1$ respectively, give the forms of the equation (1) and so the leading intermittency effects are the same in both models with leading exponents m_1 and m_2. A distinct advantage of (3) is that, somewhat remarkably, it has a closed form under composition. If the first branch function is denoted by $f_1(c_1, m_1, x)$ then the n-th iterate can be shown to satisfy $f_1^n(c_1, m_1, x) = f_1(nc_1, m_1, x)$. Similarly the second branch function f_2 satisfies $f_2^n(c_2, m_2, x) = f_2(nc_2, m_2, x)$. This enables sojourn times in the "*off*" or "*on*" regions to be calculated explicitly, by solving equations of the type $f_1^n(c_1, m_1, x) = d$, to obtain the number of iterations to the transition point.

We can use the closed iterative form above to calculate the probability of 'escape' from an intermittency region. Specifically, we consider the probability of a sequence of k-consecutive zeroes for the output y_n of an intermittency map f. We will use the closed form map in Eq. (3) and will assume a random uniform injection into the region $[0, d]$ for $x > d$, see [37].

If the orbit re-enters the interval $[0, d]$ at the point $\bar{x}$, then that determines the sequence length l of zeroes, namely,

$$l(\bar{x}) = \frac{1}{c_1}\left(\frac{1}{\bar{x}^{m-1}} - \frac{1}{d^{m-1}}\right) \tag{5}$$

Let $P(l(x))$ be the probability density for at least length l "zero" sequences. If we are assuming that the initial point density on the interval $[0, d]$ at $\{\bar{x}\}$ is $\hat{P}(\bar{x})$, then $\hat{P}(\bar{x})d\bar{x} = P(l(\bar{x}))dl$.

If we further assume the re-entry density $\hat{P}(\bar{x})$ to be uniformly random, i.e. $\hat{P}(\bar{x}) = \hat{P}$ a constant, then

$$P(l) = \hat{P}\left|\frac{d\bar{x}}{dl}\right| \sim Cl^{-\frac{m}{m-1}} \tag{6}$$

for small x with C a constant. This is not the case for re-entry to the interval $[0, d]$ when the full double intermittency is used.

3.2 Piece-wise linear map

A second model which has different mathematical advantages over the original Erramilli model is a piece-wise version. This was originally introduced by Wang [47] and has been used subsequently in applications to realise maps with given autocorrelation profiles [7]. The analogue is constructed with the use of two sequences. The piecewise linear map $p : \mathbb{I} \to \mathbb{I}$ is defined by two monotonic sequences z_i^L and z_i^R monotonic decreasing to zero, with $z_1^L = z_1^R = 0.5$. The map p is then defined by requiring that its graph be piecewise linear with nodal points defined by (z_i^L, z_{i+1}^L), and $(1-z_{i+1}^R, 1-z_i^R)$, $i = 1, 2, \ldots$. The piecewise linear map can replicate the intermittency behaviour at $x = 0, 1$ by choosing the sequences to decay to zero in an appropriate way. If we let $z_i^L \sim i^{-\alpha}$ and $z_i^R \sim i^{-\beta}$ with $\alpha, \beta > 1$, then the exponents for the different types of smooth and piecewise linear map have analogous asymptotic behaviour at the intermittency points. We identify the parameters as $\alpha = 1 + 1/m_1$, and $\beta = 1 + 1/m_2$. The piece-wise linear map has distinct advantages in that it is possible to calculate the invariant measure associated with the map, see [7].

3.3 Autocorrelation of the Map Output

We have already seen in Fig. 1 that the movement between strings of the output values '0' and '1' is rapid in trace (b) and much slower in trace (a). The intermittency in traffic maps produces increased sojourn times for the two states. The longer sojourn times are said to introduce *memory* into the output which is reflected in a higher correlation between the output binary sequence and the same sequence with a time-lag k. The *autocorrelation* vector of a sequence is the way in which the memory is measured. Let X_t be a scalar time series of the binary values $\{0, 1\}$ for $t = 0, 1, 2, \ldots$, and suppose the series is stationary. We define the autocorrelation of time lag k by

$$\gamma(k) = \frac{E(X_t X_{t+k}) - E(X_t)E(X_{t+k})}{\sqrt{(\mathrm{Var}(X_t)\mathrm{Var}(X_{t+k}))}}. \tag{7}$$

Let $\mu = E(X_t)$. Note that the values X_t are binary, then $E(X_t^2) = E(X_t) = \mu$ and so $\mathrm{Var}(X_t) = E(X_t^2) - E(X_t)^2 = \mu(1-\mu)$. Therefore, the autocorrelation can be re-written

$$\gamma(k) = \frac{E(X_t X_{t+k}) - \mu^2}{\mu(1-\mu)} \tag{8}$$

Given $0 \leq X_t X_{t+k} \leq X_t$, it follows that $\gamma(k) \leq 1$. Note also that if there is no correlation, i.e. the values X_t are independent of each other, then $E(X_t X_{t+k}) = E(X_t)E(X_{t+k})$ and $\gamma(k) = 0$. Thus, in general, we expect that the correlation coefficient $\gamma(k)$ will eventually decay to zero in some way. Two special types of decay are

(a) power-law decay, where $\gamma(k) \sim Ck^{-\beta}$, for some constant C and $\beta > 0$;
(b) exponential decay, where $\gamma(k) \sim C\alpha^{-k}$, for some constants C and $\alpha > 0$.

Bernoulli Map Decay

The piece-wise linear Bernoulli map is given by $f(x) = 2x \bmod 1$ on the interval [0, 1]. Given the current state is '0', then $0 \leq x < 0.5$. The probability of the transfer '$0 \rightarrow 0$' is 0.5, since it requires $0 \leq x < 0.25$, and similarly for '$1 \rightarrow 1$'. Thus we can calculate the autocorrelation exactly and we obtain

$$\gamma(k) = \begin{cases} 1 & \text{for } k = 0 \\ 0 & \text{for } k > 0. \end{cases} \tag{9}$$

Intermittency Map Decay

Only two exact results are known so far for the asymptotic properties of double intermittency maps considered here. Let the piecewise linear map constructed from two sequences $z_i = i^{-\alpha}$ at $x = 0$ and $w_i = 1 - i^{-\beta}$ at $x = 1$, α, $\beta > 1$, then the two intermittencies compete and it can be shown that

$$\gamma(k) \sim K\, k^{-c} \tag{10}$$

where $c = \min\{\alpha, \beta\} - 1$, K constant, [7]. Thus the correlation for the composite map is determined by the heaviest tail in the correlation decay arising from the two competing intermittencies.

A similar result is also available for the differentiable case [33]. In this case, the map is as in Eq. (1) with the extra condition that whenever f iterates across the line $x = d$, the formula is replaced by a random uniform injection. For example, the autocorrelation vector $c(n), n \in \mathbb{Z}^+$, of the output function y is known to have asymptotic behaviour $\gamma(k) \sim Ck^{-\beta}$, with C constant, as $k \rightarrow \infty$. The constant $\beta = (2-m)/(m-1) \in (0,1)$, for $m = \max\{m_1, m_2\}$, with $m_1, m_2 \in (\frac{3}{2}, 2)$, (see [22, 33, 7]). Furthermore, the Hurst parameter, H, is given by

$$H = 1 - \frac{\beta}{2} = \frac{3m-4}{2(m-1)}, \tag{11}$$

and ranges over the interval $(\frac{1}{2}, 1)$, as required. Thus $m_1, m_2 = 1.5$ corresponds to *Poisson-like* behaviour and as m_1, m_2 are increased towards 2, the behaviour is increasingly long-range dependent, see [7, 33]. If the transition simplification is removed then the result that the heaviest tail dominates in the piece-wise smooth case is still open but the auto-correlation decay is conjectured to remain as in Eq. (11).

3.4 Transmission Control Protocol Dynamics

The dynamics of the packet production model can be extended to incorporate packet window dynamics [13]. If the map is in the '*on*' state, each iteration of the map represents a packet generated. One sojourn period in the '*on*' side of the map represents a whole file. These *files* are then *windowed* using the *slow-start* algorithm, adding another dynamical layer to the system. The *slow-start* algorithm is as follows:

At a given host i in the network, and time $t = n$, there is a current state, $x_i(n)$, and a current window size, $w_i(n)$, for the number of packets that can be sent at time $t = n$. There is also a residual file size, $s_i(n)$, at node i which is given by the number of iterates of f such that $f^{s_i(n)}(x_i(n)) < d$, and $f^{s_i(n)+1}(x_i(n)) > d$. The source will send $p_i(n) = \min\{w_i(n), s_i(n)\}$ packets. The full window dynamics therefore takes the form, (see Erramilli *et al.*, [13]): For $x(n) < d$, i.e. packet generated-

$$w_i(n+1) = \begin{cases} 1, & \text{if } x_i(n-1) < d, \\ \min\{2w_i(n), w_{max}\}, & \text{otherwise,} \end{cases} \tag{12}$$

and $x_i(n+1) = f^{p_i(n)}(x_i(n))$.

For $x_i(n) > d$, i.e. no packet generated - $w_i(n+1) = 0$, and $x_i(n+1) = f(x_i(n))$.

This algorithm applies if all packets in a window are acknowledged before the retransmission timeout (RTO) limit is reached. If packets take longer than this to be acknowledged the window of packets is sent again with the RTO doubled and the window size set to zero. When the map is in the '*off*' state, the window size is zero and no packets are sent.

This initial value of RTO is calculated using the exponential averaging method [42]. This method keeps a running average of all round trip times. This average is weighted towards more recent round trips, and is used in calculating the *initial* RTO.

4 Topology and Models of Networks

Many different topologies appear in communication networks. Square lattices, toroidal lattices, meshes and hypercubes arise on multiprocessor computers (e.g. [30]), scale-free networks in the WWW [2] and the Internet [16]. The way that the elements of the network are connected to each other and the nodal degree properties have an impact on its functionality. The representation and study of the connectivity of a network is carried out using concepts from graph theory.

A communications network can be represented by a graph $\mathcal{G} = (\mathcal{V}, \mathcal{E})$, where $\mathcal{V}$ is the set of nodes (vertices) and $\mathcal{E}$ is the set of links (edges). The hosts, routers and switches are represented by nodes and the physical connections between them are represented by links. The links can have a direction, but here we are only going to consider undirected links. A node can transfer information to another node in the form of data-packets if there is a link between them. If there is no direct link between the nodes, then a path in the network is the sequence of distinct nodes visited when transferring data-packets from one node to another. We consider networks where there exists at least one path connecting any pair of nodes of the network.

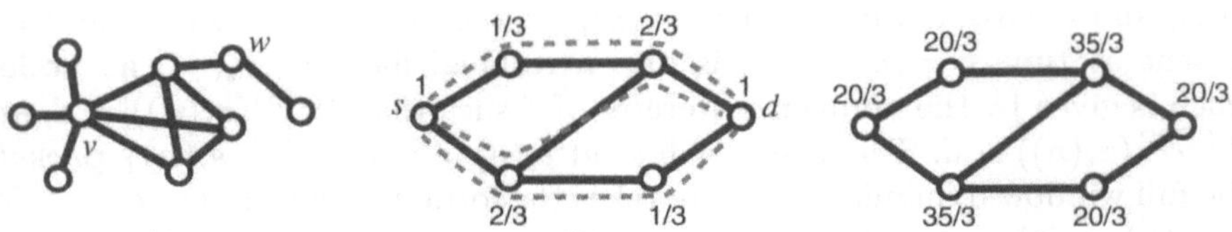

Fig. 3: (a) The node degree for node v is 6 and for w, it is 2. (b) Three different shortest paths between s and d. The length of the path is 3. The number above the nodes denotes the proportion of shortest paths $p_{s,d}(w)$ that go through that node. (c) The node medial centrality for the whole network.

The degree, k, of a node is the number of links which have the node as an end-point [49], or equivalently, the number of nearest neighbours of a node, see Fig. 3(a). The degree of a node is a local quantity. However, the node degree distribution of the entire network gives important information about the global properties of a network and can be used to characterise different network topologies.

If there is very little traffic on the network and if the journey time from one node to its neighbour is in unit time, then given any two nodes and to first approximation, the journey time will be proportional to the length of the journey, or the path length. A path that goes from source node s, to destination node d, in the smallest number of hops is called the shortest-path. The length of the shortest-path $\ell_{s,d}$ is the number of nodes visited when going from s to d. There can be more than one shortest–path between a pair of nodes. The characteristic path length

$$\bar{\ell} = \frac{1}{S(S-1)} \sum_{s \in \mathcal{V}} \sum_{d \neq s \in \mathcal{V}} \ell_{s,d}, \tag{13}$$

where S is the total number of nodes, is the average shortest-path over all pairs of nodes (see Fig. 3(b)). Sometimes $\bar{\ell}$ is referred to as the *diameter* of the network.

If there is traffic on the network the difference in the journey times of two shortest paths with the same length can be very different. Not all the journeys are equal due to the different patterns of usage of the routes. On a network, there are nodes that are more prominent because they are highly used when transferring packet-data. A way to measure this "importance" is by using the concept of *betweeness centrality* of a node. We will refer to this concept here as *medial centrality* The concept of centrality [18] was introduced in social networks to characterise the prominence of an individual in the *context of the social structure*. Given a source s, and destination d, the number of different shortest-paths is $g(s,d)$. The number of shortest-paths that contain the node w is $g(w;s,d)$. The proportion of shortest-paths, from s to d, which contain node w is

$$p_{s,d}(w) = \frac{g(w;s,d)}{g(s,d)}. \tag{14}$$

Remark: The proportion of shortest-paths and the shortest-path length are related by

$$\ell_{s,d} = \sum_{w \in \mathcal{W}} p_{s,d}(w) - 1, \tag{15}$$

where $\mathcal{W}$ is the set that contains the nodes visited by the shortest paths from s to d.

The *medial centrality* of node w is defined as [24]

$$C_B(w) = \sum_{s \in \mathcal{V}} \sum_{d \neq s \in \mathcal{V}} p_{s,d}(w) \tag{16}$$

where the sum is over all possible pairs of nodes with $s \neq d$. The medial centrality measures how many shortest paths pass a certain node (see Fig. 3(c)). A node with a large C_B is "important" because a large amount of packets flow through it, that is, it carries a large traffic load. If this node fails or gets congested, the consequences to the network traffic can be very drastic [24, 55].

4.1 Regular-Symmetric Networks

In a regular network all nodes have the same degree (see Fig. 4). By symmetry, the medial centrality is constant for all nodes. From Eq. (15) and if $C_B(w) = c$ then

$$c = \frac{1}{S} \sum_{w \in \mathcal{V}} C_B(w) = \frac{1}{S} \sum_{s,d} \ell_{s,d} - 1. \tag{17}$$

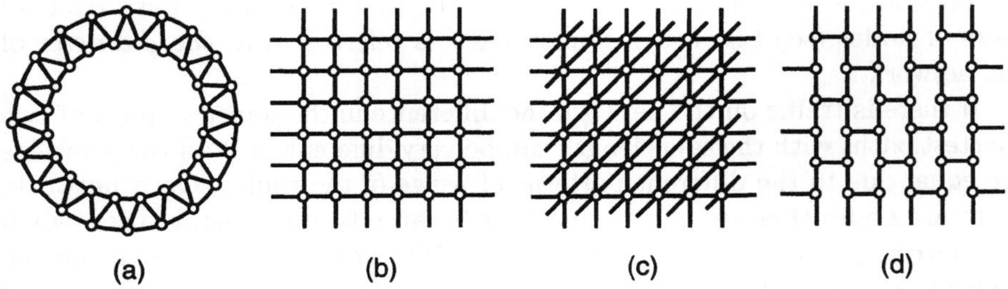

Fig. 4: Four regular-symmetric networks. (a) The ring network and (b) the rectangular toroidal network, in which the nodes on one edge of the lattice connect to nodes on the opposite edge, have degree four. (c) The triangular toroidal network has degree six and (d) the hexagonal toroidal network has degree three.

Toroidal Networks

The toroidal rectangular network ($\mathcal{H}$) is based on a square lattice of nodes in which each node has four neighbours with boundary nodes appropriately identified, [41, 19, 36, 51]. The finite rectangular lattice $\mathbb{Z}$ consists of $S = L^2$ nodes. The position of each node in the lattice is given by the coordinate vector $\mathbf{r} = (i, j)$ where i and j are integers in the range 1 to L. The network has periodic boundary conditions throughout, and so each coordinate of (i, j) is effectively reduced - mod $L + 1$ to give a toroidal topology. To measure the distance between a pair of nodes the periodic "Manhattan" metric is used, which measures the sum of vertical and horizontal displacements between two nodes.

The average shortest path can be calculated to be $\bar{\ell} = L/2$.

Useful comparative networks for traffic studies on Manhattan grid are the hexagonal and triangular networks (see Figure 4 (c & d)). Their embeddings in the plane show how we can view the three regular networks as satisfying an edge inclusion property $E(\mathcal{H}) \subset E(\mathcal{R}) \subset E(\mathcal{T})$. Comparative studies of all three types of network show consistent results [4] on the critical loads, and the onset of congestion behaviour.

4.2 Random Networks

Random networks have been used to model communications networks. The reason is that because some of the communication networks tend to have a complex topology and the interactions defining their structure are apparently random.

From a set of nodes, a random network is built by connecting every pair of nodes with probability p. If the total number of nodes is S and if $p > 1/S$ [10] then, with probability 1, the network is fully connected , or equivalently, there is at least one path connecting any pair of nodes. This is the only case we are going to consider here, as we are interested in connected networks. The degree distribution of a random graph is well approximated by a binomial distribution [10]

$$P(k) = \binom{S-1}{k} p^k (1-p)^{S-1-k}. \tag{18}$$

The degree distribution tends to be concentrated around some "typical" node degree, or average node degree $\bar{k}$, see Fig. 5(a) and [12, 9]. The characteristic path length scales with the size of the network as

$$\bar{\ell}_{\text{rand}} \approx \frac{\ln(S)}{\ln(\bar{k})} \tag{19}$$

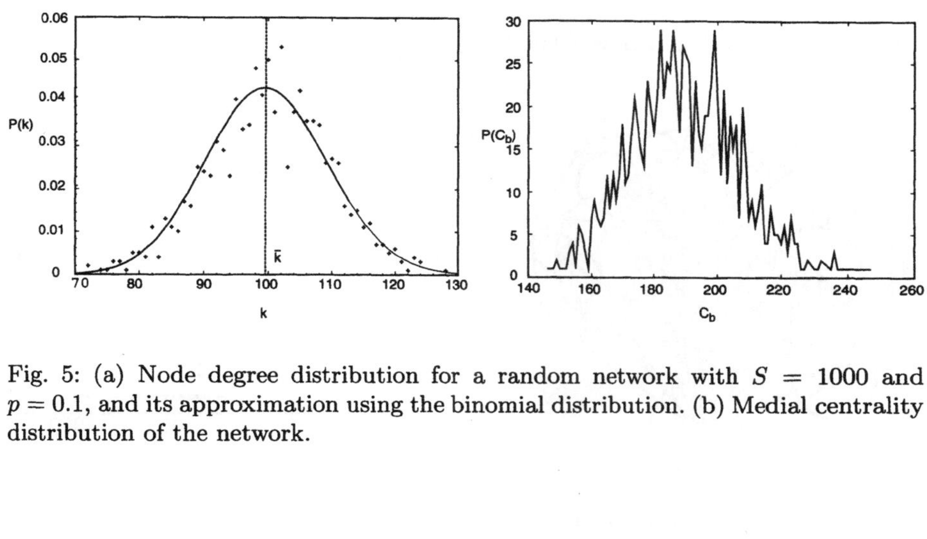

Fig. 5: (a) Node degree distribution for a random network with $S = 1000$ and $p = 0.1$, and its approximation using the binomial distribution. (b) Medial centrality distribution of the network.

4.3 Scale-free Networks

Many technological networks are not described by a random or a regular network; instead they are better described by a network where the degree distribution is described by a power law [2, 16] where

$$P(k) \sim Ck^{-\beta}, \tag{20}$$

for $\beta > 1$ and C constant. The probability that a node has k edges connected to it is given by $P(k)$. In practical terms a power law distribution means that the majority of the nodes will have very few neighbours, but there is a very small set of the nodes with a very large number of neighbours (see Fig. 6(a)). Networks with this property are known as *scale-free* because power-laws are free of a characteristic scale, that is, there is no characteristic node degree (see Fig. 6(b)).

The diameter of a scale-free network scales as $\bar{\ell} \sim C \ln(\ln S)$ [11] where S is the size of the network. This is due to the existence of "far–reaching" links which are shortcuts when going from a source to a destination. Any node that contains one of these far-reaching links is going to be highly used when transferring packet-data, that is, its medial centrality is large.

In 1999, Barabási and Albert [6] showed that it is possible to create a scale-free network by using two generic mechanisms: *growth*, the network grows by attaching a new node with m links to m different nodes present in the network; and *preferential attachment*, where new nodes are attached preferentially to nodes that are already well connected. Barabasi and Albert showed further that if the probability that a new node will be connected to node i with degree k_i is

$$\Pi(i) = \frac{k_i}{\sum_j k_j}. \tag{21}$$

then the network has a power law link distribution $P(k) \propto k^{-3}$.

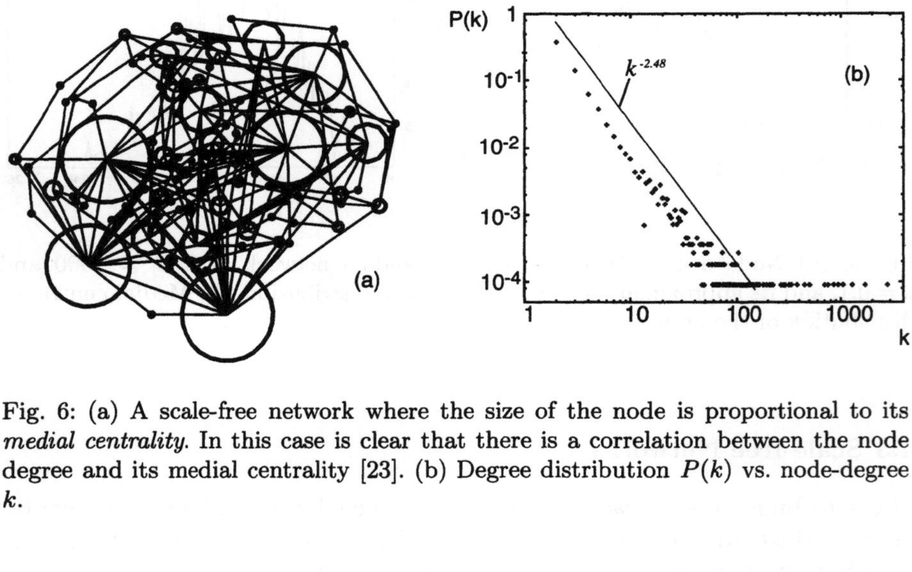

Fig. 6: (a) A scale-free network where the size of the node is proportional to its *medial centrality.* In this case is clear that there is a correlation between the node degree and its medial centrality [23]. (b) Degree distribution $P(k)$ vs. node-degree k.

The Internet

In 1999, an analysis of the Internet topology by Faloutsos *et. al* [16] suggested that the distribution of node degrees of the Internet decays as a power-law $P(k) \propto k^{-y}$, with $y = 2.22$. In 2002, Subramanian *et. al* [44] reported that the Internet features a tier structure, where at the top of this tier is the core of the network. It is well established that a realistic model of the Internet topology should generate a power law topology with a core structure. There exist network models that produce power law networks, e.g. Barabási and

Albert model [6] and Inet-3.0 [50] to mention just two of them, but they do not reproduce the core structure of the Internet [54].

A node of a network is considered "rich" if it contains a large number of links or equivalently it has a large node-degree. The core of the Internet consist of a set of nodes which have a large node degree. We refer to these nodes as the *rich-club*. In the Internet, the members of the rich-club are very well connected to each other. This means that there are a large number of alternative routing paths between the club members where the average path length inside the club is very small (1 to 2 hops). The rich-club acts as a super traffic hub and provides a large selection of shortcuts. Hence scale-free models without the rich-club structure may underestimate the efficiency and flexibility of the traffic routing in the Internet. Conversely, networks without the rich-club may over-estimate the robustness of the network to a node attack, [1, 24] where the removal of a small percentage of its richest club members can break down the network integrity.

It is possible to build network topology models which will generate a rich-club. Recently the Interactive-Growth (*IG*) model [53] was introduced as a way to generate networks that contain a rich-club. The model is a modification of the Barabási and Albert model and it reflects the evolution of the Internet, with the addition of new nodes and the addition of new links between existing nodes. The network is generated by starting with m_0 nodes connected through $m_0 - 1$ links. At each time–step, one of the following two operations is performed: 1) with probability $p \in (0, 1)$, $m < m_0$ new links are added between m pairs of nodes chosen from the existing nodes, and, 2) with probability $1 - p$, one new node is added and connected to m existing nodes.

4.4 Model of Networked Data Traffic

The model considered here has been studied by several authors [36, 19, 21, 20, 41, 45, 51, 3, 5]. The network consists of two types of nodes; router nodes that store and forward packets; host nodes that store and forward packets and are also sources and sinks of traffic. Given the network has S nodes, and a density $\rho \in [0, 1]$ of hosts then ρS is the total number of hosts. The host nodes are randomly distributed in the network.

- Traffic generation: A host creates a packet whose destination is another host. A host creates a packet using either a uniform random distribution (Poisson) or a *LRD* distribution defined by a chaotic map. Each source generates its traffic independently of the other sources.
- Queue: Each node keeps a queue of unlimited length where the packets are stored. Any packet that is generated is put at the end of the host's queue. If a packet arrives at a router it is put at the end of the hierarchical router's queue. The packet is not queued when it arrives at its destination node.
- Routing: Each node picks a packet at the head of the queue and forwards the packet to the next node. From the source/destination information that

each packet carries, the forwarding is done by one of the following routing algorithms:

1. For regular networks:
   - a neighbour closest to the destination node is selected
   - if more than one neighbour is at the minimum distance from the destination, the link to which the smallest number of packets has been forwarded is selected
   - if more than one of these link shares the same minimum number of packets forwarded then a random selection is used
2. for general networks:
   - From the state of the queues and the number of hops from source to destination, the routing is done by minimising the time–delay by considering the number of hops and the size of the queues that a packet visits when crossing the network.

The process of packet generation, hop movement, queue movement and updating of the routing table occurs at one time step.

5 Congestion

The time that elapses between the creation of a packet at source s, from its creation to its destination d, is known as the delay time $\tau_{s,d}$. The average of the delay

$$\bar{\tau}_T = \frac{1}{S(S-1)} \sum_{s,d;\tau_{s,d}\leq T} \tau_{s,d},$$

which is the average for all the packets up to time T, is an important quantity with which to assess the performance of a network. If the traffic load presented to the network is low and the queues on the nodes are empty, then, to a first approximation, the average delay is proportional to the average number of nodes that the packets visit when travelling. As the traffic load increases, the queues at the nodes start to build up, the average delay time will increase accordingly. If the traffic load increases even further, then at the critical load λ_c, the queues of some nodes will grow very quickly and the average delay time will dramatically increase or even diverge. At this critical load, we consider that the network is congested. This critical behaviour is also noticed in the network throughput. The throughput is defined as the number of packets reaching their destination per unit time per host. Starting from a low load, the throughput increases proportionally as the increase of the load, until congestion is reached. At this point the network has its maximum throughput (see Fig. 7).

If the number of packets at node i at time t is denoted by $n_i(t)$, then the total number of packets in the network is

$$N(t) = \sum_{i=1}^{S} n_i(t) = \sum_{i=1}^{S} Q_i(t). \tag{22}$$

If the network is not congested, and the average delay time is finite, then the average number of packets on the network, $\bar{N} = (1/T) \lim_{T\to\infty} N(t)$, is also finite. At the congestion point, the queues of the congested nodes increases rapidly and this implies that the total number of packets on the network continues to increase.

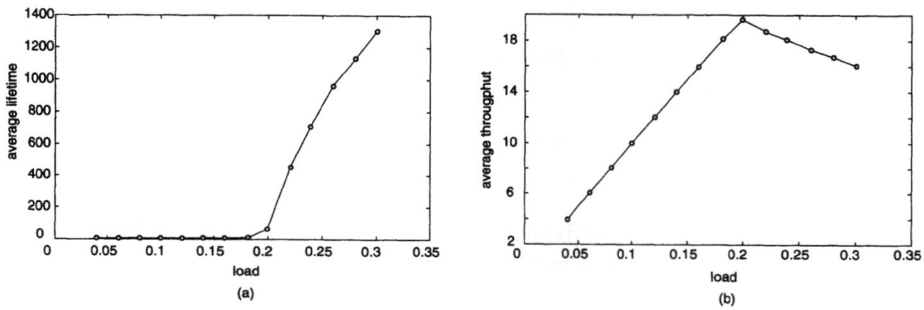

Fig. 7: (a) The average lifetime of a packet increases rapidly after the congestion point. (b) Average throughput of a Manhattan network with $S = 100$ and $\rho = 1$. As the offered load increases the average throughput increase until it reaches the critical point $\lambda_c \approx 0.2$. At this load the network is congested.

5.1 Mean Field Approximation

Total Distance Formulation

A simple way of estimating λ_c, [51], is to look for the total distance that all the packets at time t have to travel to reach their destination. In the congested phase if there are queues at all nodes, then the change in total distance is $D(N_{t+1}) - D(N_t)$, where $N(t)$ is the number of packets in the queues at time t, and $D(N_t)$ is the aggregated distance of all packets from their destination at time t. The increase in the number of packets per unit of time is $\rho\lambda S$. If $\bar{\ell}$ is the average path length, then the overall added distance is $\rho\lambda S\bar{\ell}$. By contrast, the aggregated distance is reduced by S given that every packet at the head of the queue moves one step closer to its destination. Thus the change in total distance to destination between time t and $t+1$ is

$$D(N_{t+1}) - D(N_t) = \rho\lambda\bar{\ell}S - S. \tag{23}$$

The critical load λ_c occurs when the total distance no longer decreases, $D(N_{t+1}) - D(N_t) = 0$, which implies

$$\lambda_c = \frac{1}{\rho \bar{\ell}}. \tag{24}$$

This relation refines the results of [41] and [19].

Time-Delayed Formulation

Another possible way to determine the critical load is to use Little's Law [20, 41]: *"The average number of customers in a queueing system is equal to the average arrival rate of customers to that system, times the average time spent in the system"* [28]. Little's law is a flow conservation law which can be restated as: in a steady state, the number of delivered packets is equal to the number of generated packets, or

$$\frac{dN(t)}{dt} = \rho\lambda S - \frac{N(t)}{\tau(t)}. \tag{25}$$

where $\rho\lambda S$ is the average arrival rate to the queues per unit of time, $\tau(t)$ is the average time spent in the system, and $N(t)/\tau(t)$ is the number of packets delivered per unit of time.

Remark: Little's law does not depend on the arrival distribution of packets to the queue or the service time distribution of the queues. Also it does not depend upon the number of queues in the system or upon the queueing discipline within the system. The law holds only when a steady state exists below the critical load, as shown in Fig. 8 where the load rates have been normalised.

If the load is low, the queues at the nodes tend to be empty and the average delay time is the average shortest path $\bar{\ell}$, that is $\tau \approx \bar{\ell}$. For higher loads the transit time can be approximated by the average shortest path plus the average time, $\mathcal{T}$, that a packet spends in the queues

$$\tau \approx \bar{\ell} + \mathcal{T}(N(t), \bar{\ell}). \tag{26}$$

Criticality for Regular-Symmetric Networks

If the traffic is evenly distributed in the network and the network is not congested, then there exists a steady state solution N^* for the number of packets on the network. Each queue, on average, contains N^*/S packets and the delay can be approximated by

$$\tau(t) \approx \bar{\ell}\,(1 + \bar{Q}) = \bar{\ell}\left(1 + \frac{N^*}{S}\right) \tag{27}$$

where on average, a packet visits $\bar{\ell}$ queues with average load $\bar{Q}$ and $\mathcal{T} \approx \bar{\ell}\bar{Q}$. From the steady state solution $(dN(t)/dt = 0)$ the total number of packets in the system is

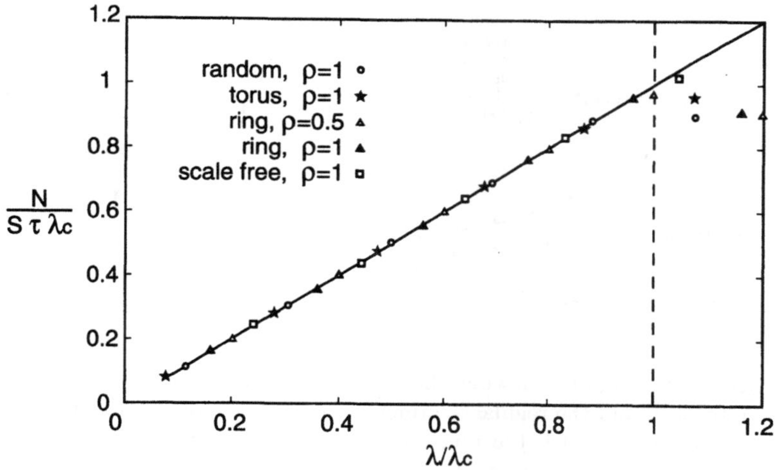

Fig. 8: Verification of Little's law for different networks. In all cases the networks have $S = 100$ nodes. The coordinates have been normalised by dividing them by the critical load of the corresponding network. For the ring network, the law was verified for two different densities of nodes. The law does not hold if the load is greater than the critical load (marked with a vertical hash line).

$$N^* = \frac{\rho\lambda\bar{\ell}S}{1-\rho\lambda\bar{\ell}} = \frac{\gamma S}{1-\gamma} \tag{28}$$

where $\gamma = \rho\lambda\bar{\ell}$ is the normalised load and N^* is the steady state solution.

The average traffic load generated at node i is

$$\lambda_i = \left(\frac{1}{\rho\bar{\ell}(1+S/N^*)}\right). \tag{29}$$

As the traffic load increases, the number of packets in the network increases accordingly. At the congestion point the number of packets on the network diverges, $N^* \to \infty$, and the critical load is

$$\lambda_c = \frac{1}{\rho\bar{\ell}}, \tag{30}$$

which is the same as Eq. (24).

General Networks

Fig. 9(a) shows, in phase space, that the transition from the free flow phase to the congested phase is well approximated by Eq. (30) for the case of regular-symmetric networks but not for scale-free networks. Equation (30) gives a good approximation to the critical load because the average queue size for all nodes is very similar ($\bar{Q} \approx N^*/S$).

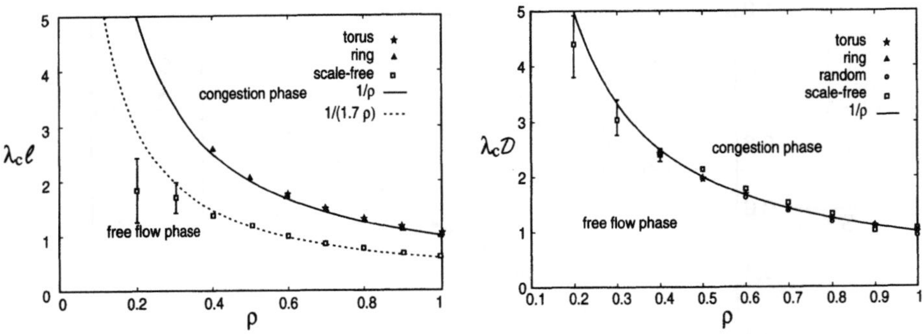

Fig. 9: (a) Phase diagram showing the change from free–flow to congestion. For the square toroidal and the regular-symmetric network the phase transition is given by $\ell_c\lambda_c = 1/\rho$. For $\rho > 0.4$, the phase transition of the scale free network is well approximated by $\ell_c\lambda_c = 1/(1.7\rho)$. (b) Renormalised phase transition for different networks topologies. The four networks have $S = 100$.

In a scale-free network, due to the disproportionate importance of some of the nodes, if these nodes get congested more readily and hence, all the network gets congested. An alternative approach is to use the medial centrality [34] to characterise the node usage. If

$$\hat{C}_B(w) = \frac{C_B(w)}{\sum_{v\in\mathcal{V}} C_B(v)} \tag{31}$$

is the normalised medial centrality then the average queue size at node i can be approximated by

$$\bar{Q}_i \approx \hat{C}_B(n_i)N^*. \tag{32}$$

An example of this approximation is shown in Fig. 10.

It is possible to approximate the typical travel time of a packet by adding the average time that the packets spends in the queues that it visits, that is

$$\tau^*(s,d) \approx \sum_{v\in\mathcal{R}(s,d)} \hat{C}_B(v)N^* \tag{33}$$

where the sum is over all nodes that the packet visits. The set $\mathcal{R}(s,d)$ is the subset of nodes obtained from the routing table. The total delay time can be approximated by considering

$$\tau(t) \approx \bar{\ell} + \frac{1}{S(S-1)} \sum_{s\in\mathcal{V}} \sum_{d\neq s\in\mathcal{V}} \sum_{v\in\mathcal{R}(s,d)} \hat{C}_B(v)\, N(t) = \bar{\ell} + \mathcal{D}N(t) \tag{34}$$

where

$$\mathcal{D} = \frac{1}{S(S-1)} \sum_{s\in\mathcal{V}} \sum_{d\neq s\in\mathcal{V}} \sum_{v\in\mathcal{R}(s,d)} \hat{C}_B(v). \tag{35}$$

The approximation to the critical load is given by

$$\lambda_c = \frac{1}{\rho S \mathcal{D}}, \tag{36}$$

(cf. Eq. (30)). As shown in Fig. 9(b), this last equation gives a very good approximation of the phase transition from free flow phase to congested phase for the regular-symmetric and scale-free networks. It is not difficult to see why

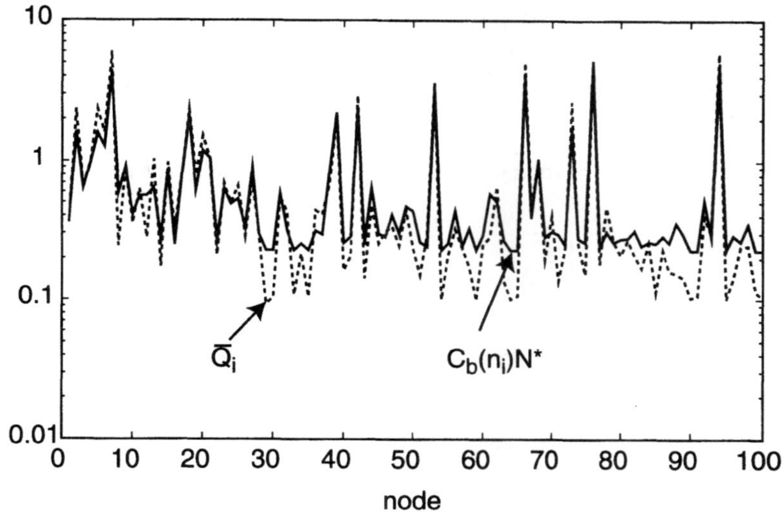

Fig. 10: Comparison between the average queue size (dotted line) and its approximation using Eq. (32) (solid line). The network is a scale-free network with $S = 100$.

Eq. (36) also works for regular-symmetric networks. If the medial centrality measures how many routes use a particular node, then all nodes are utilised to the same degree in the regular case. The constant value for the medial centrality for each node is the same (see Eq. (17)) and is given by

$$\hat{\mathcal{C}}_B(w) = \frac{\mathcal{C}_B(w)}{\sum_{v \in \mathcal{V}} \mathcal{C}_B(v)} = \frac{1}{S}. \tag{37}$$

By substituting this last expression in Eq. (36), we find $\mathcal{D} = \bar{\ell}/S$ and the expression for the critical load obtained in Eq. (30) is recovered.

Equation (36) also reproduces behaviour observed by Fuks and Lawniczak [19] in a rectangular network. The addition of few random links provides a shortcut between distant parts of the network, but can quickly become congested. A fuller addition of links such as transferring from a square to triangular lattice has the more conventional effect of increasing the critical load, [4].

5.2 *LRD* at Criticality

It is known that at the critical load the traffic statistics change from *SRD* to *LRD* [41]. The change of correlations from *SRD* (exponential) to *LRD* (power law) at criticality in phase transitions is a standard feature, and this phenomena also occurs in road traffic congestion.

The change from *Poisson* to *LRD* has been observed by analysing the time series for delays for 1 hop and 24 hop routes, and also by looking at the time series of average host and router queues lengths measured at each time tick. As mentioned before, the Hurst parameter distinguishes *SRD* traffic from *LRD*. The parameter can be obtained by considering the so-called *rescaled data*. More precisely, let X_t, $t \in \mathbb{Z}^+$ denote a discrete time series. The *adjusted range* is defined as

$$R(t,k) = \max_{0\leq i\leq k} R_i(t,k) - \min_{0\leq i\leq k} R_i(t,k), \tag{38}$$

where

$$R_i(t,k) = \sum_{l=1}^{t+i} X_l - \sum_{l=1}^{t} X_l - \frac{i}{k}\left(\sum_{l=t+1}^{t+k} X_l - \sum_{l=1}^{t} X_l\right) \tag{39}$$

The quantity $R(t,k)$ is normalized by the translated sample standard deviation

$$S(t,k) = \sqrt{k^{-1}\sum_{l=t+1}^{t+k} (X_l - \bar{X}_{t,k})^2} \tag{40}$$

where $\bar{X}_{t,k} = k^{-1}\sum_{l=t+1}^{t+k} X_l$. The R/S statisticis then defined to be

$$R/S = \frac{R(t,k)}{S(t,k)} \tag{41}$$

and it is fitted to the equation

$$\ln E[R/S] = a + H \ln k, \tag{42}$$

with H interpreted as the Hurst parameter. We considered separate R/S plots for 1 and 24 step journeys in a 32×32 network grid with 164 hosts for Poisson, and for *LRD*, sources for a range of load values. The network used had the same parameters as that used for Fig. 7 so the onset of congestion at a load of 0.3 was expected. At the smaller loads the network remains free of congestion. This means that time series of packet delays are stationary, and the original data may be used in measuring a value of H. For the higher values of the load, including $\lambda = 0.3$, congestion does occur, leading to an upward trend in delay times and queue sizes. In this case, the data are weighted to remove the trend, creating a stationary series. We observe that for $\lambda = 0.2$, $H = 0.5$ for both path lengths, indicating that very little *LRD* is present. This is corroborated by investigating the probability distributions of delays which both have the characteristic shape of exponentially decaying delay times. However, for $\lambda = 0.29$ H values are higher for both path lengths at the onset of congestion. The R/S plots show a kink with the steeper part of the curve corresponding to $H = 0.8$ in both cases. Hence the longer delays and lower frequencies do show significant *LRD*, but this is not seen at any other load value. Values for the three higher load values show very similar H values. Note that in these cases, the time series was weighted to remove the upward trend and the H values are all close to 0.5. This lack of any *LRD* seems to be caused by the phase change to the congested region above λ_c. The probability distributions for these higher values show delays shifting towards the length of the run (1×10^6 time ticks as the network becomes more and more congested). Here 1 and 24 step delays have a long tailed distribution, but this is caused by the non-stationarity of the data, not power-law autocorrelation decay.

5.3 Congestion and *LRD* Traffic

In the previous section, it was noted that *LRD* arose from interaction within the network and was not intrinsic to the traffic sources. In this section the Poisson-like sources are replaced with *LRD* sources, modelled using chaotic maps.

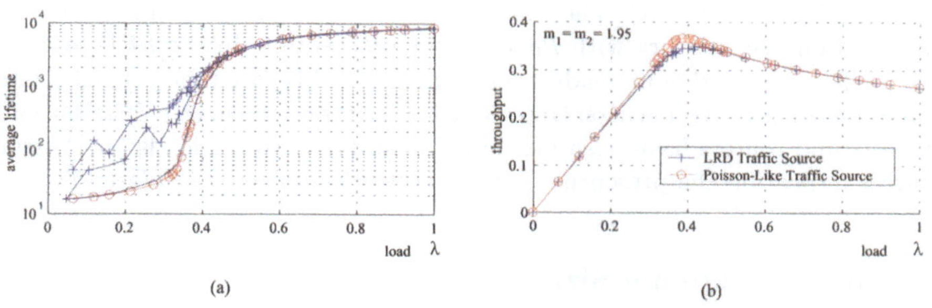

Fig. 11: (a) Average packet lifetimes are plotted as a function of the load λ for Poisson sources and also for *LRD* sources with increasing average lifetime in the pre-congestion phase as m increases through $m_1 = m_2 = 1.5, 1.8$, and 1.95. (b) Corresponding throughput for Poisson and *LRD* traffic are plotted as a function of the load λ for $m = 1.95$. Note that the lower peak value in throughput for the *LRD* traffic sources reflects the longer average lifetimes below the critical point. The peak differences diminish to zero as m is decreased to the Poisson-like value $m = 1.5$.

Fig. 11 gives a comparison of the onset of congestion in two otherwise identical Manhattan networks with host density $\rho = 0.164$, one Poisson sourced, and the other *LRD* sourced for different values of the Hurst parameter. The values of the intermittency parameters $m_1 = m_2 = m$ are kept equal in each case for simplicity. The critical load for this network is $\lambda = 0.3$.

Fig. 11(a) shows the average lifetime, or *end-to-end delay*, of a packet plotted against load λ, the average number of packets generated per host per unit time. As has been seen previously, there is a phase transition from the free phase in which lifetimes remain small to a congested phase in which lifetimes increase rapidly. Fig. 11(a) shows clear evidence of the earlier onset of congestion in the *LRD* traffic in comparison with Poisson traffic produced at the same rate. The data for Fig. 11 are shown for various values of $m_1 = m_2$. Values of $m_1 = m_2$ close to the maximum value of $m_1 = m_2 = 2$ give the highest degree of intermittency and hence the greatest contrast with Poisson traffic sources (corresponding to $m_1 = m_2 = 1.5$ in the intermittency model). The highest value used in our simulations, $H = 0.975$ has been observed in statistical investigations of real network traffic data.

Fig. 11(b) shows throughput versus load. The peak in the throughput occurs at the critical point. The network therefore reaches its peak efficiency at the critical point. The peak throughput is slightly lower for the *LRD* sources, emphasising the longer lifetimes of packets. Although the throughput is only slightly reduced, the average lifetimes increase by up to a factor of 10. This earlier onset appears to be the most important feature of *LRD* congestion within the context of the model, and has significant implications for shared backbone data network infrastructures.

The average queue sizes were shown to be closely related to the average lifetime of packets in [51]. The difference is clear for the average queue size for a rectangular network with *Poisson* and *LRD* traffic. In the *LRD* case the queues get congested more readily and as mentioned before this translates into longer lifetimes of the packets. High values of H for queue length distributions at hosts and routers have been measured at all post critical loads for Poisson sources indicating the presence of strong network-induced *LRD* (see [52]).

6 Control of Queue Sizes

The simplest way to control packet traffic is to limit the length of queues [4]. As grid bar charts of node queue size have shown by Woolf *et al.* [52], long queues in the network invariably occur at hosts.

Valverde and Solé [45] introduced a *local* control mechanism where hosts modify their rates of packet release depending on the detected *local* rate of congestion. The sources adjust their local packet release by querying how busy their neighbours are. The number of congested neighbours a node can have is

$$\zeta = \sum_{v \in \mathcal{G}} \theta(n_v(t)) \tag{43}$$

where $\mathcal{G}$ is the set containing the neighbours of node, $n_v(t)$ is the number of packets in node v at time t and θ is the Heaviside function $\theta(x > 0) = 1$ and $\theta(x \leq 0) = 0$. The control mechanism consists of adjusting the rate of packet production $\lambda_i(t)$ using

$$\lambda_i(t+1) = \begin{cases} \min\{1, \lambda_i(t) + \mu\}, & \zeta < 4 \\ 0 & \zeta \geq 4. \end{cases} \tag{44}$$

If the neighbouring nodes are not congested the load increases at a rate of μ, and drops to zero if all the neighbours are congested. This control mechanism is inspired by the "additive increase/multiplicative decrease" [43] used in *TCP* packet networks. Valverde and Solé studied this control method in a Manhattan network with *Poisson* traffic and unlimited queue capacity. It was noticed that the packet release and the density of hosts satisfies the relation $\lambda \sim \rho^{-1}$ (see Eq. 24). On average, the packet release reaches a steady critical state. This does not mean that the local traffic creation rates are all similar, as the congestion levels vary widely between nodes. An interesting observation made by Valverde and Solé is the existence of a synchronization effect in the congestion state of distinct nodes.

6.1 Control in Scale-free Networks Using *TCP*

The above model is an oversimplified packet network as the topology of a real network is, of course, not a regular network. Also, the model does not allow for packets dropped in transit (as occurs in the Internet) because of the unlimited queues on the nodes. In this section we introduce a more realistic model by considering a scale-free network generated using the *IG* algorithm with a power law decay index of 2.22. Similarly to the previous simulations, each node is designated as either a host or a router. Routers have a single routing queue that receives packets in transit across the network, and releases them back onto the network at a rate governed by the connectivity of the node. The difference with the previous examples is that for this network the simulation is of the fixed-increment time advance type rather than next-event time advance. This allows the routing queue service rate to be set as Ck^{α} packets per time tick of the simulation, where k is the degree of the router node, and C constant. This means that nodes with larger degrees produce more traffic. The index α has been chosen to be between 1 and 2 here. Hosts have identical routing queues and function in the same way, but additionally act as sources. They have transmit buffers that hold packets generated by *LRD* and Poisson traffic sources until they have been acknowledged. The exact mechanism for this is described below.

The control mechanism uses a simplified version of *TCP* Reno [43] as the network protocol. This is the predominate protocol used on the Internet at present. This version is derived from that described in Erramilli *et al* [13]. It concentrates on the slow-start mechanism because in real networks this not only affects all connections, but is also the dominant effect for most connections. It also has a more marked effect on the network dynamics than congestion avoidance.

As in our previous sections, a double intermittency map is used as the basis for each *LRD* traffic source, and a uniform random number generator for each

Poisson source. However, they are used in a slightly different way. One sojourn period in the '*on*' side represents a whole file which is then *windowed* using the *TCP* slow start algorithm: at the start of the file the window size is set to 1. Only a single packet is sent at that time tick. When a packet reaches its destination an acknowledgement is returned to the source: once this packet has been acknowledged, the window size is doubled and the next two packets are sent. When both these packets have been acknowledged, the window size is doubled again and a new window of packets is sent. The doubling process is repeated until the end of the file or the maximum window size (a fixed value). This is described mathematically in section 3.4 on *TCP* dynamics.

The routing algorithm uses a pre-calculated look-up table of shortest paths. All links between nodes are assumed to have unit length. At each time step packets are forwarded from the head of each routing queue. If an acknowledgement packet reaches its destination, this triggers the release of the next window of packets from that host.

A necessary modification of the traffic model was to make server strengths reflect the degree(and importance) of the node in the network . This has significant effects in increased throughput without overload, see [52].

Fig. 12 shows a comparison of different server strengths as α is varied. Two types of source are considered. The same *IG* network with the same pattern of hosts is used. Results from *LRD* sources and Poisson sources differ greatly. Throughputs (Figs 12(c) and 12(d)) at the lower server strengths (α equal to 1 and 1.1) are qualitatively similar: the throughput matches the load up to a threshold and then levels out. However, this threshold is more than 50% higher for the Poisson sources. If the exponent α is increased to a value of 1.5 this threshold can no longer be seen; throughputs for the two source types are similar. When servers are strongest ($\alpha = 2.0$) the situation is reversed: the network with *LRD* sources has a higher throughput, able to handle the maximum load applied to it without becoming overloaded.

Figs. 12(a) and 12(b) show average lifetimes for the two source types. At low loads behaviour is similar to that seen in our earlier work with regular lattices and open-loop protocols: lifetimes for Poisson sources are much less than for *LRD.* In the case of Poisson sources there is a similar transition from the free state to the congested state. At increasing server strengths the transition becomes less pronounced. This is due to the increased network capacity which slows the onset of congestion.

Comparisons have been made between the different topologies and packet transport algorithms to give four combinations [51, 52]:

– scale-free(produced using the *IG* model) with a closed-loop algorithm (*TCP*).

– Manhattan with an open-loop algorithm.

– scale-free(produced using the *IG* model) with an open-loop algorithm.

– Manhattan network with a closed-loop algorithm.

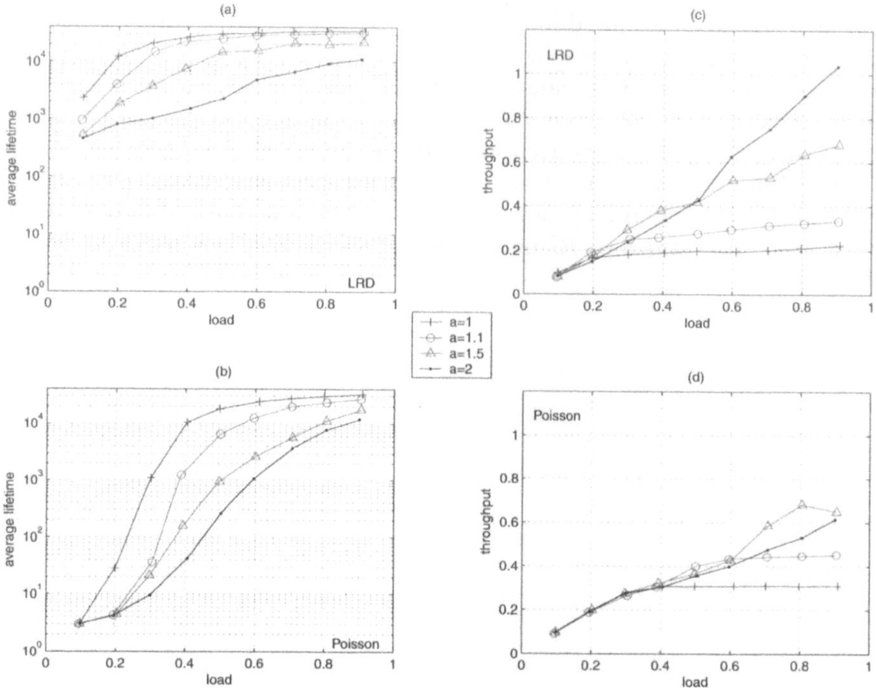

Fig. 12: The number of packets that can be served at each time instant is increased according to a power law n^α, where n is the degree of the node and $\alpha = 1,\ 1.1,\ 1.5,\ 2$.

The network had 1024 nodes in both cases and the host density was taken to be 589/1024. The index α of the server strength $0.25n^\alpha$ was fixed at 1 for all four combinations. The host distribution was random for the Manhattan network; by comparison the *IG* network connectivity 1 and 2 nodes were selected to be hosts.

The closed-loop *TCP* algorithm increases lifetimes for both regular and scale free networks. The requirement of *TCP* that packets be acknowledged before the next window of packets is sent is very restrictive in the sense that new windows cannot be sent by hosts more frequently than the round trip times (RTT) or ping times. Congestion is reacted to immediately because it causes an increase in RTT's and makes sources back off. Since file sizes are not that great and window sizes are often reset to 1 due to the round-trip time maximum RTO limit, throughput can never be that high. This results in packets being delayed in the transmit buffer and is the primary cause of increased packet lifetimes. By contrast the open-loop algorithm does not react to congestion and allows unlimited queues to build up at routers. For sufficiently high loads open-loop networks become congested and lifetimes approach those of the closed-loop networks.In fact a Manhattan network with an open-loop algorithm has longer lifetimes than an *IG* network with a closed-loop algorithm primarily because of the much shorter average path lengths in the *IG* network, see [52]. Identical networks using open-loop algorithms have higher throughputs; the *IG* network performs more efficiently for both types of algorithm. Similar behaviour is observed for Poisson sources.

Thus, regular open-loop simulations are fundamentally different to the closed-loop simulations of *IG* networks and so detailed comparisons are not very useful.

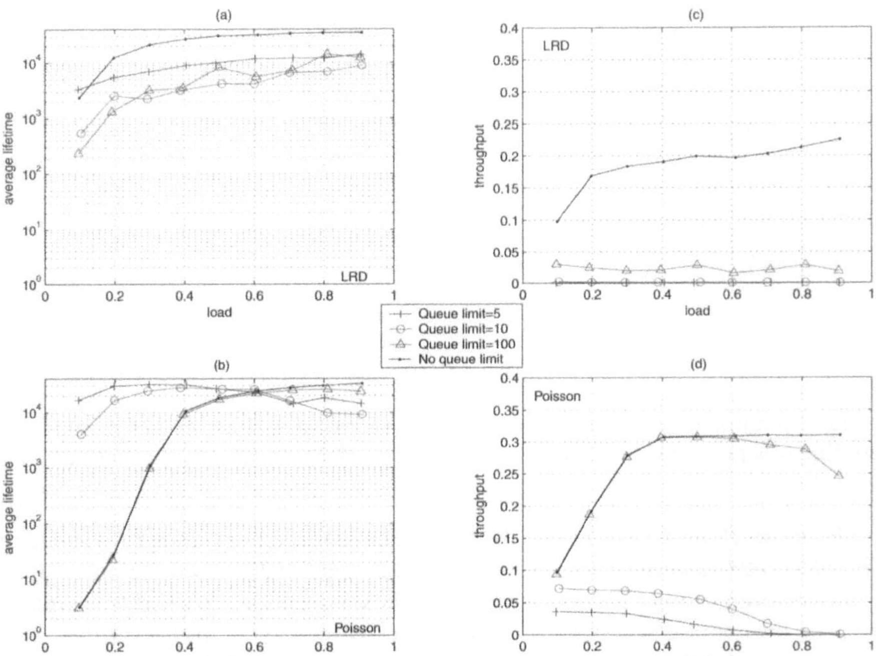

Fig. 13: Effect of varying queue length limit on average packet life time and throughput for *LRD* and Poisson sources.

In Fig. 13 packet dropping of real networks by limiting the routing queue lengths. Very severe packet loss has been modelled here in order to test the extreme situation. Real networks generally suffer much less packet loss. For *LRD* sources Fig. 13(a) average lifetimes are greatly reduced when the queue limit is decreased. When queue lengths are limited, packets are dropped at the routers and therefore re-sent more frequently. This causes shorter waits in the transmit buffer. This can be seen most clearly at the very low queue length limit. In the case of Poisson sources Fig. 13(b) average lifetimes behave quite differently. Lifetimes peak at a load of 0.3 for the queue limit of 5. This peak shifts to higher values as the queue limit increases. When there is no queue limit the lifetimes are the same in regular networks. Fig. 13(c) shows throughputs for *LRD* sources. These are greatly reduced when a queue limit is applied. At the smaller queue limits throughput is close to zero. In the case of Poisson sources (Fig. 13(d)) throughput is similar for the queue limit of 100, but also much less at the lower queue limits. However, throughputs are still much higher than for the *LRD* sources. This is caused by the shorter queues in the case of Poisson sources. Most queue lengths are less than a 100, meaning that this limit has little effect.

The same *IG* network has been used, but hosts are now selected randomly with the same density of 589/1024. Results are similar. The network with randomly placed hosts always performs slightly better than the one with hosts placed at the low degree hosts. Average lifetimes are lower and throughput slightly higher. This shows that the inclusion of a small number of rich nodes in fact makes little difference.

References

1. R. Albert, H. Jeong, and A. Barabási. Error and attack tolerance of complex networks. *Nature*, 406:378–381, 2000.
2. R. H. Albert, A. Jeong, and A.-L. Barabási. Diameter of the world wide web. *Nature*, 401:130–131, 1999.
3. D.K. Arrowsmith, R.J. Mondragon, J.M.Pitts, and M. Woolf. Internet packet traffic congestion. *Nonlinear Dynamics for Coding Theory and Network Traffic, in the IEEE Proceedings of Systems and Circuits*, 3:746–749, 2003.
4. D.K. Arrowsmith, R.J. Mondragón, J.M. Pitts, and M.Woolf. Phase transitions in packet traffic on regular networks: a comparison of source types and topologies,. *Preprint*, 2004.
5. D.K. Arrowsmith and M.Woolf. Modelling of TCP packet traffic in a large interactive growth network. *IEEE Proceedings of Systems and Circuits*, 5:477–480, 2004.
6. A. L. Barabási and R. Albert. Emergence of scaling in random networks. *Science*, 266:509–512, 1999.
7. M. Barenco and D.K. Arrowsmith. The autocorrelation of double intermittency maps and the simulation of computer packet traffic. *Jnl of Dynamical Systems*, 19(1):61–74, 2004.
8. J. Beran. Statistics of long memory processes,. *Monographs on Stats and Appl. Prob.*, 61, 1994. Chapman & Hall.
9. B. Bollobás. Degree sequences of random graphs. *Discrete Maths.*, 33:1–19, 1981.
10. B. Bollobás. *Random Graphs*. Academic Press, 1985.
11. R. Cohen and S.Havlin. Scale–free Networks are Ultrasmall. *Phys. Rev. Lett.*, 90(5):058701, 2003.
12. P. Erdös and A. Rényi. On Random Graphs I. *Publ. Math. Debrecen*, 6:290–297, 1959.

13. A. Erramilli, M. Roughan, D. Veitch, and W. Willinger. Self-similar traffic and network dynamics. *Proc. of the IEEE*, 90(5):800–819, May 2002.
14. A. Erramilli, R. P. Singh, and P. Pruthi. Chaotic Maps as Models of Packet Traffic. *In Proc. ITC 14, The Fundamental Role of Teletraffic in the Evolution of Telecommunication Networks*, pages 329–338, 1994.
15. A. Erramilli, R. P. Singh, and P. Pruthi. An application of deterministic chaotic maps to model packet traffic. *Queueing Systems*, 20:171–206, 1995.
16. M. Faloutsos, P. Faloutsos, and C. Faloutsos. On Power-Law Relationships of the Internet Topology. *Proc. ACM/SIGCOMM, Comput. Commun. Rev.*, 29:251–262, 1999.
17. H. H. Fowler and W. Leland. Local Area Network Traffic Characteristics, with Implications for Broadband Network Congestion Management. *IEEE Jour. on Sel. Areas in Comm*, 9:1139–1149, September 1991.
18. L.C. Freeman. A set of measures of Centrality based on Betweeness. *Sociometry*, 40:35–41, 1977.
19. H. Fukś and A. T. Lawniczak. Performance of data networks with random links. *Mathematics and Computers in Simulation*, 51:103–119, 1999.
20. H. Fukś, A.T. Lawniczak, and S. Volkov. Packet delay in data network models. *ACM Transactions on Modeling and Computer Simulation*, 11(3), 2001.
21. K. Fukuda, H. Takayasu, and M. Takayasu. Origin of critical behavior in Ethernet traffic. *Physica A*, pages 289–301, 2000.
22. A. Giovanardi, G. Mazzini, and R. Rovatti. Chaos based self–similar traffic generators. *Proc. NOLTA*, pages 747–750, 2000.
23. K. I. Goh, E. Oh, K. Kahng, and D. Kim. Betweenness centrality correlation in social networks. *Phys. Rev. E*, 67:017101, 2003.
24. P. Holme and B. J. Kim. Vertex Overload Breakdown in Evolving Networks. *Physical Review E*, 65:066109, 2002.
25. C. Huang, M. Devetsikiotis, I. Lambadaris, and R. Kaye. Fast Simulation for Self-Similar Traffic in ATM Networks. In *Gateway to Globalization, 1995 IEEE International Conference on Communications*, volume 1, pages 438–444, Seattle, Washington, 1995. IEEE ICC95.
26. B. E. Huberman and R. M. Lukose. Social dilemmas and Internet congestion. *Science*, 277:535–537, 1997.
27. R. Jain and S. A. Routhier. Packet trains: Measurements and a new model for computer network traffic. *IEEE Journal on Selected Areas*, 4:986–995, 1986.
28. L. Kleinrock. *Queueing Systems. v. 1. Theory.* John Wiley & Sons, 1975.
29. W. E. Leland, M. S. Taqqu, W. Willinger, and D. Wilson. On the Self-Similar Nature of Ethernet Traffic (Extended Version). *IEEE/ACM Trans on Networking*, 2, No 1(1):1–15, Feb. 1994.
30. H. Li and M. Maresca. Polymorphic-torus network. *IEEE Transs Comp.*, 38(9):1345–1351, 1989.
31. B. Mandelbrot. Self-Similar Error Clusters in Communication Systems and the Concept of Conditional Stationarity. *IEEE Trans. On Communication Technology*, COM-13:71–90, March 1965.
32. J. L. McCauley. *Chaos, Dynamics and Fractals and Algorithmic Approach to Deterministic Chaos.* Cambridge University Press, 1995.
33. R. J. Mondragón. A Model of Packet Traffic using a Random Wall Model. *Int. Jou. of Bif. and Chaos*, 9 (7):1381–1392, 1999.
34. R. J. Mondragón. Congestion and centrality. *in preparation*, 2004.

35. I. Norros. Studies on a model for connectionless traffic, based on fractional brownian motion. *Conf. On Applied Probability in Engineering, Computer and Communication Sciences*, Paris:16–18, June 1993.
36. T. Ohira and R. Sawatari. Phase Transition in Computer Network Traffic Model. *Physical Review E*, 58:193–195, 1998.
37. Y. Pomeau and P. Maneville. Intermittent transition to turbulence in dissipative dynamical systems. *Commun. Math. Phys.*, 74:189–197, 1980.
38. P. Pruthi. *An application of chaotic maps to packet traffic modelling.* PhD thesis, KTH, Stockholm, Sweden, Oct. 1995.
39. P. Pruthi and A. Erramilli. Heavy-Tailed ON/OFF Source Behaviour and Self-Similar Traffic. *ICC 95*, June 1995.
40. M. Roughan and D. Veitch. A Study of the Daily Variation in the similarity of Real Data Traffic. *Tech. Rep. 0070, SERC, Software Engineering Research Centre, Level 3, 110 Victoria St, Carlton Vic, 3053, Australia*, 1997.
41. R. V. Solé and S. Valverde. Information transfer and phase transitions in a model of internet traffic. *Physica A*, 289:595–605, 2001.
42. W. Stallings. *Data and Communications.* Prentice Hall Int. Inc., 2000.
43. W. R. Stevens. *TCP/IP Illustrated*, volume Volume 1: The Protocols. Addison-Wesley, 1994.
44. L. Subramanian, S. Agarwal, J. Rexford, and R. H. Katz. Characterizing the Internet Hierarchy from Multiple Vantage Points. *Proc. of INFOCOM 2002*, June, 2002.
45. S. Valverde and R.V. Solé. Self-organized critical traffic in parallel computer networks. *Physica A*, 312:636, 2002.
46. D. Veitch. Novel Models of Broadband Traffic. In *Proc. Globecom '93*, pages 362–368, Murray River, Australia, 1992. 7th Australian Teletraffic Research Seminar.
47. X-J Wang. Statistical physics of temporal intermittency. *Phys Rev A*, 40(11):664761, 1989.
48. W. Willinger, M. S. Taqqu, R. Sherman, and D. V. Wilson. Self-similarity through high-variability: statistical analysis of Ethernet LAN traffic at the source level. *IEEE/ACM Transactions on Networking*, 5(1):71–86, 1997.
49. R.J. Wilson. *Introduction to Graph Theory.* Longman Group Limited, Whistable Kent, Great Britain, 1972. ISBN:0-582-44762-3.
50. J. Winick and S. Jamin. Inet–3.0 Internet Topology Generator. *Tech. Report UM-CSE-TR-456-02, University of Michigan*, 2002.
51. M. Woolf, D.K. Arrowsmith, R.J. Mondragón, and J. M. Pitts. Optimization and phase transition in a choatic model of data traffic. *Physics Reviews E*, 66:056106, 2002.
52. M. Woolf, D.K. Arrowsmith, S. Zhou, R.J.Mondragón, and J.M. Pitts. Dynamical modelling of *tcp* packet traffic on scale-free networks. *submitted to Phys Rev E*, 2004.
53. S. Zhou and R. J. Mondragón. Towards Modelling the Internet Topology – The Interactive Growth model. In *Proc. of ITC18*, 2003.
54. S. Zhou and R. J. Mondragón. Accurately modelling the internet topology. *submitted to Physical Review E*, 2004.
55. S. Zhou and R. J. Mondragón. Redundancy and robustness of the *as*-level internet topology and its models. *IEE Electronic Letters*, 40(2):151–152, January 2004.

Chaos-Based Generation of Artificial Self-Similar Traffic

Gianluca Setti[1,3], Riccardo Rovatti[2,3], and Gianluca Mazzini[1,3]

[1] DI University of Ferrara - via Saragat 1, 44100 Ferrara, Italy
email: (gsetti,gmazzini)@ing.unife.it
[2] DEIS University of Bologna - viale Risorgimento 2, 40136 Bologna, Italy
e-mail: rrovatti@deis.unibo.it
[3] ARCES, University of Bologna - via Toffano 2/2, 40125 Bologna, Italy

1 Introduction

The exploitation of the unique features of chaotic dynamics for performance improvement recently became a widely followed path in several field of Electrical Engineering. This is certainly due to the understanding of the mixed deterministic/stochastic nature of a chaotic system and to the consequent adoption of tools from statistical dynamical systems theory for their analysis [1][2]. By means of such a statistical approach it was clarified that the applications where the use of chaotic dynamics can be more beneficial are those where the statistical features of the involved signals or quantities are the dominant factor. Additionally, the same tools are also of great importance to develop the *quantitative models* needed to control the statistical features of the processes generated by discrete-time chaotic systems.

Applications of such chaos-related techniques in the field of information technology range from chaos-based cryptography [3][4], noise generation [5] or suppression [6], electromagnetic interference control [7]-[11], and telecommunication, which has been so far the most fruitful area for applying chaos-related methodologies (see [12] for a fairly complete survey, as well as [13]-[21] for a complete description of the application to Direct Sequences Code Division Multiple Access communication systems).

Beside the appreciable importance of the results achieved for the above applications, the existence of a plethora of different tasks to which the chaos-based approach (and the statistical techniques in particular) has been successfully applied, highlight the existence of a general scheme, which can in principle be applied to design a dynamical systems able to optimize the performance for the application under exam. Such a procedure consists of two phases. In as a first step a model of the target application must be developed to link specifications (performance and constraints) with the statistical features of the signals entailed in the considered task. In the second phase, link will be exploited to deepen the comprehension of what improvements are achievable by the design and use, in the chosen applications framework, of a chaotic system with controllable statistical features.

Modern Information Technology (IT) offers a perfect environment for the application of such methodology. IT deals, in fact, with systems made of a population of active and often intelligent units deeply interconnected and interacting. Examples range from Ethernet-based LANs to frame-relay or ATM geographic links, from Internet and its protocols to wireless sensor networks, from distributed memory or processing hierarchies to distributed cooperative computation. The complexity of the activities carried out in those systems and the number of independent and often unpredictable entities interacting with them forces the extensive use of statistical tools for their modeling and design. Hence, future successful development of related technologies will depend on our ability of modeling stochastic processes of increased complexity as well as of designing artificial systems able to generate them.

One of the most significant example comes from digital communication networks where the experimental discovery of self-similar of fractal processes [22] [23] [24] opened a definite chasm between classical traffic and queuing analysis and the most recent trends and methods.

As discussed in [25], modeling techniques in this field were forced to evolve from classical Poisson processes, to renewal processes with finite variance of inter-arrival times, to Markov chain regulating the transition between some of those renewal processes. The present state of this evolution is the widely accepted second-order self-similar model [25][26] from which most of the recent network traffic analysis originate.

A key step in this analysis was the discovery that complex self-similar behaviors can be interpreted as the superimposition of elementary binary (ON/OFF) processes representing the network activity (f.i. presence or absence of a packet) which features heavy-tailed (i.e. very slowly decaying) distribution of ON or OFF times [27][28].

Intuitively speaking, self-similarity has to do with invariance of behavior at different time scales. More formally, we may consider a second-order stationary process $\{y_i\}$ $i = 1, 2, \ldots$ and we may "observe" it at different scales by defining aggregated quantized trajectories

$$y_i^{(A)} = \frac{1}{A} \sum_{l=0}^{A-1} y_{Ai+l}$$

where A is the aggregation factor and, by definition, the finest scale sequence remains $y_i^{(1)} = y_i$. Such (aggregated) process represents the sequence of activity (ON=1) and inactivity (OFF=0) on the network at different time scales.

Hence, the process is *second-order self-similar* if a constant $\beta \in]0, 1[$ exists such that

$$\frac{C_2^{(xA)}(0)}{C_2^{(A)}(0)} \sim x^{-\beta} \quad \frac{C_2^{(A)}(\tau)}{C_2^{(A)}(0)} \sim K\tau^{-\beta} \tag{1}$$

with $A \to \infty$ and $x > 0$, where the symbol $\sim$ means "behave asymptotically as," and where $C_2^{(A)}(\tau)$ and $C_2^{(A)}(0)$ are the second-order autocorrelation and variance of the aggregated process. The first condition (1) requires that the variance do not decay too rapidly when the aggregation scale increases (*Slowly Decaying Variance*), while the polynomial decay of the autocorrelation in the first equation (1) states that the process features *Long-Range Dependence (LRD)*. Both are peculiar features of self-similar traffic models as opposed to more classical ones (e.g., Poisson models) that have a decay x^{-1} of the variance function (*Fast Decaying Variance*) and an exponential decay of the correlation function (*Short Range Dependence*).

In the literature the parameter $\beta \in]0,1[$ of (1) is is often replaced by the Hurst parameter $H = (1 - \beta/2) \in]0.5, 1[$, that represents the degree of the process "burstiness". The closer to 1 the H, the more self-similar the process. If H is close to 0.5 the process shows a smoother (less bursty) behavior when the time aggregation increases [22] in analogy with what happens for Poisson-like processes.

Another characteristic of network traffic relates to the time distributions for which the network is in the ON or OFF state, which are *heavy-tailed*, i.e., polynomially decaying, for at least one of the two states. More formally, at least one of the ON or OFF time distributions will have a complementary distribution such that

$$F(\tau) = \Pr\{t > \tau\} \approx \tau^{-\alpha} L(\tau), 0 < \alpha < 2 \tag{2}$$

where $L(\tau)$ is a slowly varying function at infinity [29] [26]. Note that the classic Poisson models are characterized by ON and/or OFF sojourn times which are both *light-tailed*, i.e. such that the complementary distribution has the form $F(\tau) = \Pr\{t > \tau\} \approx \gamma^{\tau} L(\tau)$, $0 < \gamma < 1$ [30] [26].

The intuitive explanation of the fundamental process generating this non-trivial behavior can be found in [25] and references therein. Here, we are conversely more interested in the understanding of the modeling of an artificial mathematical structure able to reproduce them. More specifically, we will describe how such an observation can be at least partially systematized in a general framework addressing a multiplicity of quantized renewal processes with possibly infinite sojourn time variance, the transitions among which is administered by a general Markov-like meta-process. The relationship between sojourn-time statistics and self-similarity is re-discovered and partially generalized in this framework whose aim is to contribute to the set of formally grounded tools the necessity of which is widely recognized [25]. Such a new entry in our tool-box is based on a suitable generalization of the one-dimensional chaotic maps used for chaos-based communication optimization [2] and can be directly used in the already ascertained scenarios.

As an example, they will be used at the end of the chapter to address the problem of synthesizing ON/OFF process for which both the average and the strength of self-similarity are given, thus showing how it is sometimes possible to conciliate and even exploit the difference between "short-term" and "long-term" traffic features.

The organization of this chapter is as follows. In section 2 we report the formal definition of Piecewise-Affine Markov maps and show how it is possible to give a statistical description of their behavior. More specifically we briefly recall the path linking the (generalized) Perron-Frobenius operator to a finite dimensional operator called Symbolic Dynamics Tracking Tensor (SDTT) and to the tensor-based formulation of the correlation functions whenever the observables are quantized in a finite number of regions of the state space. We also give the fundamental result for the factorization of the SDTT and thus for its decomposition into smaller building blocks, as well as few elementary results on basic statistical quantities correlated with the process of the quantized trajectories. Since these maps cannot generate self-similar process we extend the class of chaotic systems we are dealing with by allowing the Markov partition to be possibly made be an infinite, but countable number of intervals; this is done in general terms in section 3, while section 4 reports the formal definition of Piecewise-Affine Pseudo-Markov systems. Then, in section 5 we briefly highlight the path from the Perron-Frobenius operator to the SDTT similar to section 2 that could be followed for achieving similar results for the more general pseudo-Markov case. Then, we develop a closed-form expression of the simple case of a 2nd-order correlation function, which is given both in the time domain and in the z-transformed domain. Finally, we concentrate on systems with 2 macro-states and apply our methods to the explicit computation of correlation functions when the sojourn times are polynomially distributed, which leads to the desired generation of a quantized process with a self-similar trends.

This discussion continues on a more applicative ground in section 6 where we address the generation of self-similar processes for the synthesis of artificial network traffic. We here show that the link between autocorrelation profiles and asymptotic trends of the countable structure of the maps can be exploited to separately adjust first- and second-order characteristics of the processes and thus designing a realistic traffic generator. Some general conclusion are finally drawn in a separate Section.

As a final note let us stress that the result presented in the chapter are mainly rearranged from [2][31][32][33] and [34]. We adopt here a more heuristic, bottom-up approach with the aim to give simpler explanations of the underlying concepts and to present a ready-to-use collection of results and references.

2 Chaotic Piecewise Affine Markov Maps and Higher Order Correlations of Quantized Trajectories

It has now been known for a long time that dynamical systems including non-linear elements can exhibit non-classical behaviors including very irregular, aperiodic, noise-like trajectories, and an extreme sensitivity to initial conditions, which can make two identical systems apparently starting at identical conditions end up with totally different outputs [1]. These behaviors are generally named as *chaos* and, intuitively, are very appealing for different kind of stochastic process generation [2].

Clearly, achieving such a goal requires specific mathematical tools for dealing with the properties of chaotic systems. While such a theory does not yet generally exist, if one sufficiently restricts the range of chaotic systems being considered, many recently introduced tools become available [2].

Following this introduction, we limit ourselves to a very specific class of nonlinear models, represented by the iteration of so called piece-wise affine Markov (PWAM) maps [2]. These are discrete-time one dimensional (1-D) systems in which the state variable x is updated as

$$x_{k+1} = M(x_k) \tag{3}$$

where $M : [0,1] \to [0,1]$ is such that an interval partition $\{X_i\}, i = 1, \ldots p$ of $[0,1]$ exists so that:

1. M is affine on each X_i;
2. either $X_i \cap M(X_j) = \emptyset$ or $X_i \subseteq M(X_j)$ for any i, j (which is equivalent to saying that partition points are mapped into partition points);

If M is properly chosen, the system exhibits chaos[4]. By setting an initial condition x_0 and by iterating the map M, one can then generate a trajectory irregularly wandering in $[0,1]$. A typical case is shown in figure 1. Plot (a) shows a chaotic map, where $M(x) = 3x$ for $0 \leq x < 1/4$, to $x + 1/2$, for $1/4 \leq x < 1/2$ and to $-2x+2$ for $1/2 \leq x < 1$, and plot (b) a typical trajectory. Note that the map satisfies properties 1 and 2 for the interval partition composed by the four intervals $X_1 = [0, 1/4)$, $X_2 = [1/4, 1/2)$, $X_3 = [1/2, 3/4)$ and $X_4 = [3/4, 1]$. If M is chaotic, changing x_0 even by an extremely small quantity leads to a completely different trajectory, which progressively (and exponentially) increases its distance form the former as k increases. Hence, even in absence of noise, the behavior of the system is intrinsically unpredictable since the limited precision with which the state is known at any time prevents long-term forecasting.

A thorough and engineering-oriented illustration of PWAM chaotic maps can be found in [2]. Here it is sufficient to observe that the classical tools of system theory — which are generally based on the observation of system trajectories over time — are insufficient to get *typical* behaviors and general system properties, precisely due to the large variety of trajectories that can be generated. To overcome this impasse, the introduction of statistical tools is required. The intuitive idea is to pretend to be observing a huge number of

[4]More precisely, one wants M to be *non-singular* and *exact* following the definitions in [1].

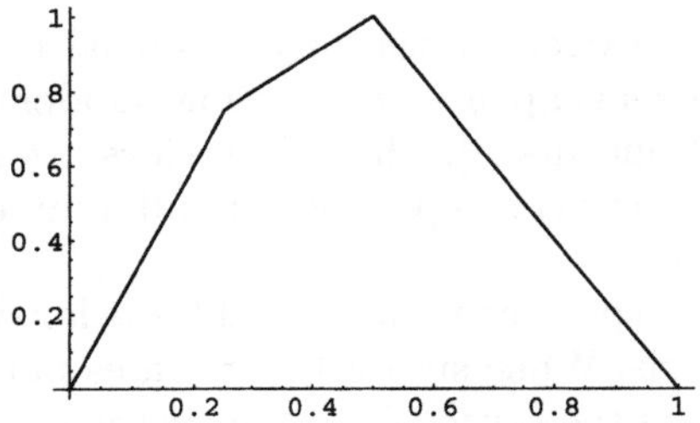

(a)

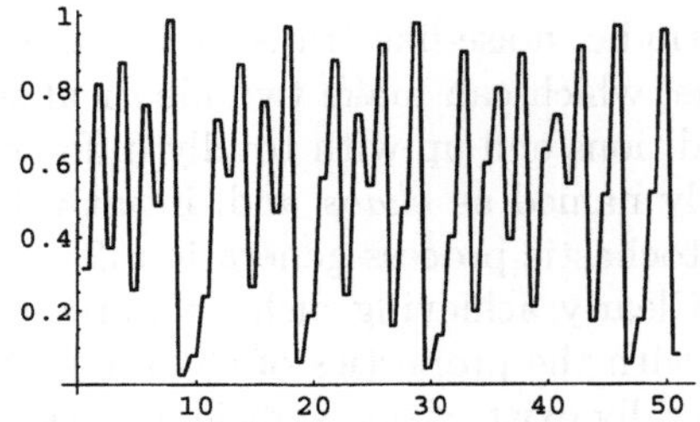

(b)

Fig. 1: Example chaotic map (a) and typical output trajectory (b).

trajectories at the same time, follow their evolution and collect statistics at every time step in order to extrapolate common features.

More formally, let us indicate with indicate with M^k its k-th iterate. Let $\mathbb{D}([0,1])$ be the set of probability densities defined on $[0,1]$ and assume that x_0 is randomly drawn according to $\rho_0 \in \mathbb{D}([0,1])$. Then, the probability density regulating the distribution of the random point $x_k = M(x_{k-1}) = M^k(x_0)$ is $\rho_k = \mathbf{P}\rho_{k-1} = \mathbf{P}^k\rho_0$, where $\mathbf{P}$ is the Perron-Frobenius Operator (PFO) [1, chap. 4]. In other terms, while M transforms points into points, $\mathbf{P}$ transforms probability densities into probability densities and we may think, instead of studying the evolution of the highly-nonlinear, upredicatable chaotic systems (3), we may turn our attention on the study of a companion dynamical system whose state variable is the PDF ρ_k of the state of system (3) and whose evolution operator is the the PFO.

Interestingly, in spite of M being non-linear, $\mathbf{P}$ does always turn out to be linear, which makes it by far more manageable from a mathematical point of view. It can also be proven that if M is exact, $\mathbf{P}$ is *constrictive*, so that starting from any ρ_0[5], ρ_k converges , with a rate $0 < r_{\text{mix}} < 1$, to a unique asymptotic density $\bar{\rho}$, which is called invariant density since $\bar{\rho} = \mathbf{P}\bar{\rho}$. The rate of convergence,that depends on the spectral portrait of $\mathbf{P}$ [35] is called the *rate of mixing* and is also an important quantity for the study of the loss of statistical dependence of trajectories generated by a chaotic map. More formally, one usually considers the quantity

$$\mathbf{E}\left[\phi(x_0)\psi(x_q)\right] = \int_{[0,1]} \phi(x)\psi(M^q(x))\bar{\rho}(x)dx \tag{4}$$

where $\phi, \psi : [0,1] \mapsto \mathbb{R}$ are two smooth functions. If one thinks of these function as *physical observables* of the system, then (4) quantifies the correlation between observing ϕ at time 0 and ψ at time q. For generic exact maps, the closed form computation of $\mathbf{E}\left[\phi(x_0)\psi(x_q)\right]$ is an almost impossible task, so

[5]Provided that ρ_0 is sufficiently well behaved [2]

that one is usually restricted to determine the rate r_{mix} of geometric convergence of (4) to its limit value when $q \to \infty$. In fact, it can be shown that a constant $\alpha > 0$ exists such that

$$\left| \mathbf{E}\left[\phi(x_0)\psi(x_q)\right] - \int_{[0,1]} \phi(x)\bar{\rho}(x)dx \int_{[0,1]} \psi\bar{\rho}(x)dx \right| \leq \alpha \sup|\phi| \sup|\psi|\, r_{\mathrm{mix}}^q \tag{5}$$

which highlights that the rate of mixing is a very informative quantity since it provides information on how quickly the system settles into a statistically regular behavior, and thus how quickly physical observables become uncorrelated.

Nevertheless, from an engineering point of view, the mere knowledge of r_{mix} and α cannot be considered as entirely satisfactory. In fact, using (5) in the equations expressing the merit figure of a specific signal processing task as a function of statistical features of the chaotic sequences would simply yield to a (hopefully tight) *performance bound* (see f.i. [13]).

On the contrary, the ability of computing the exact values of (4) would allow to analytically express the same merit figure as a function of the characteristics of M, which is surely the necessary starting point for any performance optimization procedure.

Our aim in the rest of this section is to show that this goal can be achieved if we deal with the class of PWAM chaotic maps.

Notice first that from the very definition of Markov maps one readily gets that the following theorem hold [2].

Theorem 1. *An exact Markov map M has a unique invariant density $\bar{\rho}$ is $\bar{\rho}$ is piecewise constant on the interval partition $\{X_i\}$, namely*

$$\bar{\rho} = \sum_{i=1}^{p} \bar{\rho}_i \chi_{X_i} \tag{6}$$

where χ_{X_i} is the indicator function of X_i.

To proceed further, it is interesting to observe what happens if, instead of considering any possible initial density ρ_0, one limits himself to initial probability densities which are piecewise constant in $\{X_i\}$: all subsequent probability densities $\rho_1, \ldots, \bar{\rho}$ are then compelled to be piecewise constant on the interval partition. This is noteworthy. In fact, in this case, rather than considering probability densities, one can consider the finite probability of the state variable x being at any of the intervals X_i. In other terms, at each time step k we may substitute to $\rho_k = \sum_{i=1}^{p} \rho_{ki}\chi_{X_i}$ a vector of p probabilities $\mathbf{\Pi}_k = (\Pi_{k1}, \ldots, \Pi_{kp})^T = (\rho_{k1}\mu(X_1), \ldots, \rho_{kp}\mu(X_p))^T$, where $\cdot^T$ indicates vector/matrix transposition. Similarly a finite dimension operator $\bar{\mathcal{K}}$ can be substituted for the functional operator $\mathbf{P}$. Being $\bar{\mathcal{K}}$ linear and finite-dimensional, it can be represented by a matrix, named *kneading matrix*. It has been proven that the entries of $\bar{\mathcal{K}}$ can be obtained from M as (see e.g. [2] and references therein):

$$\bar{\mathcal{K}}_{ij} = \frac{\mu(X_i \cap M^{-1}(X_j))}{\mu(X_i)} \tag{7}$$

where μ is the usual interval measure, i.e. if $X_i = [a,b]$, then $\mu(X_i) = b - a$. For instance, the kneading matrix corresponding to the map in figure 1 (a) is:

$$\bar{\mathcal{K}} = \begin{pmatrix} 1/3 & 1/3 & 1/3 & 0 \\ 0 & 0 & 0 & 1 \\ 0 & 0 & 1/2 & 1/2 \\ 1/2 & 1/2 & 0 & 0 \end{pmatrix} \quad . \tag{8}$$

As an example, suppose that $\mathbf{\Pi}_0 = (1/2, 0, 0, 1/2)$. Then $\mathbf{\Pi}_1 = \mathbf{\Pi}_0\bar{\mathcal{K}} = (5/12, 5/12, 1/6, 0)$.

It is now interesting to try to give a meaning to the individual entries in $\bar{\mathcal{K}}$. With reference to usual example map, suppose that it is know only that the initial condition x_0 falls in X_1, and nothing else. This is loosely equivalent to saying that $\mathbf{\Pi}_0 = (1,0,0,0)$. What can be said about x_1? This point can obviously fall either in X_1, or in X_2 or in X_3, but not in X_4. Furthermore, from $\mathbf{\Pi}_0\bar{\mathcal{K}}$ one can estimate that x_1 will fall in X_1 with probability $1/3$, in X_2 with probability $1/3$ and in X_3 with probability $1/3$. In other terms, the entry $\bar{\mathcal{K}}_{ij}$ of $\bar{\mathcal{K}}$ represents the probability by which a trajectory starting in X_i falls in X_j at the next step.

The process can be iterated. For instance, knowing that x_0 falls in X_1, one could build the tree-graph reported in figure 2 (a), by which the probability of finding x_k in any of the partition intervals can easily be obtained. As an example, note that the probability of finding x_2 in X_3 turns out to be $1/3 \cdot 1/3 + 1/3 \cdot 1/2 = 5/18$. It is obviously convenient to compact infinitely long

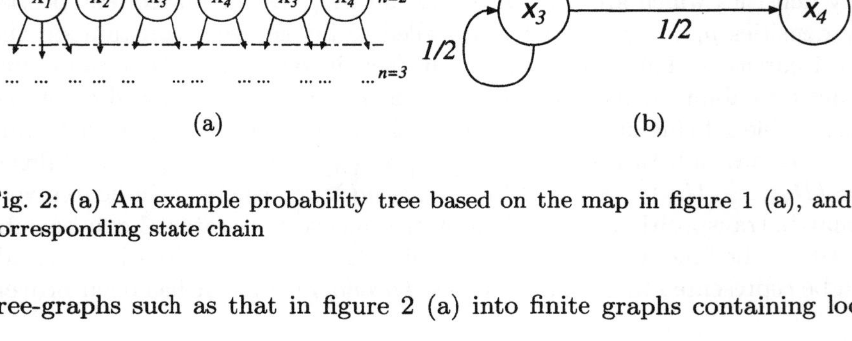

Fig. 2: (a) An example probability tree based on the map in figure 1 (a), and (b) corresponding state chain

tree-graphs such as that in figure 2 (a) into finite graphs containing loops, such as that in figure 2 (b). Note that this graph can be interpreted as a *Markov chain* [2][5] or a transition graph for a *probabilistic state machine*. The machine is in its discrete state x_i when the chaotic system has its continuous state variable x in the partition interval X_i. The weights assigned to the graph arrows represent the probabilities by which the machine travels from a state to another. The idea that a chaotic dynamics could embed a Markov chain was first conjectured by Kalman [36].

Following this approach, we are first able to easily compute $\bar{\rho}$. Since $\bar{\mathcal{K}}$ is a stochastic matrix, if we indicate with $\bar{\mathbf{\Pi}} = (\bar{\Pi}_1, \dots, \bar{\Pi}_p)^T$ the (normalized) unique eigenvector of $\bar{\mathcal{K}}$ corresponding to a unit eigenvalue, i.e. $\bar{\mathbf{\Pi}} = \bar{\mathbf{\Pi}}\bar{\mathcal{K}}$, we easily get that the coefficient defining (6) are given by $\bar{\rho}_i = \bar{\Pi}_i/\mu(X_i)$. Furthermore, we can determine closed form expressions for more sophisticated and meaningful statistical quantities such as (4). More formally, let us consider that observables functions f_i exist, which quantize the state at a given time step q_i, i.e we will assume that the function $f_i : [0,1] \mapsto \mathbb{C}$ is constant in each of p intervals of the maps Markov partition. The p-dimensional vector $\mathbf{f}_i = (f_{i1}, \dots, f_{ip})^T$ is defined so that $f_i(X_j) = f_{ij}$. A different quantization is adopted at different time steps and different trajectories are associated to different values depending on which intervals they visit. Hence it is straightforward [20][2] to prove the following theorem

Theorem 2. *Consider a PWAM map with $\bar{\rho} = 1$ and where $\mu(X_j) = 1/p$ for $j = 1, \dots, p$. Then, for trajectories generated with $\rho_0 = \bar{\rho}$ one has*

$$\mathbf{E}\left[f_1(x_{q_1}) f_2(x_{q_2})\right] = \frac{1}{p} \mathbf{f}_1^T \bar{\mathcal{K}}^{q_2 - q_1} \mathbf{f}_2 \tag{9}$$

Although important for several applications, the mere use of theorem 2 is not sufficient to give a complete characterization of the process obtained by quantizing the trajectories generated by PWAM maps. To this aim we need to compute m-order expectation with $m > 2$, i.e. terms of the kind $\mathbf{E}\left[\prod_{i=1}^m f_i(x_{q_i})\right]$, with $0 < q_1 < \cdots < q_m$.

To achieve such a goal, we need to deal with multi-index quantities, that we will call *tensors*, that generalize vectors and matrixes. Several kind of *products* can be defined among tensors and we may now define those that will be of use in the following discussion.

Given two m-order tensors $\mathcal{A}$ and $\mathcal{B}$ with identical index ranges, their *inner product* is a scalar defined as

$$\langle \mathcal{A}, \mathcal{B} \rangle = \sum_{j_1, \dots, j_m} \mathcal{A}_{j_1, \dots, j_m} \, \mathcal{B}_{j_1, \dots, j_m}$$

where every index sweeps all its range. The inner product is a commutative bilinear operator.

Given an m'-order tensor $\mathcal{A}$ and an m''-order tensor $\mathcal{B}$, their *outer product* is an $(m' + m'')$-order tensor $\mathcal{C} = \mathcal{A} \otimes \mathcal{B}$ such that

$$\mathcal{C}_{j_1,\ldots,j_{m'+m''}} = \mathcal{A}_{j_1,\ldots,j_{m'}} \mathcal{B}_{j_{m'+1},\ldots,j_{m'+m''}}$$

where every index sweeps all its range. The outer product is a non-commutative, associative and bilinear operator.

Given the m'-order tensor $\mathcal{A}$ and the m''-order tensor $\mathcal{B}$ such that the range of the last index of $\mathcal{A}$ is the same of the first index of $\mathcal{B}$, their *chain product* is an $(m' + m'' - 1)$-order tensor $\mathcal{C} = \mathcal{A} \circ \mathcal{B}$ such that

$$\mathcal{C}_{j_1,\ldots,j_{m'+m''-1}} = \mathcal{A}_{j_1,\ldots,j_{m'}} \mathcal{B}_{j_{m'},\ldots,j_{m'+m''-1}}$$

The chain product is a non-commutative associative bilinear operator.

The tensor products defined above enjoy few *distributive* properties which are recalled without proof in the following

Lemma 1. *For any four tensors $\mathcal{A}$, $\mathcal{B}$, $\mathcal{C}$ and $\mathcal{D}$ whose index ranges make the equalities below well defined, we have*

$$\mathcal{A} \otimes (\mathcal{B} \circ \mathcal{C}) = (\mathcal{A} \otimes \mathcal{B}) \circ \mathcal{C} \qquad \mathcal{A} \circ (\mathcal{B} \otimes \mathcal{C}) = (\mathcal{A} \circ \mathcal{B}) \otimes \mathcal{C}$$
$$\langle \mathcal{A} \otimes \mathcal{B}, \mathcal{C} \otimes \mathcal{D} \rangle = \langle \mathcal{A}, \mathcal{C} \rangle \langle \mathcal{B}, \mathcal{D} \rangle$$

Notice first that this newly introduced operation between multi-index quantities allows us to give a more elegant and general expression for (9) in the case of completely general PWAM maps, as formally stated in the following theorem [2][37]

Theorem 3. *For trajectories generated by a PWAM map with with $\rho_0 = \bar{\rho}$ one has*

$$\mathbf{E}\left[f_1(x_{q_1}) f_2(x_{q_2})\right] = \left\langle \mathbf{f}_1 \otimes \mathbf{f}_2, \bar{\mathbf{\Pi}} \circ \bar{\mathcal{K}}^{q_2-q_1} \right\rangle = \left\langle \mathbf{f}_1 \otimes \mathbf{f}_2, \bar{\boldsymbol{\rho}} \circ \boldsymbol{\mu} \circ \bar{\mathcal{K}}^{q_2-q_1} \right\rangle \quad (10)$$

where $\bar{\boldsymbol{\rho}} = (\bar{\rho}_1, \ldots, \bar{\rho}_p)^T$ and $\boldsymbol{\mu} = (\mu(X_1), \ldots, \mu(X_p))^T$.

Actually, the adoption of tensor-formalism allows to readily compute the m-th order expectations of the quantized trajectories [17]. In fact, let us rely on the stationarity of the process and assume $q_1 = 0$ and indicate with $\bar{\mathbf{P}}^{q_2,\ldots,q_m} = \mathbf{P}^{0,q_2,\ldots,q_m} : \mathbb{D}([0,1]) \mapsto \mathbb{D}([0,1]^m)$ an operator that is formally expressed for any $m-1$ integers $q_2 < \cdots < q_m$ as

$$\bar{\mathbf{P}}^{q_2,\ldots,q_m}[\bar{\rho}](\boldsymbol{\xi}) = \bar{\rho}(\xi_1) \prod_{i=2}^{m} \delta(\xi_i - M^{q_i-q_1}(\xi_1))$$

and which represents the joint probability density of the random vector $(x_0, x_{q_2}, \ldots, x_{q_m})$. With this we obtain

$$\mathbf{E}\left[\prod_{i=1}^{m} f_i(x_{p_i})\right] = \int_{[0,1]^m} \prod_{i=1}^{m} f_i(\xi_i) \bar{\mathbf{P}}^{p_2,\ldots,p_m}[\bar{\rho}](\boldsymbol{\xi}) d\boldsymbol{\xi} =$$
$$= \sum_{j_1,\ldots,j_m=1}^{n} \left[\prod_{i=1}^{m} f_{ij_i} \int_{\times_{i=1}^{m} X_{j_i}} \bar{\mathbf{P}}^{p_2,\ldots,p_m}[\bar{\rho}](\boldsymbol{\xi}) d\boldsymbol{\xi}\right]$$

where the integral term, which represents the probability that a trajectory starting in X_{j_1} pass trough X_{j_2} at time step q_2, trough X_{j_3} at time step q_3, etc, can be collect in a m-order tensor $\mathcal{H}(q_2, \ldots, q_m)$

$$\mathcal{H}_{j_1,\ldots,j_m}(q_2, \ldots, q_m) = \int_{\bigtimes_{i=1}^{m} X_{j_i}} \bar{\mathbf{P}}^{q_2,\ldots,q_m}[\bar{\rho}](\boldsymbol{\xi})d\boldsymbol{\xi} \tag{11}$$

which will be indicated as the SDTT.

Note now that the terms $\prod_{i=1}^{m} f_{ij_i}$ for $j_1, \ldots, j_m$ spanning the range $1, \ldots, p$, can be also compounded in the m-order tensor $\mathcal{F} = \mathbf{f}_1 \otimes \cdots \otimes \mathbf{f}_m$, and that

$$\mathbf{E}\left[\prod_{i=1}^{m} f_i(x_{q_i})\right] = \langle \mathcal{F}, \mathcal{H}(q_2, \ldots, q_m)\rangle \tag{12}$$

Since the processes with which we are dealing are stationary it is often convenient to think of the SDTTs as quantities depending on the time lags between two subsequent observation instants, i.e. on $\tau_i = p_{i+1} - p_i$. In the following we will use the two notations $\mathcal{H}(q_2, \ldots, q_m)$ and $\mathcal{H}(\tau_1, \ldots, \tau_{m-1})$ interchangeably.

It can be proved that [17][2], since we adopt PWAM maps, that the SDTT can be factorized into smaller pieces so that the expression of the m-th order moment is

$$\mathbf{E}\left[\prod_{i=1}^{m} f_i(x_{q_i})\right] = \left\langle \mathcal{F}, \bar{\boldsymbol{\rho}} \circ \boldsymbol{\mu} \circ \bigcirc_{i=2}^{m} \bar{\mathcal{K}}^{q_i - q_{i-1}} \right\rangle \tag{13}$$

where $\bar{\mathcal{K}}$ is the kneading matrix defined above and the symbols $\bigcirc$ represents the iterative application of the chain-product that is associative.

Under the assumption that let the factorization of the SDTT hold, we also know that [17][2] if $\bar{\mathcal{K}}$ is primitive (i.e. a power of $\bar{\mathcal{K}}$ exists in which no entry is null), a family of matrixes $\mathcal{A}(q)$ exists such that [17][2] $\bar{\mathcal{K}}^q = \Im + \mathcal{A}(q)$, with $\mathcal{A}(q)$ exponentially vanishing as $q \to \infty$ and $\Im = \mathbf{1} \otimes \bar{\boldsymbol{\rho}} \circ \boldsymbol{\mu}$ with $\mathbf{1} = (1, \ldots, 1)$.

With this and few tensor algebra manipulations exploiting the distributivity of the tensor products we may easily derive that any inner product of the kind $\langle \mathcal{F}, \mathcal{H}^{q_2,\ldots,q_m}\rangle$ can be expressed as a weighted sum of terms of the kind $\langle \mathcal{F}, \mathcal{A}' \circ \mathcal{A}'' \circ \ldots\rangle$. As an example for $m = 4$, we have

$$\begin{aligned}
&\mathbf{E}[f_1(x_0)f_2(x_{q_2})f_3(x_{q_3})f_4(x_{q_4})] = \\
&\langle \mathcal{F}, \bar{\boldsymbol{\rho}} \circ \boldsymbol{\mu} \circ (\Im + \mathcal{A}(q_2)) \circ (\Im + \mathcal{A}(q_3 - q_2)) \circ (\Im + \mathcal{A}(q_4 - q_3))\rangle = \\
&\langle \mathcal{F}, \bar{\boldsymbol{\rho}} \circ \boldsymbol{\mu} \otimes \bar{\boldsymbol{\rho}} \circ \boldsymbol{\mu} \otimes \bar{\boldsymbol{\rho}} \circ \boldsymbol{\mu} \otimes \bar{\boldsymbol{\rho}} \circ \boldsymbol{\mu}\rangle + \langle \mathcal{F}, \bar{\boldsymbol{\rho}} \circ \boldsymbol{\mu} \otimes \bar{\boldsymbol{\rho}} \circ \boldsymbol{\mu} \otimes \bar{\boldsymbol{\rho}} \circ \boldsymbol{\mu} \circ \mathcal{A}(q_4 - q_3)\rangle + \\
&\langle \mathcal{F}, \bar{\boldsymbol{\rho}} \circ \boldsymbol{\mu} \otimes \bar{\boldsymbol{\rho}} \circ \boldsymbol{\mu} \circ \mathcal{A}(q_3 - q_2) \otimes \bar{\boldsymbol{\rho}} \circ \boldsymbol{\mu}\rangle + \langle \mathcal{F}, \bar{\boldsymbol{\rho}} \circ \boldsymbol{\mu} \circ \mathcal{A}(p_2) \otimes \bar{\boldsymbol{\rho}} \circ \boldsymbol{\mu} \otimes \bar{\boldsymbol{\rho}} \circ \boldsymbol{\mu}\rangle + \\
&\quad \langle \mathcal{F}, \bar{\boldsymbol{\rho}} \circ \boldsymbol{\mu} \circ \mathcal{A}(p_2) \otimes \bar{\boldsymbol{\rho}} \circ \boldsymbol{\mu} \circ \mathcal{A}(p_4 - p_3)\rangle + \\
&\quad \langle \mathcal{F}, \bar{\boldsymbol{\rho}} \circ \boldsymbol{\mu} \circ \mathcal{A}(p_2) \circ \mathcal{A}(p_3 - p_2) \otimes \bar{\boldsymbol{\rho}} \circ \boldsymbol{\mu}\rangle + \\
&\quad \langle \mathcal{F}, \bar{\boldsymbol{\rho}} \circ \boldsymbol{\mu} \otimes \bar{\boldsymbol{\rho}} \circ \boldsymbol{\mu} \circ \mathcal{A}(p_3 - p_2) \circ \mathcal{A}(p_4 - p_3)\rangle +
\end{aligned}$$

$$\begin{aligned}\langle \mathcal{F}, \bar{\boldsymbol{\rho}} \circ \boldsymbol{\mu} \circ \mathcal{A}(p_2) \circ \mathcal{A}(p_3 - p_2) \circ \mathcal{A}(p_4 - p_3) \rangle = \\ \langle \mathbf{f}_1 \otimes \mathbf{f}_2, \bar{\boldsymbol{\rho}} \circ \boldsymbol{\mu} \circ \mathcal{A}(p_2) \rangle \langle \mathbf{f}_3 \otimes \mathbf{f}_4, \bar{\boldsymbol{\rho}} \circ \boldsymbol{\mu} \circ \mathcal{A}(p_4 - p_3) \rangle + \\ \langle \mathcal{F}, \bar{\boldsymbol{\rho}} \circ \boldsymbol{\mu} \circ \mathcal{A}(p_2) \circ \mathcal{A}(p_3 - p_2) \circ \mathcal{A}(p_4 - p_3) \rangle\end{aligned}$$

where the last equality holds thanks to lemma 1 and if we assume that $\mathbf{E}[f_i(x_{q_i})] = \langle \mathbf{f}_i, \bar{\boldsymbol{\rho}} \circ \boldsymbol{\mu} \rangle = 0$, $i = 1, \ldots, 4$

As a final comment, we may conclude that the asymptotic properties of 2nd and higher-order expectations depend on the spectral structure of $\bar{\mathcal{K}}$ and can be expressed for, zero-average quantization symbols, as a sum of exponentially vanishing terms. Such properties can be thus analyzed (and sometime designed) by elementary linear algebraic methods, so that when such correlation functions are the key factor in the performance of a system in which a chaotic component has been introduced (such as for DS-CDMA systems [2][13][19]), this gives us high chances of optimization by means of proper statistical chaos design.

Nevertheless, *none* of the PWAM maps introduced so far can be employed to generate self-similar processes, since terms like t is obviously not(13) cannot give rise to correlations with a polynomially vanishing trend. To solve the impasse, we need therefore to generalize the class of maps we are dealing with as formally stated in the next section.

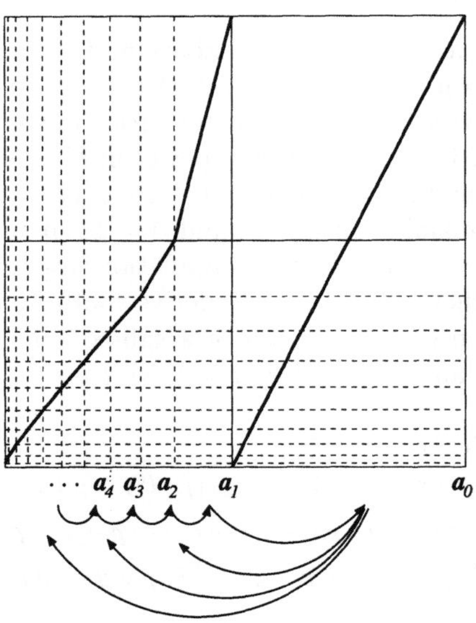

Fig. 3: A piecewise-affine Markov map with an infinite number of Markov intervals

3 A More General Model of Quantized Chaotic Trajectories

The impasse generated by the impossibility to generate self-similar process using the class of PWAM maps considered in the previous section force us to extend the set of chaotic systems we are dealing with. From now on we consider more general maps in which the Markov partition is made of a non-necessarily finite number of intervals. Despite their seeming a quite abstract exercise, these maps may be of non-negligible interest from the applicative point of view.

One of the simplest (but far from trivial) examples is a family of maps defined starting from a sequence of points $1 = a_0 > a_1 > a_2 > \ldots$ such that $\lim_{k\to\infty} a_k = 0$ and obeying

$$M(x) = \begin{cases} a_i + (a_{i-1} - a_i)\dfrac{x - a_{i+1}}{a_i - a_{i+1}} & \text{if } x \in]a_{i+1}, a_i] \text{ for } i \geq 1 \\ \dfrac{x - a_1}{a_0 - a_1} & \text{if } x \in]a_1, a_0] \end{cases} \tag{14}$$

Maps in this family are affine in each interval $]a_{i+1}, a_i]$, $M(]a_{i+1}, a_i]) =]a_i, a_{i-1}]$ for $i \geq 1$ and that $M(]a_1, a_0]) = [0, 1] = \bigcup_{i=0}^{\infty}]a_{i+1}, a_i]$. Hence, they are Piecewise-affine Markov maps with a countable Markov partition.

A possible map within this family is reported in figure 3 in which we also highlight the qualitative behavior of a typical trajectory of the system. It jumps steadily from $]a_{i+1}, a_i]$ to $]a_i, a_{i-1}]$ up to $]a_1, a_0]$, from which it moves back to a point spread over the entire $[0, 1]$. If the time spent in the region $x \leq a_1$ is statistically greater than the time spent in the region $x > a_1$ such a behavior is often indicated as "intermittency" and is a key issue in the theory of non-linear dynamical systems.

In fact, maps of this kind have been analyzed in detail (see e.g. [38] and [39]) to reveal that the discrete time signal obtained by the sequence of the map states has a peculiar autocorrelation trend and thus power-spectrum profile.

More formally, it was proved that if the trend of the a_k for $k \to \infty$ is of the kind $k^{-\zeta}$ with $\zeta > 1$ then the autocorrelation function $c(\tau)$ decays as $\tau^{-\zeta+1}$ for $\tau \to \infty$ while, if the trend of the a_k is of the kind $k^{-1}(\log k)^{-\zeta}$ with $\zeta > 1$, then the correlation decay is $(\log \tau)^{-\zeta+1}$. In terms of power spectrum, such slow correlation decays may imply a diverging power density for frequencies approaching 0. For this reason, maps of the kind (14) have been proposed to generate synthetic $1/f^{\alpha}$ noise [40] that can be of interest in the time-domain simulation of electronic circuits [41] when linearized frequency-domain analysis would hide non-negligible effects and other conventional approximated $1/f$ noise realizations are deemed insufficient.

As we will see, a variant of this kind of maps will be of use in the last section of this chapter when its quantized version will be the core for the generation of artificial non-Poisson network-traffic with features strictly related with the ones measured in real-worlds LANs.

If we decide that piecewise-affine Markov maps with a countable Markov partition cannot stay out of our scope the question of the characterization of their statistical properties arises naturally.

Note that the introduction of an assumption that forces all the observables to be constant in each of the intervals of the Markov partition does not have here the same beneficial effect that we exploited to obtain the results presented in the previous section.

In fact, when the cardinality of the Markov partition is finite, such an assumption allows to project the Perron-Frobenius operator onto a finite dimensional space and relate statistical properties to the finite spectral portrait of the kneading matrix. In our case, this is no longer possible since the kneading matrix of a map with a countable Markov partition is itself a countable matrix whose spectral portrait may be extremely difficult to characterize.

Actually, this increased complexity stems from the identification of the elements of the Markov partition with the regions in which the observables are kept constant. In fact, the observable space itself is now infinite-dimensional and forces a richer characterization of the underlying process.

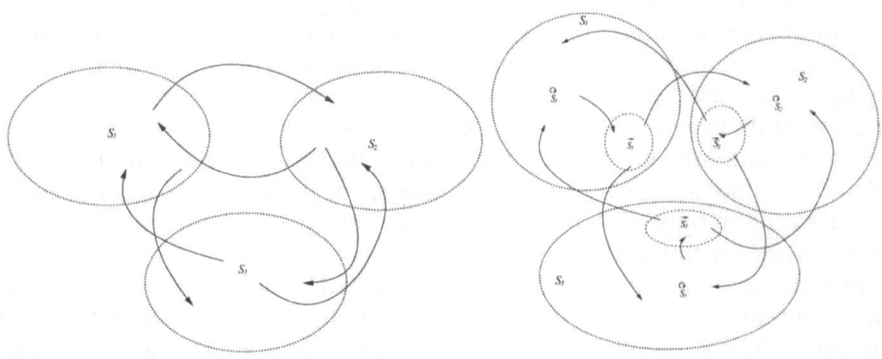

(a) (b)

Fig. 4: (a) A general model for the evolution of a system whose observables assume constant values on three regions of the state space. (b) A more refined model of the same evolution detailing the structure of each macro-state.

To avoid this, we may further limit our investigations to those observables that are constant in certain unions of Markov intervals that we will call macro-states. If the number of these unions is finite we may hope to retrieve, at least partially, the regular and finite structure that allowed the complete characterization of the SDTT for piecewise-affine Markov maps with a finite Markov partition.

With this assumption, in fact, from the observables point of view the system behaves as depicted in figure 4(a) for three macro-states, i.e. it is perceived in one of the macro-states for a certain time, then it leaves that region to land, with an assigned probability, in another region in which it sojourn for a certain amount of time, and so on.

Since each macro-state actually envelopes a possibly countable set of Markov intervals, the amount of time spent in each macro-state (the sojourn time) is a random-variable which is, in general, non-geometrically distributed. With this, the whole system cannot be characterized only by the probability of moving from one macro-state to another (the analogue of the kneading matrix $\bar{\mathcal{K}}$ introduced in the of previous section) and the characterization of the observed trajectories, i.e. the derivation of the SDTTs, must take into account both the transition probabilities between macro-states and the statistics of the sojourn times in each macro-state.

Regrettably, in general terms, local dynamics (the sojourn times statistics) and global dynamics (the transitions probabilities) are so intimately intertwined that their influence on the SDTT can be hardly separated and managed. To cope with this we need some further assumptions on how Markov intervals are grouped in macro-states. Namely we will consider as a reference a model like the one depicted in figure 4(b) in which we identify, within each macro-state S_j, a union of Markov intervals that can be aggregated in $\overset{\circlearrowleft}{S}_j \subseteq S_j$ that is mapped again into S_j (the *return* sub-macro-state) and a union of Markov intervals that can be aggregated in $\overrightarrow{S}_j \subseteq S_j$ from which the map states moves surely to another macro-state (the *exit* sub-macro-state).

Given this structure, few other technical assumptions guarantee that local and global dynamics are coupled loosely enough to recognize that the system still features a "memory-one" property, i.e. that the dependency between certain critical events in its evolution are extremely short-lasting.

These systems are called pseudo-Markov and, in analogy with what happened with the memory-one property intrinsic in purely Markov maps, we may construct for them a tensor-based mechanism leading to the expression of the z-transform of any order SDTTs.

It is worthwhile to stress that the adoption of macro-states with a detailed substructure as well as the focus on statistical properties instead of trajectory classification also somehow widens, with respect to what is done in the previous section, the gap between our approach and the classical symbolic dynamic approach (see e.g. [42]). What is highlighted and strengthened is the link between the evolution of a chaotic system and the theory of renewal processes. This idea is not new, has already led to significant results (e.g. [39][43][44][45]), and is here exploited to a larger extent to cope with quantized chaotic trajectories.

4 Piecewise-Affine Pseudo-Markov Systems

One of the possible set of assumptions supporting the idea of a memory-one property, as far as the sojourn times are concerned, is the one defining the so-called Piecewise-affine pseudo-Markov Systems (PWAPM). As it can be expected, these systems are strictly related to the PWAM maps defined in section 2 of which Pseudo-Markov systems generalize some aspects and specialize some other features.

As already discussed in the previous section, the general idea is to partition each macro-state into a *return* and *exit* subset. To guarantee that the statistics of the sojourn time in the generic macro-state $S_{j'}$ depends only on the previously visited macro-state $S_{j''}$ we must constrain the probability of moving from $S_{j''}$ to $S_{j'}$ and of staying in the latter for a certain amount of time to be independent of the path that leads from $\overset{\frown}{S}_{j''}$ to $\overrightarrow{S}_{j''}$, i.e. of the detailed evolution of the system within each macro-state.

We can do so by placing some constraints on the structure of the map. More formally, in a bottom-down approach we assume that the state space $[0,1]$ of the system (3) is partitioned into p disjoint intervals S_i that are the macro-states. Each of these intervals is further partitioned into a (possibly) countable number of subintervals X_{ij} such that

$$S_j = \bigcup_k X_{jk}$$

We will additionally assume that the map M is particularly simple when restricted to the countable set of intervals X_{jk}, i.e. that for any $X_{j'k'}$ and $X_{j''k''}$,

$$M(X_{j'k'}) \cap X_{j''k''} = \begin{cases} \emptyset \\ X_{j''k''} \end{cases} \tag{15}$$

and that M is invertible in each X_{jk} and affine in each $X_{j'k'} \cap M^{-1}(X_{j''k''}) \neq \emptyset$. Note that this is a slightly relaxed assumption with respect to the usual requirements that M is affine in each X_{jk} which is actually what our example in figure 3 does.

Nevertheless the main property of classical piecewise-affine Markov Systems is still available as confirmed by this theorem [31][32]

Theorem 4. *If the map M is exact then its unique invariant probability density $\bar{\rho}$ is constant in each X_{jk}.*

In the following we will assume that M is exact so that the concept of invariant density $\bar{\rho}$ and the associated invariant measure $\bar{\mu}$ are well defined [2].

To continue our definition of piecewise-affine Pseudo-Markov Systems we assume that for any S_j two subsets can be distinguished. They are the *returning* subset $\overset{\frown}{S}_j$ and the *exiting* subset $\overrightarrow{S}_j$. These two subsets are such that

$M(\overset{\curvearrowleft}{S}_j) \subseteq S_j$, $M(\overrightarrow{S}_j) \cap S_j = \emptyset$, $\overset{\curvearrowleft}{S}_j \cup \overrightarrow{S}_j = S_j$. Note that this implies that $\overrightarrow{S}_j$ and $\overset{\curvearrowleft}{S}_j$ are disjoint since their images through M are disjoint.

They are also union of certain intervals X_{jk} which are labeled accordingly so that

$$\overset{\curvearrowleft}{S}_j = \bigcup_k \overset{\curvearrowleft}{X}_{jk} \quad \text{and} \quad \overrightarrow{S}_j = \bigcup_k \overrightarrow{X}_{jk}$$

We will finally assume that the exiting subsets are such that for every $j' \neq j''$ and $k' \neq k''$ we have

$$\bar{\mu}(\overrightarrow{X}_{j'k'}) = \bar{\mu}(\overrightarrow{X}_{j'k''}) \tag{16}$$

$$M(\overrightarrow{X}_{j'k'}) \cap S_{j''} = M(\overrightarrow{X}_{j'k''}) \cap S_{j''} \tag{17}$$

$$\bar{\mu}(\overrightarrow{X}_{j'k'} \cap M^{-1}(X_{j''k})) = \bar{\mu}(\overrightarrow{X}_{j'k''} \cap M^{-1}(X_{j''k})) \tag{18}$$

where $\bar{\mu}$ is the invariant measure of the map. Despite their apparent technicality, equations from (16) to (18) have straightforward intuitive meaning of assessing the indistiguishability between all the subintervals making the exit part of a macro-state. In fact, (16) requires that all the subintervals are visited with the same probability, (17) requires that their images over all the macro-state are the same, and (18) requires that the transition probability to any other subinterval is always the same.

5 Statistical Characterization of PWAPM Systems

As far as the statistical characterization of quantized trajectories is concerned, we could follow a path similar to section 2 and assume to quantize the state of the map with suitable functions f_i, $i = 1, 2, \ldots, m$, corresponding to the time steps $0, q_1, \ldots, q_m$, which are such that $f_i(S_{j_i}) = f_{ij_i}$.

We could again define a SDTT $\mathcal{H}(q_2, \ldots, q_m)$ such as

$$\mathcal{H}_{j_1, j_2, \ldots, j_m}(q_2, \ldots, q_m) = \int_{\bigtimes_{i=1}^{m} S_{j_i}} \bar{\mathbf{P}}^{p_2, \ldots, p_m}[\bar{\rho}](\boldsymbol{\xi}) d\boldsymbol{\xi}$$

where the integral term is nothing but the probability that a trajectory starting in S_{j_1} falls in S_{j_2} at time step q_2, in S_{j_3} at time step q_3, etc. Hence we could write express the m-order expectation of quantized trajectories generated by a PWAPM system in the form (12), where $\mathcal{F} = \mathbf{f}_1 \otimes \cdots \otimes \mathbf{f}_m$.

It can be shown that, in analogy with what happened in section 2 for purely Markov systems, the SDTT of PWAPM systems can be decomposed into smaller and more tractable building blocks that allow its analytical computation. To keep the analytical development as simple as possible, we restrict here to the most simple case that is suitable for our needs, i.e. the generation of a self-similar process. The reader interested to the full development should refer to [33] and references therein.

Let us therefore consider a PWAPM system with two macro-states $S_1 = [0, 1/2[$ and $S_2 = [1/2, 1]$. It is worthwhile to stress that although the system structure is apparently simple, the transitions between the two macro-states S_1 and S_2 can be regulated by extremely complex laws depending on the countable Markov chains in $\overset{\circlearrowleft}{S}_j$ and $\overrightarrow{S}_j$. To exploit this complexity to our needs, let us first define the matrix $\mathcal{M}$:

$$\mathcal{M}_{j'j''}(k) = \frac{\bar{\mu}(S_{j'} \cap \bigcap_{j=1}^{k} M^{-j}(S_{j''}))}{\bar{\mu}(\overrightarrow{S}_{j'})}$$

The $j'j''$-th entry of this matrix contains the probability of moving from $S_{j'}$ to $S_{j''}$ and staying at least k time steps in $S_{j''}$. For $k = 1$, and taking into account definition (7) this joint probability can be obviously rewritten as

$$\mathcal{M}_{j'j''}(1) = \frac{\bar{\mu}(S_{j'} \cap M^{-1}(S_{j''}))}{\bar{\mu}(S_{j'})} \Big/ \frac{\bar{\mu}(\overrightarrow{S}_{j'})}{\bar{\mu}(S_{j'})} = \bar{\mathcal{K}}_{j'j''} \Big/ \frac{\bar{\mu}(\overrightarrow{S}_{j'})}{\bar{\mu}(S_{j'})}$$

Obviously, the matrix $\mathcal{M}(1)$ collapses into the kneading matrix (7) when the states S_i are made of just one Markov interval X_{ij}. In fact, in this case, our assumptions will prevent the system from assuming the same state for more than 1 time step, hence forcing $S_j = \overrightarrow{S}_j$.

To express the correlation we are aiming at it is now convenient to define two other quantities, namely

$$\mathcal{L}_{j'j''}(k) = \mathcal{M}_{j'j''}(k) - \mathcal{M}_{j'j''}(k+1) \tag{19}$$

$$j_{j'}(k) = \sum_{j} \sum_{i=k+1}^{\infty} \bar{\mu}(\overrightarrow{S}_j)\mathcal{L}_{jj'}(i) = \sum_{j} \bar{\mu}(\overrightarrow{S}_j)\mathcal{M}_{jj'}(k+1) \tag{20}$$

where $\mathcal{L}_{j'j''}$ is the probability of moving from $S_{j'}$ to $S_{j''}$ and staying in the latter exactly k time steps, and $j_{j'}(k)$ is nothing but the probability of staying at least $k+1$ time steps in $S_{j'}$ and then change to any other state so that $j_{j'}(0) = \bar{\mu}(\overrightarrow{S}_{j'})$.

To compute the entry of the matrix $\mathcal{H}(k)$, notice first that its generic entry $\mathcal{H}_{j'j''}(k)$ is obviously the probability of observing the system in the macro state $S_{j'}$ at a certain time step and observing the system in $S_{j''}$ k time steps after that. As a general remark note that, the system remains in the macro-state in which we have observed it for a certain amount of time. Then, it performs a certain number of transition sojourning in each of the intermediate macro state. Finally, it lands in the macro-state in which we observe it at the end of the time lag remaining there at least up to the observation instant.

To write an expression for $\mathcal{H}(k)$ we may formalize this remark considering that in the k time steps between the two observations the system may exhibit $0, 1, 2, \ldots$ macro-state transitions and writing

$$\mathcal{H}_{j'j''}(k) = \sum_{l=k}^{\infty} \mathrm{j}_{j'}(l)\mathcal{I}_{j'j''} + \sum_{\substack{l_1 \geq 0 \\ l_2 > 0 \\ l_1+l_2=k}} \mathrm{j}_{j'}(l_1)\mathcal{M}_{j'j''}(l_2) +$$

$$\sum_{\substack{l_1 \geq 0 \\ l_2, l_3 > 0 \\ l_1+l_2+l_3=k}} \sum_{i_1} \mathrm{j}_{j'}(l_1)\mathcal{L}_{j'i_1}(l_2)\mathcal{M}_{i_1 j''}(l_3) +$$

$$\sum_{\substack{l_1 \geq 0 \\ l_2, l_3, l_4 > 0 \\ l_1+l_2+l_3+l_4=k}} \sum_{i_1 i_2} \mathrm{j}_{j'}(l_1)\mathcal{L}_{j'i_1}(l_2)\mathcal{L}_{i_1 i_2}(l_3)\mathcal{M}_{i_2 j''}(l_4) + \ldots \tag{21}$$

where $\mathcal{I}$ is the identity matrix and the vector term $\sum_{l=k}^{\infty} \mathrm{j}_{j'}(l)$ accounts for the probability of beings observed in the same state after k time steps when no state transition happens, the following terms accounts for 1 state transition, the following for 2 state transitions, and so on.

We may now define an inner product between vector and matrixes as in:

$$\{\mathcal{A} * \mathcal{B}\}_{j'j''}(k) = \sum_{i_1} \sum_{j=-\infty}^{\infty} \mathcal{A}_{j'i_1}(j)\mathcal{B}_{i_1 j''}(k-j)$$

where the usual product between scalar has been replaced by sequence convolution. For a square matrix function we also define $\mathcal{A}^{*p}(k) = \mathcal{A}(k) * \mathcal{A}(k) * \cdots * \mathcal{A}(k)$ p times, and $\mathcal{A}^{*0}_{j'j''}(k) = 1$ if $j' = j''$ and $k = 1$, and zero otherwise.

With these definitions, the expression of $\mathcal{H}$ can be easily rewritten as:

$$\mathcal{H}(k) = \mathrm{diag}\left(\sum_{l=k}^{\infty} \mathrm{j}(l)\right) + \left\{\mathrm{diag}\,\mathrm{j} * \left[\sum_{j=1}^{\infty} \mathcal{L}^{*(j-1)}\right] * \mathcal{M}\right\}(k) \tag{22}$$

where the diag$(\cdot)$ function generates a diagonal matrix whose diagonal coincides with the argument vector. Note also that we used convolution instead of finite sums assuming that all the matrix functions vanish for all negative arguments and that $\mathcal{L}_{j'j''}(0) = \mathcal{M}_{j'j''}(0) = 0$ to take into account the conditions $l_j > 0$ of (21).

We further assume that all the matrix functions we defined are summable so that the generic z-transform:

$$\tilde{\mathcal{A}}(z) = \sum_{j=-\infty}^{\infty} \mathcal{A}(j) z^{-j}$$

converges for $|z| > 1$. With this we may now write the z-transform of $\mathcal{H}(k)$ as

$$\tilde{\mathcal{H}}(z) = \frac{\mathrm{diag}[\tilde{\mathrm{j}}(z) - z\tilde{\mathrm{j}}(1)]}{1-z} + \mathrm{diag}\,\tilde{\mathrm{j}}(z)\left[\sum_{j=0}^{\infty} \tilde{\mathcal{L}}^{j}(z)\right]\tilde{\mathcal{M}}(z)$$

and hence

$$\tilde{\mathcal{H}}(z) = \frac{\mathrm{diag}[\tilde{\mathbf{j}}(z) - z\tilde{\mathbf{j}}(1)]}{1-z} + \mathrm{diag}\,\tilde{\mathbf{j}}(z)\left[\mathcal{I} - \tilde{\mathcal{L}}(z)\right]^{-1}\tilde{\mathcal{M}}(z) \tag{23}$$

Note now that from (19) and (20) we get

$$\tilde{\mathcal{M}}(z) = \frac{\tilde{\mathcal{L}}(z) - \tilde{\mathcal{L}}(1)}{1-z} \qquad \tilde{\mathbf{j}}(z) = z\mathbf{j}(0)\frac{\tilde{\mathcal{L}}(z) - \tilde{\mathcal{L}}(1)}{1-z}$$

so that

$$\tilde{\mathcal{H}}(z) = \mathrm{diag}\left[-\frac{z}{1-z}\tilde{\mathbf{j}}(1) + \frac{z\mathbf{j}(0)}{(1-z)^2}(\tilde{\mathcal{L}}(z) - \tilde{\mathcal{L}}(1))\right] \times \left[\mathcal{I} + (\mathcal{I} - \tilde{\mathcal{L}}(z))^{-1}(\tilde{\mathcal{L}}(z) - \tilde{\mathcal{L}}(1))\right]$$

To proceed further we assume that M is completely defined by the two sequences of points

$$a_{1j} = \frac{1}{2}\sum_{l=j+1}^{\infty} \Delta_1(l) \qquad a_{2j} = 1 - \frac{1}{2}\sum_{l=j+1}^{\infty} \Delta_2(l)$$

depending only on the two probability functions Δ_1 and Δ_2 whose significance will be soon clarified.

We also set $X_{1j} = [a_{1j+1}, a_{1j}]$, $X_{2j} = [a_{2j}, a_{2j+1}]$ and impose M affine in each X_{ij} and such that $M(X_{ij}) = X_{ij-1}$ for $j > 1$ and $f(X_{10}) = S_2$ while $M(X_{20}) = S_1$. A sketch of the map structure is shown in figure 5

Since M is Markov and affine in each interval X_{ij} we obviously have that the invariant density is uniform in each of those intervals. From this, from the map construction, from the fact that $\overrightarrow{S}_i = X_{i0}$ and from the piecewise-affinity of M we also get that, after a state transition, the state is uniformly distributed in the new S_i. Note now that as long as $x \in S_i$ it passes from X_{ij} to X_{ij-1} at each time step until it reaches $\overrightarrow{S}_i$. Hence, the probability of staying exactly k time steps in S_i is equal to the probability of landing in X_{ik-1} after the state transition. Yet, by construction an noting that $\sum_{l=1}^{\infty} \Delta_i(l) = 1$, such a probability is nothing but $(a_{ik-1} - a_{ik})/(a_{i0} - a_{i\infty}) = \Delta_i(k)$.

We may therefore restrict our attention to systems in which

$$\mathcal{L}(k) = \begin{bmatrix} 0 & \Delta_2(k) \\ \Delta_1(k) & 0 \end{bmatrix} \tag{24}$$

where $\Delta_i(k)$ has now the significance of probability to stay in the state i exactly for k time steps. Additionally, by using again the fact that $\sum_{l=1}^{\infty} \Delta_i(l) = 1$, from (19) we also have:

$$\mathcal{M}(1) = \begin{bmatrix} 0 & 1 \\ 1 & 0 \end{bmatrix}$$

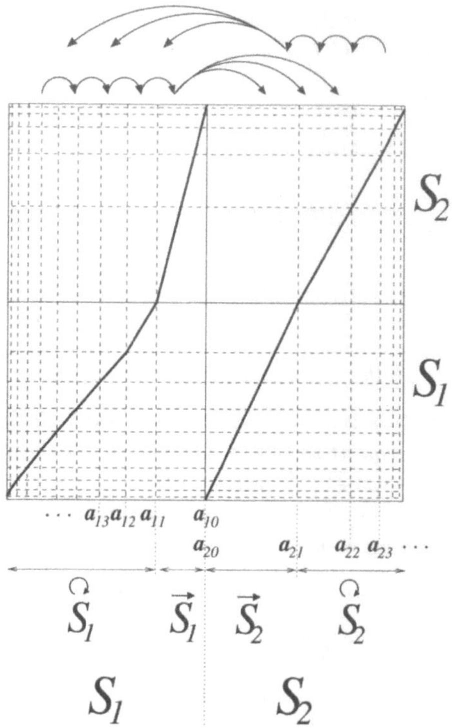

Fig. 5: A chaotic map with two macro-states whose sojourn time statistic can be adjusted at will

and thus, from (19) and defining $T_i = \sum_{l=1}^{\infty} l\Delta_i(l)$, we get:

$$\mathsf{j}(0) = \frac{(1,1)^T}{T_1 + T_2} \qquad \tilde{\mathsf{j}}(1) = \frac{(T_1, T_2)^T}{T_1 + T_2}$$

while simple calculations give

$$(\mathcal{I} - \tilde{\mathcal{L}}(z))^{-1} = \frac{1}{1 - \tilde{\Delta}_1(z)\tilde{\Delta}_2(z)} \begin{bmatrix} 1 & \tilde{\Delta}_2(z) \\ \tilde{\Delta}_1(z) & 1 \end{bmatrix}$$

The process for which we want to get self-similar features is the one constructed by considering the quantization $y_i = f(x_i)$ of trajectories $x_{i+1} = M(x_i)$ by means of a real function f such that $f(S_1) = 1$ and $f(S_2) = 0$. We will obviously deal with the process as well as with its aggregates $y_i^{(A)}$. that represents the sequence of activity (ON=1) and inactivity (OFF=0) on the network at different time scales. An additional parameter which is very important in the characterization of a self-similar traffic source is the activity index $P_{\text{ON}} = E[y_i] = E[y_i^{(A)}] = T_1/(T_1+T_2)$ [31], which gives the activity time fraction. To proceed, let us define the vector of the quantization values $\mathbf{f} = (1,0)^T$

and note that, using the generalization of (10) for the case of non-zero average process, we have $C(k) = \mathbf{f}^T\mathcal{H}(k)\mathbf{f} - P_{\mathrm{ON}}^2 = (\mathbf{f} - P_{\mathrm{ON}})^T\mathcal{H}(k)(\mathbf{f} - P_{\mathrm{ON}})$ and thus

$$\tilde{C}(z) = z(z-1)^{-2}\left(\tilde{\Delta}_1(z)\tilde{\Delta}_2(z) - 1\right)^{-1}(T_1 + T_2)^{-2}\times \tag{25}$$
$$\left[(\tilde{\Delta}_1(z) - 1)(\tilde{\Delta}_2(z) - 1)(T_1 + T_2) + (z-1)(\tilde{\Delta}_1(z)\tilde{\Delta}_2(z) - 1)T_1T_2\right]$$

To investigate the asymptotic behavior, let us review the following Tauberian result [46]:

Theorem 5. *Let $\tilde{x}(z)$ be the z transform of the sequence x_k.*

- *If a continuous function $x : \mathbb{R}^+ \mapsto \mathbb{R}^+$ exists such that $x_k = x(k)$ and $x(t)$ is asymptotically equivalent to t^{2H-2} ($H \in]0.5, 1[$) then $\tilde{x}(e^\epsilon)$ converges for $\epsilon > 0$ while it diverges as ϵ^{1-2H} for $\epsilon \to 0^+$.*
- *If $\tilde{x}(e^\epsilon) \sim \epsilon^{1-2H}$ for $\epsilon \to 0^+$ and x_k is non-negative and eventually monotonic decreasing then $x_k \sim k^{2H-2}$ for $k \to \infty$.*

Proof. Note first that for $\epsilon > 0$ we have $z^{-1} = e^{-\epsilon} < 1$ so that $\tilde{x}(e^\epsilon) = \sum_{k=0}^{\infty} x_k \left(z^{-1}\right)^k$ surely converges.

Define now the following subset of the real line $A(s) = \{\xi = k\epsilon | 0 \leq \xi \leq s\}$ and note that we can rewrite the z-transform as

$$\tilde{x}(e^\epsilon) = \frac{1}{\epsilon} \lim_{s\to\infty} \sum_{\xi \in A(s)} x_{\xi/\epsilon} e^{-\xi}\epsilon$$

which is asymptotically equivalent to the Riemann's sum of the corresponding integral with step ϵ. Hence the sum itself can be rewritten in the limit as

$$\tilde{x}(e^\epsilon) \sim \frac{1}{\epsilon}\int_0^\infty x\left(\frac{\xi}{\epsilon}\right) e^{-\xi} d\xi$$

We may now assume that $x(t)$ is negligibly different from the asymptotic behavior Xt^{2H-2} for $t > \bar{t}$. With this, the above asymptotic equivalence can be recast into

$$\tilde{x}(e^\epsilon) \sim \frac{1}{\epsilon}\int_0^{\bar{t}\epsilon} x\left(\frac{\xi}{\epsilon}\right) e^{-\xi} d\xi + \frac{X}{\epsilon}\int_{\bar{t}\epsilon}^\infty \left(\frac{\xi}{\epsilon}\right)^{2H-2} e^{-\xi} d\xi =$$
$$\int_0^{\bar{t}} x(t) e^{-t\epsilon} dt - X\epsilon^{1-2H}\left[\xi^{2H-1}\mathrm{E}_{2-2H}(\xi)\right]_{\bar{t}\epsilon}^\infty$$

where the exponential integral function is defined as $\mathrm{E}_{2-2H}(\xi) = \int_1^\infty e^{-\xi a} a^{2H-2} da$. One may now check that the exponential integral function is such that $\xi^{2H-1}\mathrm{E}_{2-2H}(\xi) \to 0$ for $\xi \to \infty$ while $(\bar{t}\epsilon)^{2H-1}\mathrm{E}_{2-2H}(\bar{t}\epsilon) = \bar{t}^{2H-1}\Gamma(2H-1)$ for $\epsilon \to 0$, $\Gamma(\cdot)$ being the conventional gamma function.

Hence, taking the limit of the above expression for $\epsilon \to 0$ one finally obtains

$$\tilde{x}(e^\epsilon) \sim \int_0^{\bar{t}} x(t)dt + X\epsilon^{1-2H}\bar{t}^{2H-1}\Gamma(2H-1) \sim \epsilon^{1-2H}$$

Conversely, we may exploit basic Tauberian theory (see e.g. [46, Theorem 8.7]) to obtain that, under our assumptions with the exception of the eventually decrease of x_k

$$\sum_{i=0}^{n} x_i \sim \frac{n^{2H-1}L(n)}{\Gamma(2H)}$$

where $L(n)$ is as slowly varying function for $n \to \infty$.

By adding the eventually decreasing property we know that $L(n)$ accounts for no oscillation around the asymptotic trend and are allowed to write

$$x_k \sim \sum_{i=0}^{k} x_i - \sum_{i=0}^{k-1} x_i \sim \frac{k^{2H-2}}{\Gamma(2H-1)}$$

Hence, we analyze the z-transform of Δ_i in the special case $z = e^\epsilon$ and $\epsilon \to 0$. In that neighborhood we may obtain an expansion of a generic $\tilde{\Delta}_i(z)$ noting that $\tilde{\Delta}(1) = 1$, that $\tilde{\Delta}_i'(1) = -T_i$ and that:

$$\tilde{\Delta}_i''(z) = z^{-2}\sum_{k=1}^{\infty} k(k-1)\Delta_i(k)z^{-k}$$

Hence, the behavior of $\tilde{\Delta}_i''(z)$ for $\epsilon \to 0$ depends on the asymptotic trend of $k^2\Delta_i(k)$.

If we assume a polynomially vanishing $\Delta_i(k) \sim k^{2H_i-4}$, with $H_i \in]0.5, 1[$, we have $k^2\Delta_i(k) \sim k^{2H_i-2}$ and thus, from Theorem 5, for $z = e^\epsilon$ and $\epsilon \to 0$ we have $\tilde{\Delta}_i''(z) \sim \epsilon^{1-2H_i}$ which accounts for an expansion of the kind:

$$\tilde{\Delta}_i(z) \sim 1 - T_i(z-1) + U_i(z-1)^{3-2H_i}$$

for some constant U_i. With the aim of comparison, let us observe that in the case of exponentially vanishing function, the same expression hold with $H_i = 0.5$.

Note that, if $H_i \in]0.5, 1[$ then $3 - 2H_i \in]1, 2[$ so that we may characterize the behavior of the z-transform or either exponentially or polynomially decaying $\Delta_i(k)$ with the three parameters T_i, U_i and $\alpha_i \in]1, 2]$ such that, when $z = e^\epsilon$ and $\epsilon \to 0$:

$$\tilde{\Delta}_i(z) \sim 1 - T_i(z-1) + U_i(z-1)^{\alpha_i} = 1 - T_i\epsilon + U_i\epsilon^{\alpha_i}$$

where we exploited also the asymptotic equivalence $e^\epsilon - 1 \sim \epsilon$. With these expansions (25) may be recast to:

$$\tilde{C}(z) \sim \frac{T_2^2 U_1 \epsilon^{\alpha_1 - 2} + T_1^2 U_2 \epsilon^{\alpha_2 - 2}}{(T_1 + T_2)^2}$$

With this, we may set $\alpha = \min_i\{\alpha_i\}$ to obtain:

$$\tilde{C}(z) \sim \epsilon^{\alpha-2} \frac{1}{(T_1 + T_2)^2} \sum_{\alpha_i = \alpha} U_i T_{3-i}^2$$

Note finally that, if $\alpha < 2$, and the autocovariance is eventually positive and monotonic, Theorem 5 implies that it obeys:

$$C(k) \sim k^{1-\alpha} = k^{2H-2}$$

where H is related to the slowest asymptotic decay in sojourn time probabilities. Since the autocovariance function has the desired expression (1), we may conclude that when at least one state has a polynomially decaying sojourn time probability the PWAPM system is able to produce second-order self-similar trajectories. The scaling parameter of such trajectories is controlled by the slowest asymptotic decay in sojourn time probabilities.

6 Self-Similarity & Network traffic

The above results on pseudo-Markov maps with two macro-states can be used to design a computer network traffic generator. The aim is to have a tool to generate traffic traces, characterized by synthetic and adjustable parameters, which can be used to evaluate off-line the performance of queue systems (e.g., switches and routers) and network protocols (e.g., protocols for medium access control or routing). Following the considerations reported in the introduction, we will model the network activity by means of an ON/OFF discrete-time process modeled by using a PWAPM map of the kind in figure 5 in which we associate S_1 with the ON condition and S_2 with the OFF condition.

As we have shown in section 5 by selecting at least one of the probabilities Δ_1 and Δ_2 as polynomially vanishing, the 2nd-order self-similarity condition reported in the second of (1) holds. As a consequence, to capture the general network traffic trend, it is sufficient to design a binary PWAPM systems with at least one macro-state featuring polynomial decay of the sojourn time probability. Furthermore, the exponent of such a polynomial decay is controlled by the slowest asymptotic decay in sojourn time probability.

In order to explain the map design procedure let us rename $\varepsilon_\tau = \Delta_1(\tau)$ and $\nu_\tau = \Delta_2(\tau)$ the probabilities of remaining in the ON and in the OFF states for τ steps.

The parameters of the target map are set by means of a Lagrangian-based iterative technique aiming at minimizing the difference between the desired sojourn distribution (at least one of the probabilities ε_τ and ν_τ must be polynomially vanishing) and those implied by the structure of the map itself.

In particular we select two cases for the nominal decays:

- $\tilde{\varepsilon}_\tau \sim A\gamma^\tau$ and $\tilde{\nu}_\tau \sim B\tau^{2H-4}$ with $A, B > 0$, $0 < \gamma < 1$, and H the Hurst parameter;
- $\tilde{\varepsilon}_\tau \sim A\tau^{2H_1-4}$ and $\tilde{\nu}_\tau \sim B\tau^{2H_2-4}$ with $A, B > 0$ and $H = \max(H_1, H_2)$ the Hurst parameter.

These two cases permit to design two different kinds of self-similar maps with two ON/OFF sojourn time distributions: *light/heavy-tailed* and *heavy/heavy-tailed.*

We want to assign $P_{\mathrm{ON}} = E[y_i]$ minimizing the deviation of ε_τ and ν_τ from the nominal decays $\tilde{\varepsilon}_\tau$ and $\tilde{\nu}_\tau$. To do so we note that $P_{\mathrm{ON}} = T_1/(T_1+T_2)$ with the ON and OFF average times being $T_1 = \sum_{\tau=1}^{\infty} \tau\varepsilon_\tau$ and $T_2 = \sum_{\tau=1}^{\infty} \tau\nu_\tau$, respectively. Hence, we must solve the following minimization problem:

$$
\begin{aligned}
&\min \sum_{\tau=1}^{\infty} \left(\frac{\varepsilon_\tau}{\tilde{\varepsilon}_\tau} - 1\right)^2 + \left(\frac{\nu_\tau}{\tilde{\nu}_\tau} - 1\right)^2 \\
&\text{s.t. } (P_{\mathrm{ON}} - 1)\sum_{\tau=1}^{\infty} \tau\varepsilon_\tau + P_{\mathrm{ON}} \sum_{\tau=1}^{\infty} \tau\nu_\tau = 0 \\
&\text{s.t. } \sum_{\tau=1}^{\infty} \varepsilon_\tau = 1\ ,\ \sum_{\tau=1}^{\infty} \nu_\tau = 1\ ,\ \varepsilon_\tau \geq 0\ ,\ \nu_\tau \geq 0
\end{aligned}
\tag{26}
$$

To this aim, note first that, since when inequality constraints are active they set the corresponding probability to be zero, the functional form of the solution of (26) can be obtained by considering only the equality constraints. With this we obtain that the optimal probabilities $\overset{*}{\varepsilon}_\tau$ and $\overset{*}{\nu}_\tau$ may have only two different functional forms, namely:

$$
\overset{*}{\varepsilon}_\tau = \begin{cases} \hat{\varepsilon}_\tau = \tilde{\varepsilon}_\tau + \dfrac{\lambda_1(P_{\mathrm{ON}} - 1)\tau}{2}\tilde{\varepsilon}_\tau^2 + \dfrac{\lambda_2}{2}\tilde{\varepsilon}_\tau^2 & \text{if } \hat{\varepsilon}_\tau \geq 0 \\ 0 & \text{otherwise} \end{cases}
$$

$$
\overset{*}{\nu}_\tau = \begin{cases} \hat{\nu}_\tau = \tilde{\nu}_\tau + \dfrac{\lambda_1 P_{\mathrm{ON}}\tau}{2}\tilde{\nu}_\tau^2 + \dfrac{\lambda_3}{2}\tilde{\nu}_\tau^2 & \text{if } \hat{\nu}_\tau \geq 0 \\ 0 & \text{otherwise} \end{cases}
$$

where λ_1 is the Lagrange multiplier corresponding to the P_{ON} constraint, λ_2 is the Lagrange multiplier corresponding to the normalization of the ε_τ and λ_3 is the Lagrange multiplier corresponding to the normalization of the ν_τ.

Regrettably, the values of λ_1, λ_2 and λ_3 depend on the indexes for which $\hat{\varepsilon}_\tau < 0$ and $\hat{\nu}_\tau < 0$ and a suitable procedure must be devised to solve the problem.

To this aim note first that, given the asymptotic positivity of $\hat{\varepsilon}_\tau$ and $\hat{\nu}_\tau$, only a finite number of vanishing entries exist in $\overset{*}{\varepsilon}_\tau$ and $\overset{*}{\nu}_\tau$.

Moreover, any generic minimization problem in a sequence space with equality and positivity constraints may benefit from the following theorem.

Theorem 6. *Let the sequences $\{u_\tau\}_{\tau=1}^{\infty}$ and $\{a_{i\tau}\}_{\tau=1}^{\infty}$ for $i = 1, \ldots, n$ be given along with the real numbers b_i for $i = 1, \ldots, n$. Assume that the solution $\overset{*}{\underline{u}} = \{\overset{*}{u}_\tau\}_{\tau=1}^{\infty}$ of the minimization problem:*

$$\min \sum_{\tau=1}^{\infty} (u_\tau - 1)^2$$

$$\text{s.t.} \sum_{\tau=1}^{\infty} a_{i\tau} u_\tau - b_i = 0 \quad \forall i \ , \ u_\tau \geq 0$$

exist. Let also $\hat{\underline{u}} = \{\hat{u}_\tau\}_{\tau=1}^{\infty}$ be the solution of the relaxation, without the constraint $u_\tau \geq 0$, which surely exists. If $\hat{\underline{u}} \neq \overset{*}{\underline{u}}$ then when $\hat{u}_\tau < 0$ we have $\overset{*}{u}_\tau = 0$.

Note that we may rewrite (26) to fit the assumptions of theorem 6 if we set $u_{2\tau-1} = \varepsilon_\tau / \tilde{\varepsilon}_\tau$ and $u_{2\tau} = \nu_\tau / \tilde{\nu}_\tau$ that leave the positivity constraints unchanged.

We may now address the solution of (26) and solve the relaxed problem where the Lagrange's multipliers λ_1, λ_2, and λ_3 are determined only by the satisfaction of the three equality constraints in (26). This procedure must be iterated until the solution of the relaxed problem has no negative components. Note that termination is guaranteed from the finiteness of the number of vanishing probabilities in the solution of (26) and from the fact that, when the solution of the relaxed problem has no negative components then it coincides with the solution of the non-relaxed problem.

Once that the two probabilities distributions $\overset{*}{\varepsilon}_\tau$ and $\overset{*}{\nu}_\tau$ are known we may construct a chaotic map M as described in section 5 whose iteration causes the state $x \in [0, 1]$ to switch between the ON condition $x \in [0, 1/2]$ and the OFF condition $x \in]1/2, 1]$ with the given statistics for the sojourn times.

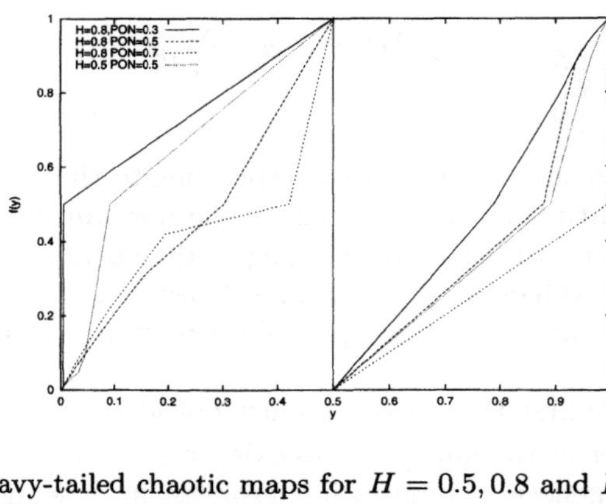

Fig. 6: Light/heavy-tailed chaotic maps for $H = 0.5, 0.8$ and $P_{\text{ON}} = 0.3, 0.5, 0.8$.

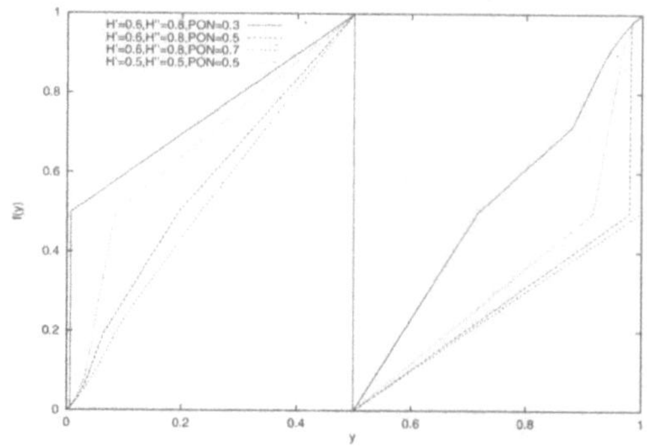

Fig. 7: Heavy/heavy-tailed chaotic maps for $H = 0.5, 0.8$ and $P_{\text{ON}} = 0.3, 0.5, 0.8$.

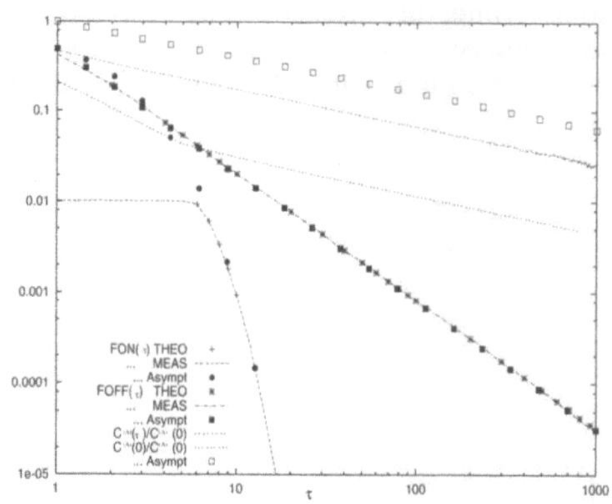

Fig. 8: ON/OFF sojourn time distributions (theoretical, measured and asymptotic), correlation and variance functions (measured and asymptotic) relative to the light/heavy-tailed chaotic map, with $H = 0.8$ and $P_{\text{ON}} = 0.3$

In particular, to obtain a chaotic map with light/heavy-tailed sojourn profiles we set: $\tilde{\varepsilon}_\tau \sim A\gamma^\tau$ and $\tilde{\nu}_\tau \sim B\tau^{2H-4}$. Thus, by following the described map design criterion, we obtain the maps in figure 6, for $H = 0.5, 0.8$ and $P_{\text{ON}} = 0.3, 0.5, 0.8$.

To obtain a chaotic map with heavy/heavy-tailed sojourn profiles we set: $\tilde{\varepsilon}_\tau \sim A\tau^{2H_1-4}$ and $\tilde{\nu}_\tau \sim B\tau^{2H_2-4}$. Thus, by following the described map design criterion, we obtain the maps in figure 7, for $H = 0.5, 0.8$ and $P_{\text{ON}} = 0.3, 0.5, 0.8$.

In order to verify the behavior of the ON/OFF sojourn distributions of the proposed maps, in figures 8 (light/heavy-tailed) and 9 (heavy/heavy-tailed), the complementary distributions of the ON time, $F_{ON}(\tau)$ and of the OFF time, $F_{OFF}(\tau)$ are reported, for $H = 0.8$ and $P_{\text{ON}} = 0.3$. In particular, the theoretical trends corresponding to $\overset{*}{\varepsilon}_\tau$ and $\overset{*}{\nu}_\tau$ (i.e., derived by means

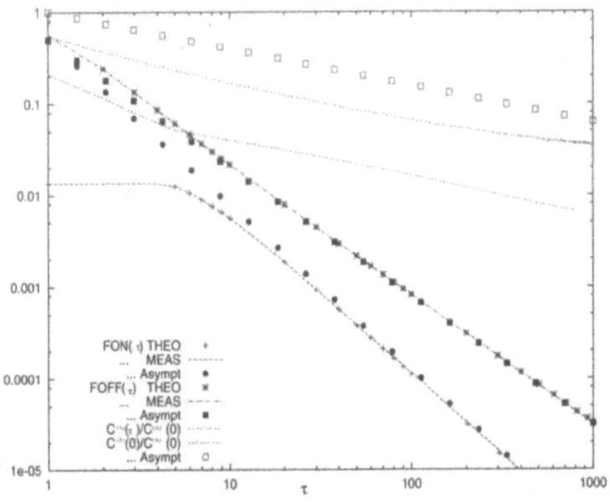

Fig. 9: ON/OFF sojourn time distributions (theoretical, measured and asymptotic), correlation and variance functions (measured and asymptotic) relative to the heavy/heavy-tailed chaotic map, with $H_1 = 0.6, H_2 = 0.8$ and $P_{\mathrm{ON}} = 0.3$

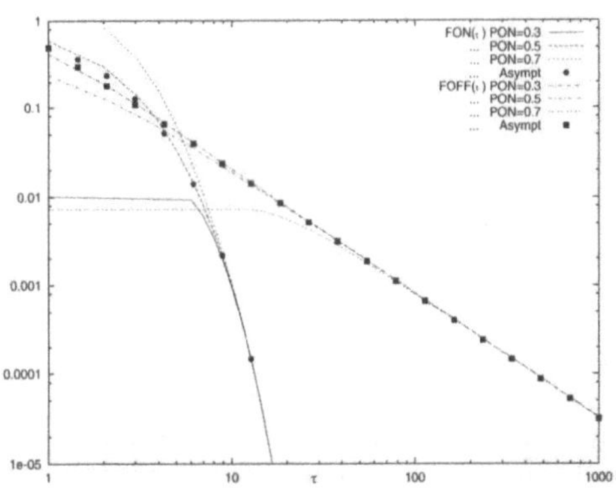

Fig. 10: ON/OFF sojourn time distributions (theoretical and asymptotic) relative to the light/heavy-tailed chaotic map, with $H = 0.8$ and $P_{\mathrm{ON}} = 0.3, 0.5, 0.7$

of the Lagrange-based iterative procedure described above); the asymptotic ones corresponding to $\tilde{\varepsilon}_\tau$ and $\tilde{\nu}_\tau$ (i.e., the target behavior for $\tau \gg 1$), and the measured ones (i.e., obtained by iterating the map) are reported for both systems. The good match between the curves can be verified.

In the same figures also the correlation and variance functions, $C_2^{(A)}(\tau)/C_2^{(A)}(0)$ and $C_2^{(\tau A)}(0)/C_2^{(A)}(0)$ (x has been substituted by τ to include this function in the same graph), obtained by iterating the maps, are reported. Note that the simulated trends match with the theoretical asymptotic decay $\tau^{-\beta} = \tau^{2H-2}$.

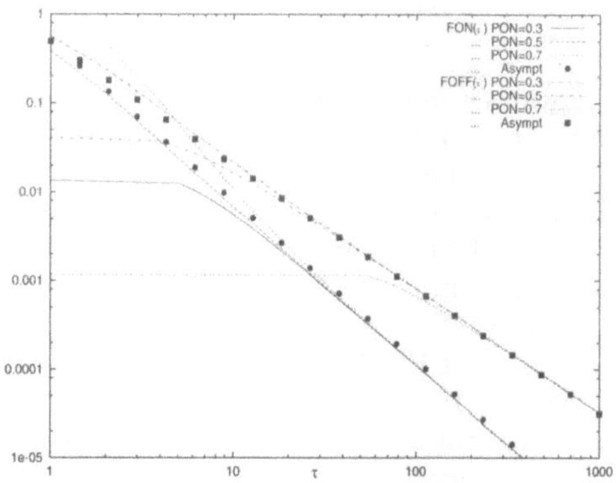

Fig. 11: ON/OFF sojourn time distributions (theoretical and asymptotic) relative to the heavy/heavy-tailed chaotic map, with $H_1 = 0.6, H_2 = 0.8$ and $P_{\mathrm{ON}} = 0.3, 0.5, 0.7$

In figures 10 and 11 only the ON/OFF sojourn distributions have been reported (theoretical and asymptotic trend), by considering $H = 0.8$ and different $P_{\mathrm{ON}} = 0.3, 0.5, 0.7$.

As a final remark, note that, since the described procedure allows to set P_{ON} and H independently, we are able to generate synthetic traffic processes corresponding to most of the known scenarios.

7 Conclusion

This chapter deals with the development of the formal tools needed to cope with determination of the statistical features of the process generated by Piecewise Affine Pseudo Markov systems, that are a suitable generalization of well-know PWAM maps able to produce self-similar process. The price we have to pay is that such maps can be characterized by an infinite but countable number of Markov intervals.

To maintain a perspective that is coherent with what was developed for PWAM maps, we decided that in the transition from a finite Markov partition to a countable one, the dimensionality of the space of observable functions must be kept finite.

As an example of use we applied the methods described in the body of the paper to the characterization of the self-similar behaviors that may be exhibited by quantized trajectories of suitably defined chaotic maps. We have also shown that this characterization can be exploited to give design guidelines yielding chaos-based synthetic LAN traffic generators that can mimic most of the traffic conditions of interest.

References

1. A. Lasota, M.C. Mackey, *Chaos, Fractals, and Noise*, Springer-Verlag, 1994
2. G. Setti, G. Mazzini, R. Rovatti, S. Callegari, "Statistical Modeling of Discrete-Time Chaotic Processes: Basic Finite-Dimensional Tools and Applications," Proceedings of the IEEE, vol. 90, pp. 662-690, May 2002
3. L. Kocarev, "Chaos-based Cryptography: a Brief Overview," IEEE Circuits and Systems Magazine, vol. 1, n. 3, pp. 6-21, 2001
4. G. Jakimoski, L. Kocarev, "Chaos and Cryptography: Block Encryption Ciphers Based on Chaotic Maps," IEEE Transactions on Circuits and Systems - Part I, vol. 48, n. 2, pp. 163-169, 2001
5. T. Kohda, A. Tsuneda, Statistics of Chaotic Binary Sequences, IEEE Transactions on Information Theory, vol. 43, pp. 104-112, 1997
6. T. Schimming, H. Dedieu, M. Hasler, M. Ogorzałek, "Noise Filtering in Chaos-Based Communication," chap. 8 in *Chaotic Electronics for Telecommunications*, M. P. Kennedy, R. Rovatti, G. Setti (Eds.), CRC Press, Boca Raton, 2000
7. S. Callegari, R. Rovatti, G. Setti, "Chaotic Modulations can Outperform Random Ones in EMI Reduction Tasks," IEE Electronics Letters, vol. 38, n. 12, pp. 543-544, June 2002
8. S. Callegari, R. Rovatti, G. Setti, "Spectral Properties of Chaos-Based FM Signals: Theory and Simulation Results," IEEE Transactions on Circuits and Systems- Part I, vol. 50, pp. 3-15, 2003
9. S. Callegari, R. Rovatti, G. Setti, "Chaos-Based FM Signals: Application and Implementation Issues," IEEE Transactions on Circuits and Systems- Part I, vol. 50, pp. 1141-1147, 2003
10. G. Setti, R. Rovatti, S. Callegari, M. Balestra, "Control of Chaos Statistics for the Generation of Timing Signals with Improved EMC", in *Chaos Control: Theory and Applications*, G. Chen, D. J. Hill, X. Yu (Editors), pp. 341-369, Springer-Verlag, 2003
11. M. Balestra, A. Bellini, S. Callegari, R. Rovatti, G. Setti, "Chaos-Based Generation of PWM-Like Signal for Low-EMI Induction Motor Drives: Analysis and Experimental Results," IEICE Transactions on Electronics, vol. 87-C, n. 1, pp. 66-75, 2004
12. M. P. Kennedy, R. Rovatti, G. Setti (eds.), *Chaotic Electronics in Telecommunications*, CRC Press, Boca Raton, 2000
13. G. Mazzini, G. Setti, R. Rovatti, "Chaotic Complex Spreading Sequences for Asynchronous CDMA - Part I: System Modeling and Results," IEEE Transactions on Circuits and Systems - Part I, vol. 44, n. 10, pp. 937–947, 1997
14. R. Rovatti, G. Setti, G. Mazzini, "Chaotic Complex Spreading Sequences for Asynchronous CDMA - Part II: Some Theoretical Performance Bounds," IEEE Transactions on Circuits and Systems - Part I, vol. 45, n. 4, pp. 496–506, 1998
15. R. Rovatti, G. Mazzini, G. Setti, "Interference Bounds for DS-CDMA Systems Based on Chaotic Piecewise-Affine Markov Maps," IEEE Transactions on Circuits and Systems- Part I, vol. 47, pp. 885-896, June 2000
16. G. Mazzini, R. Rovatti, G. Setti, "Interference Minimization by Autocorrelation Shaping in Asynchronous DS-CDMA Systems: Chaos-Based Spreading is Nearly Optimal," Electronics Letters, vol. 35, pp. 1054-1055, 1999
17. R. Rovatti, G. Mazzini, G. Setti, "A Tensor Approach to Higher-Order Expectations of Quantized Chaotic Trajectories - Part I: General Theory and Special-

ization to Piecewise Affine Markov Systems," IEEE Transactions on Circuits and Systems - Part I, pp. 1571-1583, November 2000

18. G. Mazzini, R. Rovatti, G. Setti, "A Tensor Approach to Higher-Order Expectations of Quantized Chaotic Trajectories - Part II: Application to Chaos-Based DS-CDMA in Multipath Environments," IEEE Transactions on Circuits and Systems - Part I, pp. 1584-1597, November 2000
19. R. Rovatti, G. Mazzini, G. Setti "Enhanced Rake Receivers for Chaos-Based DS-CDMA" IEEE Transactions on Circuits and Systems- Part I, vol. 48, pp. 818-829, 2001
20. G. Setti, R. Rovatti, G. Mazzini "Tensor-Based Theory for Quantized Piecewise-Affine Markov Systems: Analysis of Some Map Families," IEICE Transactions on Fundamentals, vol. 84, n. 9, pp. 2090-2100, 2001.
21. G. Mazzini, R. Rovatti, G. Setti, "Chaos-Based Asynchronous DS-CDMA Systems and Enhanced Rake Receivers: Measuring the Improvements," IEEE Transactions on Circuits and Systems- Part I, vol. 48, pp. 1445-1453, 2001
22. W.E. Leland, M.S. Taqqu, W. Willinger, D.V. Wilson, "On the Self-Similar Nature of Ethernet Traffic,", IEEE Transactions on Networking, vol. 2, pp. 1-15, 1994
23. W.E. Leland, D.V. Wilson, "High-time Resolution Measurements and Analysis of LAN Traffic: Implications for LAN Interconnections," in Proc. of INFOCOM'91, 1991.
24. A. Erramilli, P. Pruthi, W. Willinger, "Self-Similarity in High-Speed Network Traffic Measurements: Fact or Artifact?, " 12th Nordic Teletraffic Seminar-NTS 12 (VTT Symposium 154), 1995.
25. A. Erramilli, M. Roughan, D. Veitch, W. Willinger, "Self-Similar Traffic and Network Dynamics," Proceedings of the IEEE, vol. 90, n. 5, pp. 800-819, 2002
26. K. Park, W. Willinger, *Self-Similar Network Traffic and Performance Evaluation* Wiley & Sons, 2000.
27. B.B. Mandelbrot, "Long-Run Linearity, Locally Gaussian Processes, H-Spectra and Infinite Variances," International Economic Review, vol 10, pp. 82-113, 1969
28. D.R. Cox, "Long-Range Dependency: a Review," in *Statistics: an Appraisal* H.A. David ad H.T. David eds., pp. 55-74, Iowa State University Press, Ames, Iowa, 1984
29. E. Erramilli, P. Singh, "An application of deterministic chaotic maps to model packet traffic," Queuing Systems, vol. 20, pp. 171-206, 1995.
30. S. Molnar, I. Maricza, "Source Characterization in Broadband Networks," Cost 257, 2000.
31. R. Rovatti, G. Mazzini, "Tensor Function Analysis of Quantized Chaotic Piecewise-Affine Pseudo-Markov Systems - Part I: Higher Order Correlations and Self-Similarity," IEEE Transactions on Circuits and Systems - Part I, vol. 49, n. 2, pp. 137-149, 2002
32. R. Rovatti, G. Mazzini, "Tensor Function Analysis of Quantized Chaotic Piecewise-Affine Pseudo-Markov Systems - Part II: Higher Order Correlations and Self-Similarity," IEEE Transactions on Circuits and Systems - Part I, vol. 49, n. 2, pp. 150-162, 2002
33. R. Rovatti, G. Mazzini, G. Setti, A. Giovanardi, "Statistical Modelling and Design of Discrete-Time Chaotic Process: Advanced Finite-Dimesiional Tools and Applications," Proceedings of the IEEE, vol. 90, n. 5, pp. 820-841, 2002

34. A. Giovanardi, G. Mazzini, R. Rovatti, "Criteria to Design Chaotic Self-Similar Traffic Generators," IEICE Transactions on Fundamentals, vol. E84-A, n. 9, pp. 2155-2164, 2001
35. G. Keller, "On the Rate of Convergence to Equilibrium in One-Dimensional Systems," Communications in Mathematical Physics, vol. 96, pp. 181-193, 1984
36. R. E. Kalman, "Nonlinear aspects of sampled-data control systems," Proceedings of the Symposium on Nonlinear Circuit Analysis, 1956, vol. VI.
37. A. N. Shiryaev, *Probability*, 2nd edition, Springer-Verlag, 1995
38. M. Mori, "Fredholm Determinant for Piecewise Linear Transformations," Osaka Journal of Mathematics, vol. 27, pp. 81-116, 1990
39. M. Mori, "On the Intermittency of a Piecewise Linear Map (Takahashi Model)," Tokyo Journal of Mathematics, vol. 16, pp. 411-428, 1993
40. S.B. Lowen, "Fractal Renewal Processes generate $1/f$ noise," Physical Review E, vol. 47, pp. 992-1001, 1993
41. G. Setti, R. Rovatti, "Chaos-Based Generation $1/f^{[0,1[}$ Noise for Circuit Simulation", 1999 International Symposium on Nonlinear Theory and its Applications (NOLTA'99), 1999, Hawaii, USA, pp. 565-568
42. S. Wiggins, *Introduction to Applied Nonlinear Dynamical Systems and Chaos*, Springer-Verlag, 1996
43. S. Isola, "Renewal Sequences and Intermittency", Journal of Statistical Physics, vol. 97, pp. 263-280, 1999
44. S. Isola, "Return Times and Mixing Properties", Invited Keynote Speech at IEEE Workshop on Nonlinear Dynamics of Electronic Systems, 2000
45. A.L. Baranowski, W. Schwarz, "Statistical Analysis and Design of Continuous-Discrete Chaos Generators," IEICE Transactions on Fundamental, vol. E82-A, pp.1762-1768, 1999
46. A.M. Odlyzko, "Asymptotic Enumeration Methods," in *Handbook of Combinatorics*, R.L. Graham, M. Grötschel, L. Lovász (eds), vol. 2, pp. 1063-1229, Elsevier - MIT Press, 1995
47. H.S. Wilf, "Generatingfunctionology", Academic Press, San Diego, 1994
48. S. Wolfram, *The Mathematica Book*, 4th edition, Cambridge University Press, Cambridge, 1999

Macroscopic Dynamics in Large-Scale Data Networks

Jian Yuan[1,2] and Kevin Mills[1]

[1] National Institute of Standards and Technology USA kmills@nist.gov
[2] Tshingua University Beijing, China jyuan@tsinghua.edu.cn

1 Introduction

Modern society grows increasingly reliant on the Internet, a network of global reach that supports many services and clients. However, in such a large-scale distributed network, meeting quality-of-service requirements presents a difficult challenge because hotspots of network load move around and traffic anomalies arise unpredictably in space and time. In this chapter, we will demonstrate that observing network dynamics at a macroscopic level is likely to contribute to better network engineering and management.

The Internet is an enormous network of networks without central control or administration. Millions of computers around the world attach to the Internet through many autonomous regional networks of routers, which interconnect through backbone networks of routers in a distributed, hierarchical fashion. Internet computers exchange data with each other in units called packets, where each packet is accepted by a router, stored temporarily, and then forwarded on to a next router. This accept-store-and-forward cycle begins when a source computer transmits a packet to an entry router and continues until the packet is forwarded to its intended destination by an exit router. The Internet's design is guided by the end-to-end principle [1], which allocates simple functionality to routers, while pushing complexities of specific applications and of congestion avoidance mechanisms outside the network and into attached computers. The implication of this design principle is that Internet routers see no relationship among individual packets, while "end-to-end" protocols implemented in Internet-attached computers manage all state associated with data exchanges. The basic communication protocol of the Internet is called TCP/IP (Transmission-Control Protocol/Internet Protocol) [2]. IP is a "hop-by-hop" protocol used by source computers to inject packets into the Internet, and used by Internet routers to store-and-forward packets among multiple routers along a path, and then finally to forward the packet to its destination computer. TCP is an end-to-end protocol operating on logical connections between pairs of computers.

TCP includes a congestion-control algorithm to ensure that a sender does not transmit more data than the network can handle. The TCP congestion-control algorithm exhibits a self-organizing property: when a large number of logical connections share the Internet, underlying interactions among the connections avoid router congestion simultaneously over varying spatial extent; however, the network-wide effects created by such interactions are difficult to determine. The spatial-temporal dynamics of Internet traffic is also difficult to characterize due to highly variable user demands and to unpredictable resource availability. Further, not every client using the Internet is honest or co-operative. For example, distributed denial-of-service (DDoS) attacks, which arise when large numbers of compromised computers send traffic simultaneously toward a victim (e.g., a web server or a router) [3], may intermittently disturb the normal operating condition of the Internet. All these sources of variability inhibit easy characterization of Internet-wide traffic dynamics.

We suspect that, where many globally distributed data flows simultaneously transit a large network, the self-organizing properties of the TCP congestion-control algorithm might lead to the emergence of collective behavior, as in other complex systems [4]. Collective emergent phenomena often can be identified when the behavior of an entire system appears more coherent and directed than the behavior of individual parts of the system. In this way, any single data flow across a large network would not face a totally random condition, but more likely would adapt itself to a steady collective state, in which the flow could make little change. If such an emergent collective property occurs in large networks, and if we can describe and visualize the associated patterns, then perhaps such knowledge can be used to improve global network performance and to increase resistance to subtly engineered DDoS attacks.

Since emergent coherent behavior exhibits a spatial-temporal dependence among collective data flows over a whole network, correlation might be key to describing emergent patterns. A number of empirical studies on traffic measurements have convincingly demonstrated that actual Internet traffic exhibits long-range dependence (LRD) [5, 6, 7], which implies the existence of nontrivial correlation structure at large timescales. However, in these studies, the LRD found in Internet traffic was not attributed to an emergent, spontaneous order at the macroscopic (whole network) level. Instead, these studies attributed LRD to the linear multiplexing of a large number of highly variable traffic sources [8]. This explanation apparently ignores any nonlinear relationships that might arise as collective flows compete for network resources (router buffers and link capacity) over space and time. To understand the potential collective effect in large-scale networks, we conducted our own studies to identify the reasons behind LRD traffic phenomena [9, 10, 11]. We found that network size has greater influence than other factors—e.g., high variability in traffic sources and choice of transport mechanism—on the *temporal* dynamics of network congestion. Our findings suggesting the importance of network size in generating emergent collective behavior led us to consider how we might examine both the spatial and temporal dynamics of network congestion.

Recently, graph wavelets have been proposed for *spatial* traffic analysis given knowledge of aggregate traffic measurements extracted at intervals over all links [12]. This method can provide a highly summarized view of traffic load throughout an entire network. There seems to be no stringent time limit for producing a snapshot of network-wide load with such spatial traffic analysis; however, spatial-temporal traffic analysis, which reveals the time-varying nature of spatial traffic, may have to perform in a timely manner. Currently, spatial-temporal traffic analysis presents practical difficulties, not only because large-scale distributed networks exhibit high-dimensional traffic data, but also because mining large amounts of data may strain memory and computation resources in even the most advanced generation of desktop computers. Moreover, routers may be heavily utilized, and thus fail to collect and transfer data, often when the routers are of most interest (due to their congested nature). Given these practical constraints, it would be appealing to reduce the amount of data to transfer and process, while retaining the ability to observe spatial-temporal traffic dynamics. We believe that the emergence of collective behavior (with its associated global order) could be exploited to concisely capture spatial-temporal patterns with sparse observation points. In other words, if emergent behavior arises in a large network, then traffic will be correlated over wide space-time and, thus, might be characterized by sampling a small number of points. On the other hand, if network traffic exhibits little space-time correlation, then sampling a small number of points would not prove particularly revealing.

A recent study of correlations among data flows in a French scientific network, *Renater* [13], detected the signature of collective behavior. The *Renater* study uses methods from random matrix theory (RMT) to analyze cross-correlations between network flows. In essence, RMT compares a random correlation matrix—a correlation matrix constructed from mutually uncorrelated time series—against a correlation matrix for the data under investigation. Deviations between properties of the cross-correlation matrix from the investigation data and the correlations in the random data convey information about "genuine" correlations. In the case of the *Renater* study, the most remarkable deviations arise about the largest eigenvalue and its corresponding eigenvector. The largest eigenvalue is approximately a hundred times larger than the maximum eigenvalue predicted for uncorrelated time series. The largest eigenvalue appears to be associated with a strong correlation over the whole network. In addition, the eigenvector component distribution of the largest eigenvalue deviates significantly from the Gaussian distribution predicted by RMT. Further, the *Renater* study reveals that all components of the eigenvector corresponding to the largest eigenvalue are positive, which implies their collective contribution to the strong correlation. Since all network data flows contribute to the eigenvector, the eigenvector can be viewed as the signature of a collective behavior for which all flows are correlated.

In fact, the eigenvector corresponding to the largest eigenvalue provides an important clue, which led us to a novel method for observing spatial-temporal dynamics at the macroscopic level [14]. As the macroscopic pattern emerges from all adaptive behaviors of data flows in various directions, hotspots should be exposed through their correlation information, as the joining points of significantly correlated data flows. Note that the details of the components of the eigenvector of the largest eigenvalue reveal this information, with the larger components corresponding to the more correlated flows. Therefore, we define a weight vector by grouping eigenvector components corresponding to a destination routing domain together to build up information about the influence of the domain over the whole network. Contrasting weights of the weight vector against each other in space and time, we not only can summarize a network-wide view of traffic load, but can also locate hot spots, and can even observe how spatial traffic patterns change from one time period to the next.

Using this macroscopic-level method inevitably encounters issues of scale, that is, gathering data from numerous distributed measurement points, and consuming computation time and memory when analyzing data. The *Renater* study assumes complete information from all network connection points, which proves feasible because the *Renater* network contains only about 30 interconnected routers. We have figured out how to scale down the coverage problem by exploiting an emergent collective phenomenon, called the correlation increase [14]. Correlation increases arise from *collective response* of the entire network to changes in traffic. This effect has already been observed in the framework of stock correlations, where cross-correlations become more pronounced during volatile periods as compared to calm periods [15]. Indeed, higher values of the largest eigenvalue occur during periods of high market volatility, which suggests strong collective behavior accompanies high volatility. This connection has value in our analysis because Internet traffic behavior appears to be nonstationary [16]. An increase in cross-correlation allows us to infer a shift in the spatial-temporal traffic pattern of large areas of interest outside those few areas where measurements are made. This approach could significantly reduce requirements for data, perhaps to the point where analysis could occur in real time.

In this chapter, we use simulation results to show how this innovation could succeed in a large TCP/IP network. We apply our technique to identify network hotspots and to expose large-scale DDoS attacks in our simulation environment. The rest of this chapter is structured as five sections. Section 2 delineates a simulation model we developed recently to study space-time characteristics of congestion in large networks, and to analyze system behavior as a coherent whole. In Section 3, we describe our technique for spatial-temporal traffic analysis. In Section 4, we show how our technique captures network-wide patterns shifting over time. Section 5 demonstrates the macroscopic effect of DDoS flooding attacks, and shows how our technique could provide significant information to detect and defend against such attacks. We present concluding remarks in Section 6.

2 Modeling a Large-scale TCP/IP Network

Network simulation plays a key role in building an understanding of network behavior. Choosing a proper level of abstraction for a model depends very much on the objective. Studying collective phenomena seems to require simulating networks with a large spatial extent. Appropriate models for such studies should also include substantial detail representing protocol mechanisms across several layers of functionality (e.g., application, transport, network, and link), yet must also be restricted in space and time in order to prove computationally tractable. Previously, we adopted a two-tier modeling approach that maintains the individual identity of packets, producing a full-duplex "ripple effect" at the packet level, and that also accommodates spatial correlations in a regular network structure [10, 11]. While our two-tier model has been applied successfully to qualitatively understand some traffic characteristics in large-scale networks [11, 14], some doubts exist about the realism inherent in the regular network structure of such a model. In this chapter, we retain the individual identity of packets but we replace the regular network structure of our previous two-tier model with a large-scale irregular topology chosen to resemble the topology of a real network.

2.1 Topology

Here, we transform our regular two-tier model into an irregular four-tier topology, as shown in Figure 1. (The host-computer tier is not shown in Figure 1.) While the network core becomes heterogeneous and hierarchical, (tier-four) host-computer behavior remains homogeneous at the edge of the network. The (tier-one) backbone topology, including eleven (backbone) routers (A, B, ...K), resembles the original Abilene network, as described elsewhere [12]. Links between backbone routers have varying delays. For example, the longest link between backbone routers D and F has a 20-ms propagation delay; the shortest propagation delay (3 ms) exists on the link between backbone routers J and K. Forty (tier-two) subnets connect to the backbone through subnet routers, represented by designators such as A1 and B2. Each subnet contains a variable number of (tier-three) leaf routers, such as A1a and B2b. Each leaf router supports an equal number (200 in this chapter) of (tier-four) source hosts, and a variable number ($\leq$ 800 in this chapter) of (tier-four) receivers, activated on demand. Link capacities gradually increase from host (tier four) to backbone, with (tier-one) backbone links being hundreds of times faster than links to (tier-four) hosts.

2.2 Traffic Sources

There are a total of 22,000 sources in our model, which operates at the packet level. Each source models traffic generation as an ON/OFF process, which alternates between wake and sleep periods with average durations λ_{on} and

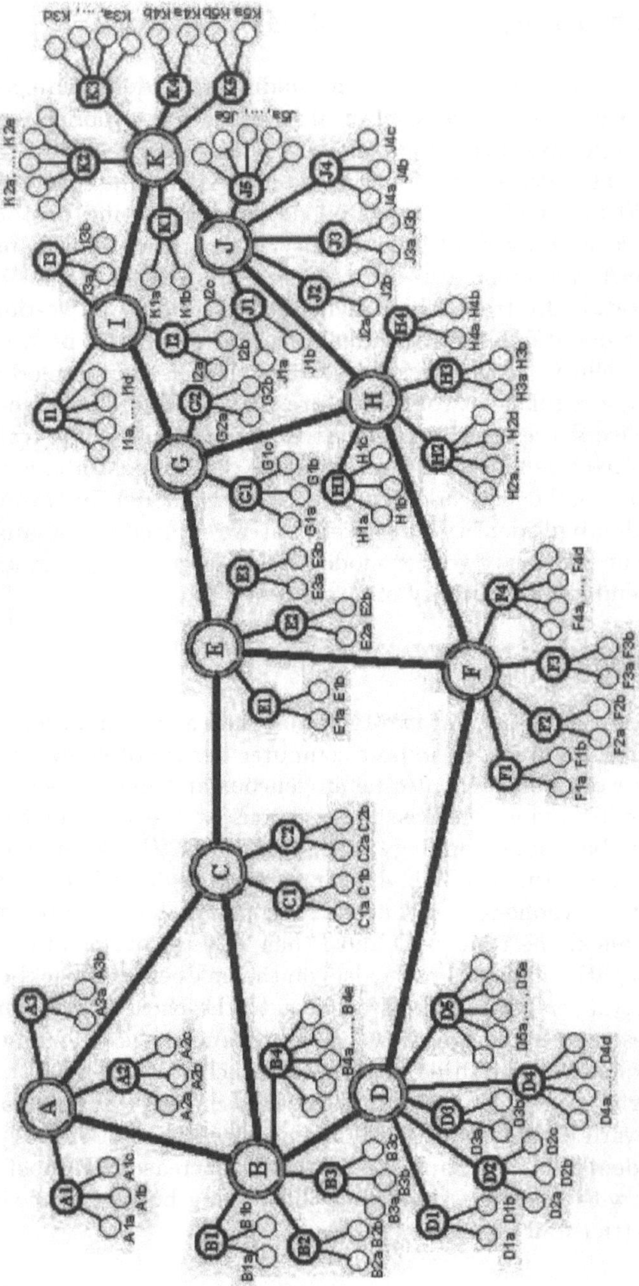

Fig. 1: Four-tier simulation model with 11 (tier-one) backbone routers, 40 (tier-two) subnet routers, and 110 (tier-three) leaf routers. The 22,000 (tier-four) source hosts and the up to 88,000 (tier-four) receivers are not shown.

λ_{off}, respectively. When awake, a source may send, subject to any restrictions imposed by TCP, one packet at each time-step to the source's attached leaf router. The packet will be placed at the end of the router's queue. At the beginning of each ON period, a destination receiver is chosen randomly from among leaf routers other than the local leaf routers, i.e., all flows must transit through at least one backbone link. When sleeping, the source does not generate new packets at each time-step. ON/OFF sources provide a convenient model of user behavior.

Empirical measurements on the Internet observe a heavy-tailed distribution of transferred file sizes [7]. Here, we use the Pareto distribution for both ON and OFF processes with the same shape parameter α [11]. In this chapter, $\lambda_{on} = 50$, $\lambda_{off} = 5000$ and $\alpha = 1.5$.

2.3 Routers

Packets, the basic unit of transmission on TCP/IP networks, contain destination addresses used by routers to correctly forward and source addresses used by receivers to identity the destination address for reply packets. To store and forward packets, which in our model travel a constant, shortest path between a source-destination pair for each flow, all routers maintain a queue of limited length (160 packets/router here), where arriving packets are stored until they can be processed: first-in, first-out. For convenience, in this chapter we assume that every discrete simulation time-step is 1 millisecond. However, each leaf router, subnet router, or backbone router can in turn forward 5, 20, or 160 packets during one millisecond. This simulates capacity differences among various link classes from leaf-access to backbone in a hierarchically structured network. With such parameter settings, simulated backbone links are very lightly loaded.

We select several subnet routers as observation points, e.g., B4, D5, F4, I1, and J5, which record all outbound flows to destination leaf routers. In this chapter, we assume that a central collector reliably receives a continuous stream of measured data from observation points in time to perform analysis for our various experiments.

3 Representing Macroscopic-Level Traffic Dynamics

In this section, we discuss briefly our approach to represent traffic dynamics at a macroscopic-level. First, we describe how we represent network flow data. Second, we outline our use of cross-correlation analysis. Finally, we depict our technique to summarize network-wide traffic load using a weight vector.

3.1 Representing Network Flow Data

Assume that there are N leaf routers, interconnecting through subnet routers and backbone routers to form a large-scale distributed network, where L subnet routers are deployed as observation points to log outbound traffic. First, we need to represent packets flowing between distinct source-destination pairs at each sampling interval. Let $\boldsymbol{x} = (\boldsymbol{x}_1, \boldsymbol{x}_2, \ldots, \boldsymbol{x}_N)^T$ denote the flow vector of corresponding packet counts, observed in L subnets during a given time interval. Each element of this flow vector is itself a vector defining the number of packets flowing into the corresponding leaf router from each of the observation subnets in the distributed network. The method to obtain all flow variables in this vector is to first enumerate all the destination leaf routers and then the observation posts by 1 to L, and group these indices by leaf router: the subnets sending to the first leaf router in the first block, $\boldsymbol{x}_1$, and those sending to the second leaf router in the second block, $\boldsymbol{x}_2$, and so forth. Thus, we form $\boldsymbol{x}$ with subvectors in the order $\boldsymbol{x}_1 = (x_{11}, x_{21}, \ldots, x_{L1})^T$, $\boldsymbol{x}_2 = (x_{12}, x_{22}, \ldots, x_{L2})^T, \ldots, \boldsymbol{x}_N = (x_{1N}, x_{2N}, \ldots, x_{LN})^T$, where x_{ij} represents packet flow from the ith observation point (i =1, 2, ..., L) to the jth leaf router (j =1, 2, ..., N). Each flow variable x_{ij} is normalized as f_{ij} by its mean m_{ij} and standard deviation σ_{ij},

$$f_{ij} = (x_{ij} - m_{ij})/\sigma_{ij}. \tag{1}$$

Then, the normalized flow vector $\boldsymbol{f}$, corresponding to $\boldsymbol{x}$, comprises N normalized subvectors, $\boldsymbol{f}_k$ (k =1, 2, ..., N), where each subvector is formed from normalized flow variables f_{ik}($i \leq L$ and $k \leq N$).

3.2 Cross-Correlation Analysis

Cross-correlation analysis is a tool commonly used to analyze multiple time series. We can compute the equal-time cross-correlation matrix $\boldsymbol{C}$ with elements

$$C_{(ij)(kl)} = \langle f_{ij}(t) f_{kl}(t) \rangle, \tag{2}$$

which measures the correlation between f_{ij} and f_{kl}, where $\langle \cdots \rangle$ denotes a time average over the period studied. The cross-correlation matrix is real and symmetric, with each element falling between –1 and 1. Positive values indicate positive correlation, while negative values indicate an inverse correlation. A zero value denotes lack of correlation.

We can further analyze the correlation matrix $\boldsymbol{C}$ through eigenanalysis [17]. The equation

$$\boldsymbol{Cw} = \lambda \boldsymbol{w} \tag{3}$$

defines eigenvalues and eigenvectors, where λ is a scalar, called the eigenvalue. If $\boldsymbol{C}$ is a square K-by-K matrix, e.g., $K = L(N-1)$ here, then $\boldsymbol{w}$

is the eigenvector, a nonzero K by 1 vector (a column vector). Eigenvalues and eigenvectors always come in pairs that correspond to each other. This eigenvalue problem has K real eigenvalues, some of which may repeat. An eigenvector is a special kind of vector for the matrix it is associated with, because the action of the underlying operator represented by the matrix takes a particularly simple form on the eigenvector input: namely, simple rescaling by a real number multiple. The eigenvector $\boldsymbol{w}^1$ corresponding to the largest eigenvalue λ_1 often has special significance for many applications. There are various algorithms for the computation of eigenvalues and eigenvectors [17]. Here, we exploit the MATLAB **eig** command, which uses the QR algorithm to obtain solutions [18].

3.3 Defining the Weight Vector

The cross-correlation matrix contains within itself information about underlying interactions among various flows. The components of the eigenvector $\boldsymbol{w}^1$ of the largest eigenvalue λ_1 represent the corresponding flows' influences on macroscopic behavior, abstracted from the matrix $\boldsymbol{C}$ into the pair (λ_1, $\boldsymbol{w}^1$). The eigenvector $\boldsymbol{w}^1$ comprises N subvectors, i.e., $\boldsymbol{w}^1 = (\boldsymbol{w}^1{}_1, \boldsymbol{w}^1{}_2, \ldots, \boldsymbol{w}^1{}_N)^T$. The kth subvector $\boldsymbol{w}^1{}_k$, corresponding to the kth leaf router, is formed from components $w^1{}_{ik}$ ($i \leq L$ and $k \leq N$) representing the ith obsevation point's contribution to the kth leaf router. We consider the square of each component, $(w^1{}_{ik})^2$, instead of $w^1{}_{ik}$ itself because $\sum_{i,k}(w^1_{ik})^2 = 1$ [19].

We define the weight S_k ($k = 1, 2, \ldots, N$) to be the sum of all $(w^1{}_{ik})^2$ in the kth subvector $\boldsymbol{w}^1{}_k$, i.e.,

$$S_k = \sum_i^L (w^1_{ik})^2. \tag{4}$$

S_k represents the relative strength of the contributions of the flows towards the kth leaf router. Thus, the knowledge of weight vector $\boldsymbol{S} = (S_1, S_2, \ldots, S_N)$ across varying k constitutes one summary view of network-wide traffic load. The evolving pattern of spatial-temporal correlation might allow us to infer where and when network congestion emerges.

4 Capturing Shifting Spatial-temporal Patterns

Internet access is never evenly distributed. Flash-crowds are quite common. Hot spots might develop and break up more quickly than the network could be re-provisioned to respond. However, capturing the movement of hot spots seems very difficult. Here, we try to use our technique to observe the macroscopic dynamics of such phenomena.

To deliberately induce congestion, we let one selected leaf router have an additional five percent probability for selection as the destination domain. This is a natural way to change the network-wide traffic demand at longer timescale. Figure 2 depicts a change in congestion in leaf routers. The vertical axis represents the congested location within 11 backbone-router zones, each of which denotes the subset of leaf routers therein. At first, leaf router H4b is congested (up until time, t, is 400 s). From $t = 400$ s, C2b is selected as a new location to induce congestion. This congestion-induction technique offers an easily interpreted framework to analyze spatial-temporal pattern shifts driven by varying traffic demand.

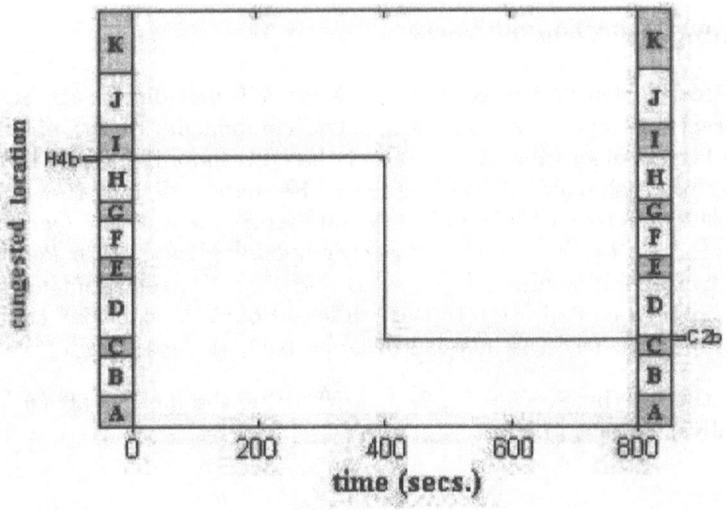

Fig. 2: Congested location changing over time from leaf router H4b to leaf router C2b.

4.1 Timescale of Interest

When focusing on network-wide behavior, the timescale of interest should not be fine-grained. The microscopic fluctuations observed at shorter timescales usually reflect local details, while the driving force of traffic demand seems to vary over much longer timescales. The timescale of interest in our experiments appears at a middle range, similar to the concept of a critical timescale beyond which traffic fluctuation is supposed to exhibit greater influence than microscopic fluctuations [20]. At this middle timescale, macroscopic (coherent) behavior emerges as a connecting link between short-range microscopic fluctuations and the longer-range driving force of variations in traffic demand. This coherence is expected to emerge as a result of adaptive behaviors among data flows in different directions, but then to continue to shift its spatial-temporal pattern under the force of traffic demand.

We first form an observation system of eleven points ($L = 11$) by selecting one subnet router in each backbone-router zone, instead of observing all subnets in all backbone-router zones. We observe, at a granularity of 200ms, every fine-grain flow from these subnet routers to every leaf router ($N = 110$). (In a subsequent section, we will try to further reduce observation points.)

Now, we calculate the weight vector $\boldsymbol{S}$ with M data points ($M = 200$ in this chapter), which span a first period ($M/2$ points) and a second period ($M/2$ points). Selecting an appropriate data length for analysis might be largely considered a trial-and-error process (or the subject of future work). Here, we selected $M = 200$, which seemed to work fine. We tried $M = 100$ and 300, which confirmed a data length of 200 more suitable for our experiments. Two weight vectors are calculated at the aggregated levels $T = 0.4$ s and $T = 2$ s, and shown respectively in Figure 3(a) and 3(b). The weight vector with $T = 2$ s shows two prominent weights at leaf routers C2b and H4b (S_{C2b} and S_{H4b}), revealing the network-wide pattern of congestion arising in these two domains. However, the pattern does not appear when $T = 0.4$ s. To clarify the role of timescale here, we further show the sum of S_{C2b} and S_{H4b} at different aggregated levels in Figure 3(c). We find that the sum of S_{C2b} and S_{H4b} gradually increases as T increases, up until about $T = 2$ s.

4.2 Increased Correlation

Figure 4(a) shows the sum of S_{C2b} and S_{H4b}, which is calculated with $T = 1.6$ s and with the time window, MT ($= 200 \times 1.6$ s $= 320$ s), sliding ahead every 16 s. The corresponding λ_1 shows in Figure 4(b). The time axis indicates the end of the moving time window. The sum of S_{C2b} and S_{H4b}, and the largest eigenvalue λ_1 undulate almost in the same way, reaching higher values during the period of pattern shifting than during calm periods. The increased correlation in the simulation data emerges gradually after the second period starts, spreading the varying traffic demand to the entire network. During this transient period, flows in different directions have to adapt their behaviors to the changing congestion, and the flows continue to react to each other until they reach collectively a new coherent pattern.

With the measurement and analysis method, as outlined above in Section 3, applied at the appropriate timescale, as cross-correlations become more pronounced, traffic patterns over the whole system become more visible. In the remaining experiments, described below, we focus on macroscopic dynamics at the timescale $T = 2$ s.

4.3 Spatial-Temporal Pattern

It might prove feasible to design sample-based techniques suitable to identify network-wide patterns that remain invariant for a long time. However,

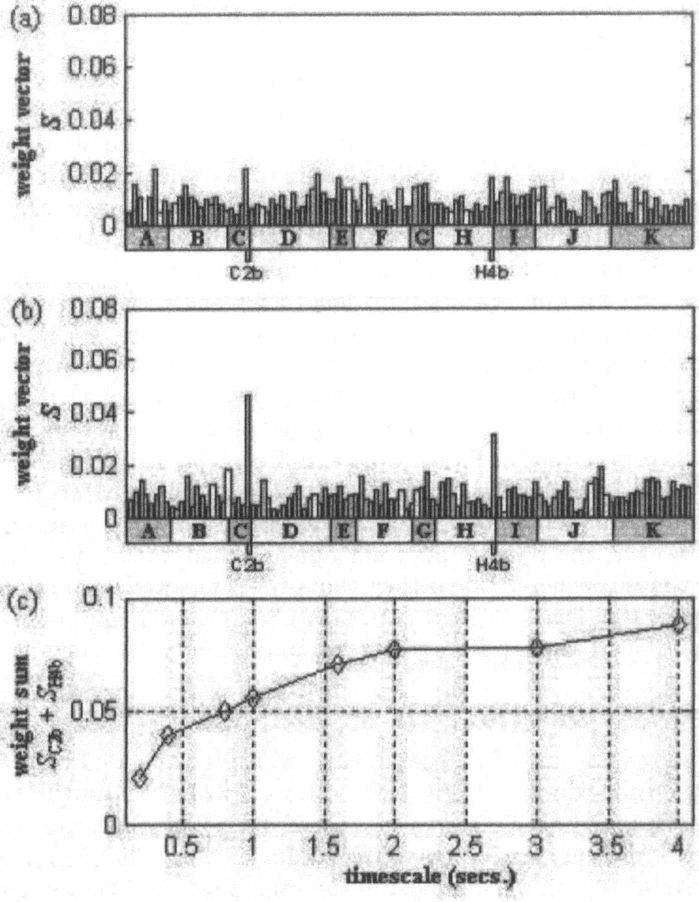

Fig. 3: Two weight vectors at T = 0.4 s (a) and T = 2 s (b), and (c) the sum of S_{C2b} and S_{H4b} changing at different timescales with $L = 11$.

when traffic demands vary over a large dynamic space-time range, these same techniques could fail to detect the more quickly changing patterns. By taking advantage of increased correlation arising over volatile periods, we might be able to use a sample-based version of our proposed method to identify shifting network-wide congestion patterns. In the following, we use only measurements from five (i.e., $L = 5$) subnet routers (B4, D5, F4, I1, and J5) to perform our analysis.

To show how the spatial traffic pattern changes, we calculate the weight vector $\boldsymbol{S}$ using M data points within a moving time window MT from one time period to the next. Figure 5 shows the weight vector $\boldsymbol{S}$ evolving with T = 2 s and with the time window MT (= 200 × 2 s = 400 s) sliding ahead

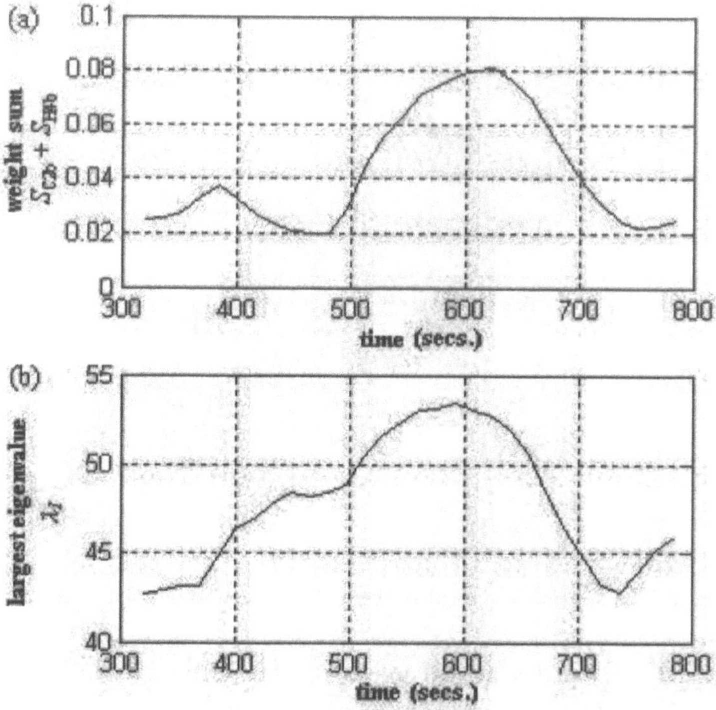

Fig. 4: (a) the sum of S_{C2b} and S_{H4b}, and (b) the largest eigenvalue λ_1 with $T =$ 1.6 s and $L = 11$.

every 10 s. The time axis indicates the end of the moving time window. We can see the enhanced weights of C2b and H4b in the shifting spatial-temporal pattern. While the new congestion appears at C2b, the existing congestion at H4b, which was indistinguishable during the previous calm period, also exposes itself to the weight vector. The five observation points, which are not near C2b or H4b, really "sense" by themselves the gentle load fluctuation of these two leaf routers. The load wave seems to bring about a collective response in the entire network. This indicates that network-wide traffic appears correlated, and that spatial-temporal dynamics evolves as a coherent whole at some appropriate timescale. Therefore, macroscopic-level observation appears to provide significant information that could be exploited to achieve better network engineering and management.

With fewer observation points, the increased correlation during transient periods is very helpful for capturing the network-wide (spatial) pattern of traffic shifting over time. While both S_{C2b} and S_{H4b} become enhanced during periods of shifting pattern, we know, as shown in Figure 2, that the congestion on C2b will persist, and that H4b will gradually recover its normal

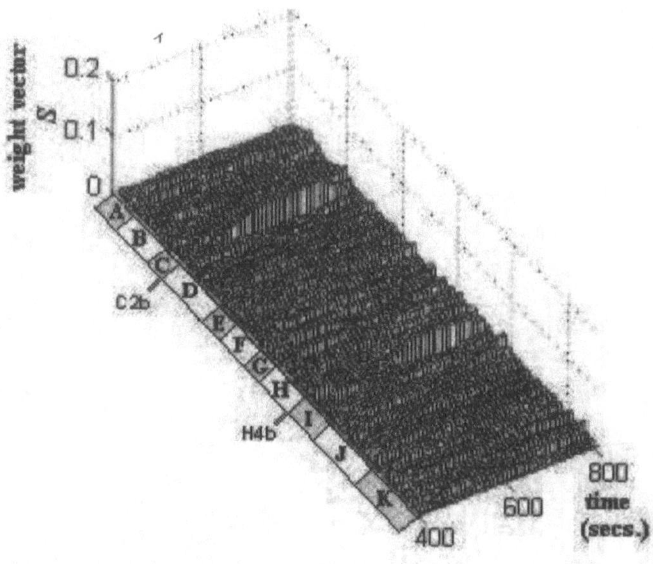

Fig. 5: The spatial-temporal pattern evolving with $T = 2$ s and $L = 5$.

condition. If we need distinguish among routers with increasing and diminishing congestion, then other techniques, such as active probing for bandwidth or delay, might be applied to specific targets identified by our passive method of network-wide observation.

In larger networks, such as the Internet, it is very difficult, if not impossible, to observe the spatial-temporal pattern of congestion over the whole top tier, which encompasses on the order of 10,000 autonomous systems. As discussed in the next section, spatial aggregation, e.g., from the leaf-router to subnet-router level, can help to implement a coarser space and time observation. First, however, we try to observe only a selected subset of the top network tier for the case of shifting congestion illustrated in Figure 2, while still including leaf-router details. We use measurements from the same five routers (subnets B4, D5, F4, I1, and J5) as before to form a spatial-temporal pattern over only three backbone-router zones of C, D, and E (comprising 26 leaf routers). Figure 6 shows the spatial-temporal pattern of the three regions, and reveals the congestion arising in C2b. This result suggests that our technique might provide a useful means to observe spatial-temporal dynamics in selected networks in a timely manner.

5 Monitoring DDoS Flooding Attacks

Distributed denial of service (DDoS) attacks present a very serious threat to the stability of the Internet. By simply exploiting the tremendous asymmetry

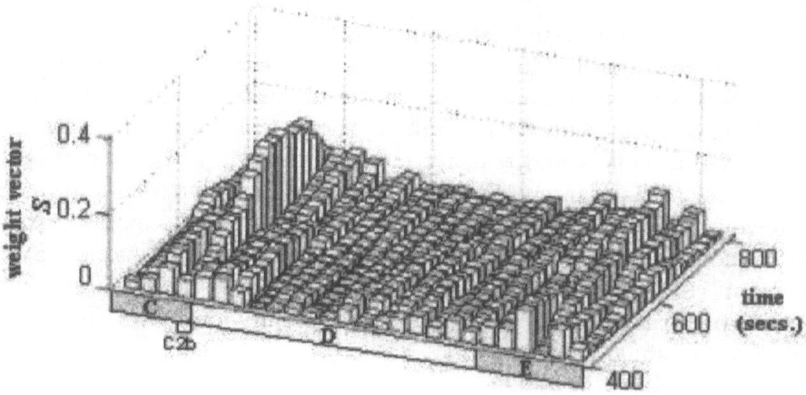

Fig. 6: The spatial-temporal pattern observed in three backbone-router zones, C, D and E, with $L = 5$.

between the large-scale distributed network resources and local capacities at the victim, a flooding-based DDoS attack can build up an intended congestion very quickly near an attacked target. DDoS attacks use forged source addresses and other techniques [21] to conceal the locations of the true attack sources; thus DDoS attacks are among the most difficult to detect and stop. Today's Internet infrastructure is extremely vulnerable to such large-scale coordinated attacks, which may easily and effectively remove an attack victim from the Internet, even without exploiting any particular vulnerabilities in network protocols or weaknesses in system design, implementation, or configuration.

To avoid congestion in the Internet, all flows under end-to-end controls adapt themselves in a self-organized, distributed manner. This adaptive behavior of flows in different directions plays a crucial role in keeping the Internet stable and in forming macroscopic traffic patterns. During a DDoS attack, the attack sources do not honor the normal end-to-end congestion control algorithms; rather, they overwhelm the intended victim, causing legitimate, well-behaved flows to back off, and then ultimately to starve. In addition, large-scale DDoS attacks also impair transit traffic flows, which happen to share a portion of the congested network. Such network-wide phenomena might show themselves in shifting patterns of spatial-temporal traffic.

5.1 Modeling DDoS Attacks

To observe the macroscopic effect of DDoS attacks, we arrange 50 attack sources in our simulation model, which are distributed uniformly throughout the network. We enable our attack sources to launch constant-rate attacks collectively or using a subgroup technique (described further below). In our experiments, there are a total of 22,000 source nodes, and more than 10,000 simultaneously active TCP connections; thus, DDoS flows cannot be easily identified from the legitimate background traffic.

Usually, DDoS attacks directed against the network infrastructure can lead to more widespread damage than those directed against individual web servers. Here, one leaf router (I1a) will be the attack target. Routers under attack may fail to collect and transfer measurement data. Usually, it is difficult to monitor areas of interest without obtaining measurements from those areas. However, our analysis technique provides the ability to monitor areas of interest without such local measurements. We assume in our experiments that the attack on I1a disables the observation point deployed at the subnet-router I1; thus, we perform our analysis using data from only four observation points (B4, D5, F4 and J5; $L = 4$).

5.2 Constant Rate Attack

Constant rate, the simplest attack technique, is typical of known DDoS attacks. We arrange for all the 50 attack sources to launch constant-rate attacks collectively (that is, simultaneously). Here, we do not have the attack sources generate attack packets with full force [22], so that they cannot be easily identified through attack intensity at the source or in intermediate networks. We assume that the variable H represents the intensity of an attack source. Since sources can only create one packet every millisecond, the maximum attack rate is one packet per millisecond, i.e., $H \leq 1$ (packet/ms). We experiment with a constant-rate DDoS attack where $H = 1/10$, that is, each attack source creates one attack packet for every 10 milliseconds beginning from $t_0 = 500$ s.

Figure 7 shows the weight vector $\boldsymbol{S}$ evolving with $T = 2$ s and with the time window MT (= 200 × 2 s = 400 s) sliding ahead every 10 s. We find that the attack really leads to a network-wide shift of spatial-temporal correlation, and the congestion on the victim (I1a) reveals itself at the enhanced weight of I1a. Since we observe this phenomenon and get the time and location of the attack without any help from the suffering victim, the network-wide monitoring could be used to activate specific detection and filtering mechanisms to isolate and stop the attack flows.

We also can observe the spatial-temporal pattern of the constant-rate attack at the subnet level by spatially aggregating the destinations of network flow at the subnet level from current measurements at the leaf-router level. Figure 8 shows such a coarser observation, where the weight vector $\boldsymbol{S}$ evolves with $T = 2$ s and with the time window MT (= 200 × 2 s = 400 s) sliding ahead every 20 s. Here, we can find that the constant-rate DDoS attack against I1a also results in the congestion on the subnet I1. With a lower computing time requirement, the coarser observation at this upper level still reveals a very useful picture of spatial-temporal dynamics.

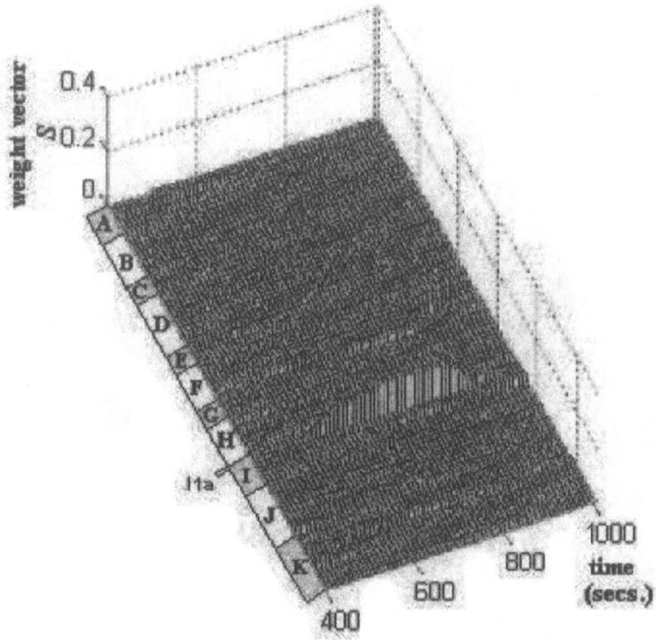

Fig. 7: The spatial-temporal pattern of the constant-rate attack with $H = 1/10$ and $L = 4$.

5.3 Subgroup Attack

Attackers constantly modify attack dynamics to evade detection. Attack dynamics can be made very sophisticated should an attacker desire. For example, next we divide the 50 attack sources into three subgroups, which are distributed separately in the left, the middle and the right parts of the larger network. Once the attack starts at $t_0 = 500$ s, one of the three subgroups is always active so that the victim experiences continuous denial of service [21]. Given the dynamic nature of such a coordinated attack, it is extremely hard to detect where attack packets originate, and to stop them at intermediate or source networks to reduce overall congestion and increase resources available to legitimate traffic.

Figure 9 shows the weight vector $\boldsymbol{S}$ evolving with $T = 2$ s and with the time window MT ($= 200 \times 2$ s $= 400$ s) sliding ahead every 10 s. We find that the subgroup attack reveals itself in the shifting spatial-temporal pattern. Comparing Figures 7 and 9, we find that for our analysis technique the dynamic nature of the subgroup attack seems advantageous, because the increased correlation induced by shifts in attack traffic keeps the weight of the victim I1a salient over a longer time range.

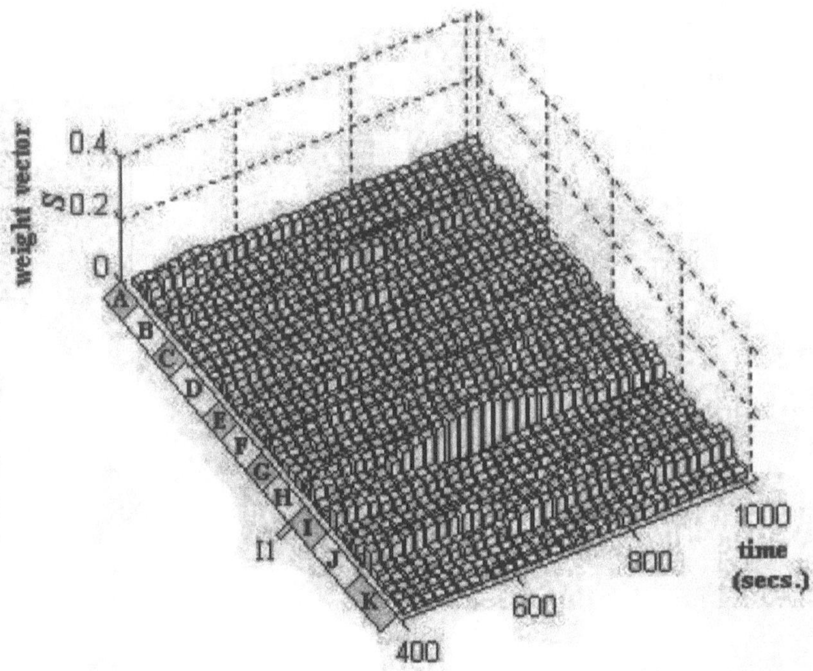

Fig. 8: The spatial-temporal pattern of $H = 1/10$ constant-rate attack at the subnet level with $L = 4$.

During the subgroup attack, we also observed a smaller portion of the larger distributed network, aggregated at the subnet level. Figure 10 shows the spatial-temporal pattern of five backbone-router zones (from G to K), where the weight vector $\boldsymbol{S}$ evolves with $T = 2$ s and with the time window MT ($= 200 \times 2$ s $= 400$ s) sliding ahead every 20 s, revealing the congestion arising in the subnet I1. The effects of the subgroup attack remain evident, while the aggregated, subnet-level observation of only a portion of the network requires less computing time than for the case of Figure 8.

6 Concluding Remarks

In large-scale networks, such as the Internet, spatial-temporal correlations emerge from interactions among adaptive transport connections and from variations in user demands. By exploring the collective dynamics of large-scale networks, we seek ways to understand spatial-temporal correlations. We realize that capturing macroscopic patterns in correlations over time may help us to understand shifting traffic patterns, to identify operating conditions, and to reveal traffic anomalies.

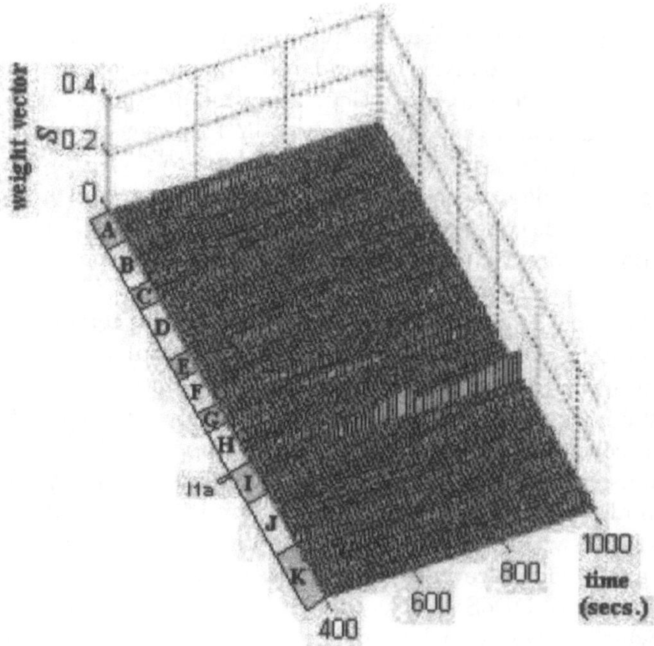

Fig. 9: The spatial-temporal pattern of the subgroup attack with $H = 1/5$ and $L = 4$.

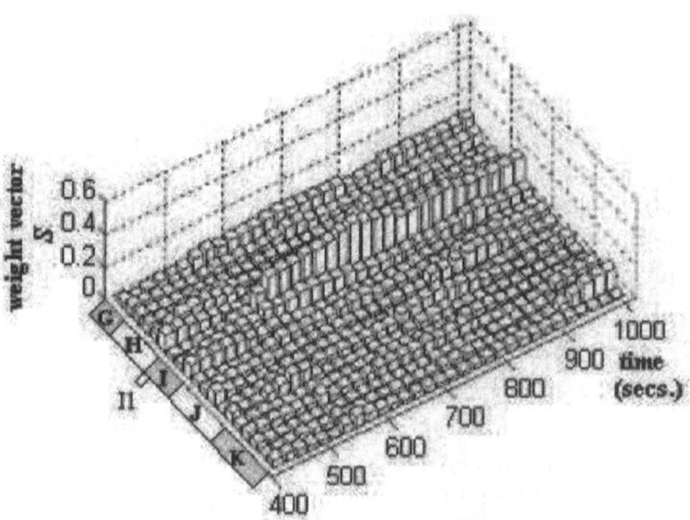

Fig. 10: The spatial-temporal pattern of the subgroup attack, observed in five zones of backbone routers from G to K with $L = 4$.

Analyzing spatial-temporal characteristics of traffic in large-scale networks requires both a suitable analysis method and a means to reduce the amount of data that must be collected. In particular, routers may be heavily utilized or under DDoS attack, and thus fail to collect and transfer data, but often also happen to be the parts of interest to monitor (due to their congested nature). In this chapter, we describe a novel technique that provides a useful way to observe network-wide congestion patterns shifting over time. To illustrate this technique and its potential promise, we reported results from some simulation experiments.

We applied this technique successfully to identify network hotspots induced deliberately in a large-scale network. In particular, the effect of transient periods helped us to capture the network-wide traffic pattern shifting over time. We indicated that the spatial-temporal dynamics of network traffic appears as a coherent whole at an appropriate timescale.

We demonstrated how to use this novel technique to expose large-scale distributed denial-of-service (DDoS) attacks. We find that DDoS flooding attacks lead to a network-wide shift in spatial-temporal correlation, and that congestion on the attack victim reveals itself in these spatial-temporal patterns. The macroscopic effect of DDoS attacks can provide significant information about where and when a DDoS attack might be underway, and could trigger further detection and filtering without any information from the attack victim. In particular, we find that the dynamic nature of the (more stealthy) subgroup attack seems to be an advantage in revealing the victim's plight, because increased variation in traffic patterns lead to increased correlation, which is exploited by our analysis technique.

Since observing the whole Internet in detail is impractical, we suggested a means to efficiently observe selective portions in detail, or to apply spatial aggregation to observe larger-scale networks with less detail. In either case, our analysis method lowers computing time requirements, while revealing shifting traffic patterns over both space and time. If proven successful when applied to real network measurement data, our proposed technique could become a powerful tool to monitor spatial-temporal behavior network-wide in real time, and could ultimately contribute to improvements in network engineering and management.

References

1. Saltzer J, Reed D, Clark D (1984) End-to-end arguments in system design, ACM Trans. Computer System, 2(4), pp. 277-288
2. Stevens W R (1994) TCP/IP Illustrated, Vol. 1, Addison-Wesley Pub. Co., Reading, MA
3. Moore D, Voelker G, Savage S (2001) Inferring Internet denial of service activity, In: Proceedings of the USENIX Security Symposium, Washington, DC, USA, August

4. Bar-Yam Y (1997) The Dynamics of Complex Systems (Studies in Nonlinearity). Perseus Books, ISBN 0-201-55748-7, August
5. Leland W E, Taqqu M S, Willinger W, Wilson D V (1993) On the self-similar nature of Ethernet traffic, In: Proc. ACM SIGCOMM '93, 183-193
6. Paxson V, Floyd S (1994) Wide-area traffic: The failure of Poisson modeling, In: Proc. ACM SIGCOMM '94, 257-268
7. Crovella M E, Bestavros A (1996) Self-similarity in world wide web traffic: Evidence and possible causes, In: Proceedings of the 1996 ACM SIGMETRICS International Conference on Measurement and Modeling of Computer Systems, May
8. Willinger W, Taqqu M S, Sherman R, Wilson D V, (1995) Self-similarity through high-variability: statistical analysis of Ethernet LAN traffic at the source level, In: Proc. ACM SIGCOMM '95, 100-113
9. Yuan J, Ren Y, Shan X (2000) Self-organized Criticality in a Computer Network Model, Physical Review E, 61(2), 1067-1071
10. Yuan J, Mills K (2002) Exploring Collective Dynamics in Communication Networks, Journal of Research of the National Institute of Standards and Technology, 107 (2), 179-191
11. Yuan J, Mills K (2003) Implication of Internet Traffic Characteristics for Network-Adaptive Distributed Systems, submitted to Performance Evaluation, December
12. Crovella M, Kolaczyk E (2003) Graph Wavelets for Spatial Traffic Analysis, In: Proceedings of IEEE Infocom 2003,San Francisco, CA, USA, April
13. Barthelemy M, Gondran B, Guichard E (2002) Large scale cross-correlations In: Internet traffic, Physical Review E 66 (2002) 056110
14. Yuan J, Mills K (2004) A cross-correlation based method for spatial-temporal traffic analysis, accepted by Performance Evaluation
15. Plerou V, et al. (2002) Random matrix approach to cross correlations in financial data, Physical Review E 65 066126
16. Thompson K, Miller G J, Wilder R (1997) Wide-Area Internet Traffic Patterns and Characteristics, IEEE Network 11 (6) 10-23
17. Bai Z, Demmel J, Dongarra J, Ruhe A, van der Vorst H (2000) Templates for the Solution of Algebraic Eigenvalue Problems: A Practical Guide, Society for Industrial and Applied Mathematics, Philadelphia, PA
18. Many authors (1998) MATLAB Ueser's Guide, The MathWorks, Inc., Natick, MA, USA
19. Goh K I, Kahng B, Kim D (2001) Spectra and eigenvectors of scale-free networks, Physical Review E 64 051903
20. Grossglauser M, Tse D (1999) A time-scale decomposition approach to measurement-based admission control, In: Proceedings of IEEE Infocom '99, pp. 1539-1547, New York, NY, March
21. Yuan J, Mills K (2004) Monitoring the macroscopic effect of DDoS flooding attacks, sumbitted to IEEE Transactions on Secure and Dependable Computing
22. Mirkovic J, Prier G, Reiher P (2002) Attacking DDoS at the Source, Proceedings of ICNP 2002, pp. 312-321, Paris, France, November

Modelling the Complex Internet Topology

Guanrong Chen[1], Zhengping Fan[1], and Xiang Li[2]

[1] Centre for Chaos Control and Synchronization, City University of Hong Kong, Kowloon, Hong Kong SAR, P. R. China
[2] Department of Automation, Shanghai Jiao Tong University, Shanghai, 200030, P. R. China

1 Introduction

Internet is one of the most impressive creatures of our civilization; it has literally changed the ways we do business, communication, education, entertainment, and many other activities today. For example, one can conveniently contact friends by email and even video conversation based on the Internet. You may wonder how the Internet delivers emails from one end to the other end, and how the Internet makes a connection between two computers at the ends? When you are searching some useful information through the Internet by using a search engine such as Google, have you thought of how it works? When the so-called Worm virus spreads throughout the Internet, have you ever thought of how a virus propagates from a few end-hosts to the entire huge network? Technically, all these problems rely on the Internet route protocols [1], searching algorithms [2], etc., and the virus spreading depends heavily on the Internet topology [3].

1.1 Historical Background

At the early stage of the Internet development, solving these kinds of problems was relatively easy because the size of the Internet was small and its structure was simple. However, as the Internet continues to grow explosively, in particular during the period of the mid-1990s, it has become more and more difficult if not impossible to deal with such complicated problems. An alternative approach is to study the problems based on an abstract model of the actual structure of the Internet. In this attempt, the Internet topology is critical and has significant impact on the achieved results. For example, it was observed [4] that the efficiency of the dynamic multi-casting algorithm is considerably reduced when it is applied to the Internet model using the random graph structure as compared to the hierarchical graph structure. It has also been found [5] that multi-cast resource reservation styles are quite different in the linear, tree, and star-shape Internet topologies.

In the past few years, graph-based methodologies have become a common tool for analyzing the Internet structure, where the nodes represent autonomous systems or routers and the links denote the interactions among the nodes. The earliest Internet model was a stochastic model, where a set of nodes are distributed in a plane uniformly at random, and then a link is added between each pair of nodes with a constant or a varying probability. The most popular stochastic model for the generation of an Internet-like structure is the Waxman model [6], where the probability to connect two nodes from u to v is given by

$$P(u,v) = \alpha e^{-d/(\beta L)}$$

where $0 < \alpha,\ \beta \leq 1$, d is the Euclidean distance from node u to node v, and L is the maximum distance between any two nodes in the network.

After the Waxman model, regular graphs such as ring, star and grid structures were also proposed. An obvious benefit to use a deterministic model is that it makes the analytic studies tractable when comparing the performances of different algorithms on a network.

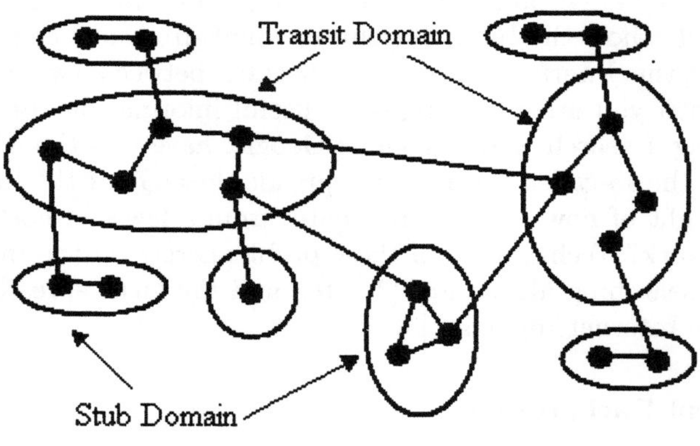

Fig. 1: Transit-Stub model of the Internet

Recently, a so-called Transit-Stub model was proposed [7], where three levels corresponding to transit domains, stub domains, and local area network (LAN) domains, which are attached to the stub nodes, are constructed in order to reflect the hierarchical domain structure and locality presented in the Internet. This model begins at the top (Transit domain/WAN) level and proceeds down to the lowest (LAN) level of the hierarchy. The nodes within the same level are placed in a rectangular region of the plane and their scales can change depending on which level they belong to. After a connected subgraph is generated, where each node represents a Transit domain, a certain number of connected subgraphs in which each node represents a Stub domain are generated for each node in every Transit domain. Eventually, some extra links are added between pairs of nodes, where one node comes from a Transit domain and the other from a Stub domain. The generated graph is illustrated by Fig. 1.

However, the recent discovery on the Internet structure identifies the real Internet topology being neither completely random nor completely regular (hierarchical), but shows a prominent characteristic of self-organization, with which the degree distribution of nodes follows a power-law form. To mimic such a scale-invariant feature of the Internet topology, Barabasi and Albert first proposed a scale-free network model [8], the BA model for brevity. The BA model responsible for the self-organization characteristic in real networks mainly involves the following three construction steps:

The initial network consists of m_0 isolated nodes, and one of the following operations is performed at each time step:

(1) **Adding new links between the existing nodes**
With probability p, m ($m \leq m_0$) new links are added into the network: one end of a link is chosen at random, and the other end is selected with probability

$$\prod(k_i) = \frac{k_i + 1}{\sum_l (k_l + 1)} \tag{1}$$

where k_i is the degree of node i.

(2) **Re-wiring**
With probability q, m links are rewired: First, a node i and a link l_{ij} attached to node i are selected at random, and then this link is replaced with a new link $l_{ij'}$ that connects node i to node j' which is chosen with probability $\Pi(k_{j'})$ given by (1).

(3) **Incremental growth**
With probability $1 - p - q$, a new node is added into the network: the new node has m new links connected to the already existing nodes in the network with probability $\Pi(k_i)$.

In this model, the probabilities satisfy $0 \leq p < 1$, $0 \leq q < 1 - p$, and $0 \leq p + q < 1$.

Using the mean-field theory [8][11], one can find that the degree k_i of a node i changes over time at the rate

$$\frac{\partial k_i}{\partial t} = (p - q)m\frac{1}{N(t)} + m\frac{k_i + 1}{\sum_l (k_l + 1)} \tag{2}$$

where the network size is $N(t) = m_0 + (1 - p - q)t$, and the total number of links are $\sum_l k_l = 2(1 - q)mt - m$.

At the initial time t_i, the number of links of node i is $k_i(t_i) = m$, thus $k_i(t)$ can be expressed in the form

$$k_i(t) = [A(p,q,m) + m + 1] \left(\frac{t}{t_i}\right)^{1/B(p,q,m)} - A(p,q,m) - 1 \quad (3)$$

where

$$\begin{aligned} A(p,q,m) &= (p-q)\left(\tfrac{2m(1-q)}{1-p-q} + 1\right) \\ B(p,q,m) &= \tfrac{2m(1-q)+1-p-q}{m} \end{aligned} \quad (4)$$

If the parameters satisfy $q < \min(1-p, (1-p+m)/(1+2m))$, then the connectivity distribution of node i is in a power-law form [8]:

$$P(k) \propto (k + A(p,q,m) + 1)^{-\gamma}$$

where $\gamma = 1 + B$.

In this model, the mechanism responsible for the emergence of the scale-free topology of the Internet globally works in terms of the preferential attachment. That is, the probability that an existing node receives new links depends on the total number of links in the whole network. However, the localization property of real networks cannot be reflected in this model. In particular, in the Internet, a router favors a connection of shortest distance when placing new links, which results in that those routers within the same region have more connections but those in different regions have less links. Consequently, the routers have larger clustering coefficient within the same region but have smaller values in different regions. Another example is the World Trade Web (WTW). In the WTW, it is reported [9] that the globally preferential attachment mechanism does not work for those countries that have less than 20 trade connections with other countries; yet many countries are accelerating their economy cooperations in various regional economy-cooperative organizations such as EU, ASEAN, and NAFTA. This indicates that preferential attachment mechanisms only exist within local economy regions of the WTW.

1.2 A Local-World Model

To capture the localization properties of these real networks, a local-world model for dynamically evolving networks was proposed in [10], which is generated by the following algorithm:

The initial network has m_0 nodes and e_0 links, and then the following two steps are performed:

(1) M nodes are selected from the existing network, which are considered as the "local world" for the forthcoming nodes.
(2) A new node is added into the network at each time step, which connects to m nodes in its local world determined by Eq. (1), with the probability

$$\Pi_{local}(k_i) = \frac{M}{m_0 + t} \frac{k_i}{\sum_j {}_{local} k_j}$$

In this model, if $m < M < m_0 + t$, it represents a transition between power-law and exponential scaling networks. In particular, the original BA model [11] is a special case of this local-world evolving network model.

Motivated by this promising work, in this chapter, the local-world concept is further developed to model the inter-domain of the Internet. A multi-local-world (MLW)model with a localization property is proposed for a better description of the Internet. Clearly, the Internet can be considered as a collection of many interconnected subnetworks. If a subnetwork in the Internet is viewed as a "local-world," then the Internet consists of several interconnected "local-worlds." This observation thereby leads to the novel concept and notion of MLW.

The rest of this chapter is organized as follows: Section 2 first introduces some experimental results on the router-level and the AS-level of the Internet topology. The MLW model is then described for modelling the Internet in Section 3. This new mode is based on a carefully study of the Internet AS graphs, with a comparison to the BA model. Finally, Section 4 concludes the chapter.

2 Real Map of the Internet

The structure of the Internet can be considered as a loose coalition of autonomous administration domains. Each domain in the Internet operates with its own policies, services, and prices. While within each Autonomous Systems (ASs) domain, It consists of lots of routers. In order to route information within an AS, an interior routing protocol (i.e., Interior Gateway Protocol or IGP) is used, and an inter-domain routing protocol (i.e., Exterior Gateway Protocol or EGP) is used for the purpose of routing information between different ASs. The Internet structure is visualized by Fig. 2.

2.1 Different Internet Topologies

Typically, the Internet is studied at two different granularities: (1) at the router level, routers are represented by nodes and links are the physical connections among them; (2) at the AS level, since each AS is approximately mapped to an Internet Service Provider (ISP), each AS can be represented by a single node and two ASs are connected if there exists a BGP peer connection between them.

Router-level Internet topology: A common tool to map the router-level Internet topology is traceroute (Unix traceroute or Windows NT tracert.exe), which uses hop-limited probe, consisting of a hop-limited IP packet and the corresponding ICMP response, to probe every possible IP address and record every touched router and the corresponding links. A pioneering work [12]

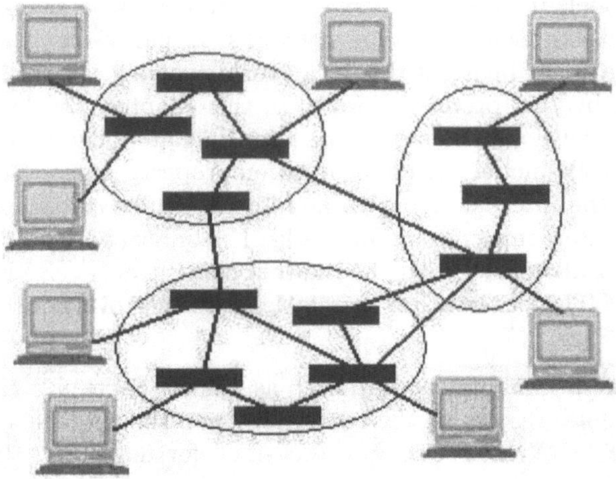

Fig. 2: Illustration of the Internet topology

is to use traceroute to trace 5000 hosts, which were selected in a network accounting database in 1995. After the destinations had been determined, 11 hosts among the 5000, were used as the new sources of routes to trace the remaining destinations, which eventually produced a graph consisting of 3888 nodes and 4857 edges by ignoring some routes that could not be traced due to transient routing or other technical problems. The analytical results of the obtained Internet topology shows that more than 70% of the nodes have degree 1 or 2 and they belong to class Leaf or Relay. However, a limitation of this method is that it needs to choose a certain number of destinations representing a subset of the Internet structure to obtain the routing information, leaving the completeness of the resulting map to a significant dependence on the chosen destinations before probing.

A technique called intelligent heuristic [13] can overcome this drawback, which uses a heuristic to decide whether to map the network from a single node and does not require an initial database of targets for exploring the network topology. Based on some careful analysis of the collected data, which consists of nearly 150,000 interfaces and nearly 200,000 links, it was found [13] that the degree distribution of nodes with connectivity less than 30 has a power-law form. However, the distribution of nodes with degree larger than 30 is significantly different: it has a faster cut-off than power-law, indicating that there may exist another law governing the distribution of higher-degree nodes in the network. In addition, the distribution of numbers of pairs of nodes within at most d hops of each other in the network follows neither exponential growth [14] nor a power-law [15]. Recently, there are some analysis [16] on data collected during October and November 1999, which shows that the hierarchical characteristic almost does not exist in the router-level of the

Internet topology. It was also found that the degree distribution of nodes has a power-law behavior, which however is smoothed by a clear exponential cut-off. Therefore, the Weibull distribution, instead of the power-law distribution, can better fit the collected data, agreeing with the result reported in [13]. However, this approach could not give a complete map of the Internet topology since it fails to map the details of the Stub networks, although it can capture the map of Transit portion of the Internet. It is then suggested [17] that probing from a large number of sources may be able to improve the performance regarding the completeness of the traceroute-style probes.

Recently, BGP has been used for routing tables to determine the destinations of traceroutes [18]. A directed probing technique [19] is to interpret BGP tables thereby identifying relevant traceroutes and pruning the remainders. The path reduction technique can be used to identify redundant traceroutes, so as to map the route-level Internet topology. An advantage of using these two techniques is that it can significantly reduce the number of required traces without sacrificing the accuracy. Actually, comparing to the brute-force all-to-all approach, this method of combining directed probing technique with path reduction technique can reduce the number of required traces significantly by three orders of magnitude. Some analytical results on the collected data during December 2001 and January 2002 have shown that the Weibull distribution can better fit the complementary cumulative distribution function of router outdegree than the Pareto (power-law) distribution.

However, because ISPs generally regard their router-level topologies as confidential, and there also exist some technical problems such as multiple alias interfaces, it is still a challenging task today to map a relatively complete router-level Internet topology.

AS-level Internet topology: The BGP routing tables of each AS contains a spanning tree from that node to every reachable AS. Therefore, the addition and deletion of an ISP in the network can be reflected by the BGP routing tables. It indicates that the AS-level Internet topology can be constructed by investigating the BGP routing tables. This method is used by the National Laboratory for Applied Network Research (NLANR) to collect BGP-related information [20]. Starting from November 1997, NLANR has collected 1253 daily instances of the BGP routing tables from the Oregon route server.

Based on the collected data during 1997 and 1999, there is a study [21] of some statistical properties of the Internet topology such as average connectivity, average clustering coefficient, and average path-length, suggesting that the "betweenness" of a network may be a unique invariant variable when the network suffers from evolving dynamics.

It has also been observed [22] that the Internet topology exhibits power-laws of the general form $y = x^a$ for the following four relationships:

(1) The out-degree d_v of a node v is proportional to the rank of the node, r_v, to the power of a constant R, that is, $d_v \propto r_v^R$.

(2) The frequency of nodes, f_d, with an out-degree d, is proportional to the out-degree to the power of a constant o, that is, $d_v \propto r_v^o$.
(3) The total number of pairs of nodes, $P(h)$, within h hops, is proportional to the number of hops to the power of a constant H, that is, $P(h) \propto h^H$, $h \ll \delta$, where δ is the diameter of the network.
(4) The eigenvalues λ_i of the network are proportional, to the order i, to the power of a constant ε, that is, $\lambda_i \propto i^{\varepsilon}$.

These four factors for different Internet topology models have been examined [23]. It was found that the factors (1) and (2) are the key contributors. It indicates that a model that can generate Internet-like topology should involve at least two key mechanisms: one is the preferential attachment rule and the other is the incremental growth mechanism.

In another study [24] of the four factors, it led to a proposal of a new Internet topology generator: Inet. This scheme can generate a topology that well approximate the actual Internet AS-level topology.

Considering the mechanism of BGP routing works, an instantaneous snapshot of the BGP routing table may not discover links belonging to less preferred or non-advertised paths. Therefore, the AS-level Internet graph obtained from the data collected by Oregon route server may provide a very incomplete pattern of the physical connectivity that exist in the actual Internet. By merging the data collected by NLANR and the BGP peering relationship information from 70 different ASs that maintain *Looking Glass* sites [25], a more complete Internet AS-level map was suggested [26]. Based on the enriched dataset, it was found that the actual Internet does not follow a strict power-law, and the event of "death of links" during the evolvement of the Internet cannot be neglected in the consideration of the actual Internet AS topology. It was confirmed [16] that the actual Internet indeed deviates from the strict power-law behavior. Lately, it was recommended [27] that the power-law should be a necessary but not a sufficient condition for a topology to be realistic for the Internet.

It should be noticed that the method of analyzing the BGP tables to construct the AS-level Internet structure suffers from several limitations. BGP connectivity does not reflect the redundancy of different parts of the network. Also, it does not capture public or private exchange points within the infrastructure or short-term AS path variation. On the other hand, the current used BGP, evolving from earlier distance vector routing algorithms, suffers from a potential routing table oscillations. For example, it was reported [28] that the event of losing connectivity in the Internet occurs frequently. Therefore, the existing constructed topology maps using BGP data are incomplete. Recently, the *skitter* tool was used to explore the Internet topology at the IP address granularity [29]. The *skitter* monitor uses active probing techniques to measure the Internet path to a destination by sending a sequence of Internet Control Message Protocol (ICMP), similar to the *traceroute* utility. To obtain a more complete Internet map, the destinations are chosen from different lists such as DNS clients, Ipv4 space, and web servers. Combining the BGP routing tables with the IP address database, IP address paths can be converted to the AS paths, yielding an AS core structure. Figure 3 shows an example of the AS graph during the first week of August 2001.

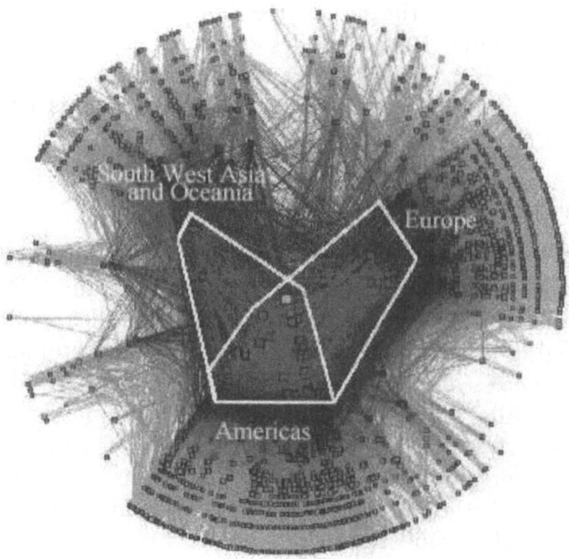

Fig. 3: AS core graph of the Internet during the first week of August 2001 [29]

It can be seen from Fig. 3 that most links of large ISPs in Asia and Europe were within their own continents or connected to those in the America, and the links between ISPs within Asia and Europe are very few. It strongly indicates that the AS-level topology of the Internet has a prominent localization property.

3 MODELLING THE INTERNET

The previous section has reviewed the actual Internet map at the router level and the AS level. In this section, a multi-local-world (MLW) model is developed to better describe the Internet topology on the basis of AS level granularities. The factors discussed in the following five subsections have been carefully studied based on the collected Internet AS-graph data.

On the Birth and Death of ASs

Define an AS as being "born" when a new ISP joins the Internet and as being "dead" when it never appears again after it disappeared. According to the data collected by the Oregon router server, the number of ASs in the Internet increased from 4320 in November 1998 to 9520 in November 2000. It implies that the Internet on the basis of AS-level granularity does grow. During this period, although there were so many dead ASs at each time unit (e.g., day, month, or year), the number of dead ASs was always small as compared with the number of born ASs at the same time. For example, the born AS number is about 200 in November 1998 while less than 50 ASs died at the same time. It indicates that the birth rate of ASs are larger than the death rate at the same instant. For simplicity, the event of death of ASs is not considered in the new MLW model.

On the Birth and Death of Links

When a new AS is added into the network, it creates a certain number of links to the existing nodes in the network. On the other hand, there also appear new interconnections between the already existing nodes, which is called the "birth" of emerging links. A reason could be that a customer usually wants to have an extra access service for fault tolerance, or to avoid possible traffic congestion. Meanwhile, a link between two nodes may be disconnected, which is called the "death" of the deleted link, if it never appears again after it is deleted. On the other hand, the end of the deleted link with lower connectivities has no links with the other existing nodes within a short time duration. Thus, the number of "dead" links does not include those disappeared links that have an end with lower connectivities.

The analysis on the Internet AS topology data shows that the event of "death" of links should not be neglected, as did in the BA model, because the death rate of links is comparable with its birth rate. In fact, for the actual Internet, in some cases the number of dead links can even be larger than that of the born links. For example, in November 2000, the number of dead links is about 1300 which was 800 more than the born links in that month. For this reason, both "birth" and "death" of links will be included in the MLW model.

On the Re-wiring Mechanism

An ISP in the Internet may rewire one of its links to connect the nodes with higher connectivity in order to gain more benefits, for example, reducing the distance from it to other nodes in the network. However, recent research shows [20] that the re-wiring mechanism may not be a significant factor in the evolution of the Internet AS topology. Therefore, the MLW model will not consider this mechanism.

On the Preferential Attachment Rules

When a new node joins the network, an existing node i in the network receives a link from the new node with a certain probability, which may depend linearly on the connectivity of node i, in the form

$$\prod(k_i) = \frac{k_i}{\sum_l k_l} \tag{5}$$

where k_i is the degree of node i at the time step t_i, and $\sum_l k_l$ is the total degree of all nodes in the network.

The probability $\prod(k_i)$ can also be a nonlinear function of the degree of node i, in the form

$$\prod(k_i) = \frac{k_i^p}{\sum_l k_i^p} \tag{6}$$

where the parameter $p > 0$.

An investigation of the actual Internet AS topology data shows [21] that the newly added nodes actually create new links by a linear preferential attachment rule, which was lately confirmed by another study [30]. Therefore, in the MLW model, the linear preferential attachment rule is used to quantify the probability of a node receiving a new link from a newly added node or from an existing one.

On the other hand, the links attached to a node with lower degrees may be more likely being removed, because ISPs always tend to delete those infrequently used links so as to reduce the maintenance cost. Combining with the linear preferential attachment rule when adding a link between two nodes, the probability of a link attached to node i being deleted may be expressed as

$$\Pi'(k_i) = \frac{1}{N(t) - 1}(1 - \Pi(k_i)) \tag{7}$$

where $N(t)$ is the number of nodes in the network at the time step t. This term will be normalized such that $\sum_i \Pi'(k_i) = 1$. Note that this form has been used in [31] to model networks with the scale-free feature.

On the Localization Effect in the Internet

At the AS level, the Internet hierarchy can be schematically divided into international connections, national backbones, regional networks, and local area networks. The nodes in the regional networks are tightly connected, leading to a high clustering coefficient within the networks. These highly clustered regional networks are then interconnected sparsely by national backbones or international connections.

When a new node j is being determined to join a regional network, the nodes in other regional networks, even those with very large degrees, will have very little impact on the decision of this newly added node. In other words, the ability that a node i in this regional network can capture a new link from the newly added node may depend primarily on its position relative to the other nodes within the same regional network, but not to the entire multi-regional Internet. This regional network is called a "local-world," while the entire Internet can be considered as a collection of many local-worlds. In the MLW model, the probability with which a node i in a local-world Ω receives a new link from the newly added node is described by

$$\prod(k_i) = \frac{k_i + \alpha}{\sum\limits_{j\in\Omega} (k_j + \alpha)} \tag{8}$$

where Ω means the Ωth local-world in which the node i locates, and the parameter $\alpha > 0$ represents the "attractiveness" of node i, which is used to govern the probability for those "young" nodes to get new links.

Similarly, the probability of a link attached to node i being deleted can be rewritten as

$$\Pi'(k_i) = \frac{1}{N_\Omega(t) - 1}(1 - \Pi(k_i)) \tag{9}$$

where $N_\Omega(t)$ represents the number of nodes within the Ωth local-world in the network, and $\Pi(k_i)$ is determined by Eq. (8).

On the other hand, the local-worlds in the whole network are connected sparsely. As a result, the deletion of links between different local-worlds may render some local-worlds being isolated. More importantly, those links removed from the network may most likely come from the nodes within the same local-world. Therefore, only links among nodes within the same local-world are allowed to be removed in the MLW model.

3.1 The MLW Model for the Internet

The proposed MLW model is generated by the following scheme:

Start with m isolated local-worlds, in which suppose that there are m_0 nodes and e_0 links equally in each local-world. At each step, then, perform one of the following five operations at random:

(i) With probability p, a new local-world is created, which contains m_0 nodes and e_0 links. Meanwhile, a unique identifier is generated to identify this new local-world.

(ii) With probability q, a new node is added to an existing local-world, which has m_1 links to the nodes within the same local-world. To do this, first, a local-world Ω is selected at random, and then a node in the local-world Ω is chosen with a probability given by (8), with which the new node connects to the selected nodes. This process is repeated m_1 times.

(iii) With probability r, m_2 links are added to a chosen local-world. To do this, first, a local-world Ω is selected at random, and then one end of a link is chosen randomly while the other end of the link is selected with a probability given by (8). This process is repeated m_2 times.

(iv) With probability s, m_3 links are deleted within a chosen local-world. To do this, first, a local-world Ω is selected at random, and then one end of a link is chosen randomly while the other end of the link is selected with a probability given by (9). This process is repeated m_3 times.

(v) With probability u, a selected local-world has m_4 links to the other existing local-worlds. To do this, randomly select a local-world and a node in this local-world with a probability given by (8), which acts as one end of a link. Then, another node of the link, which is in another local-world chosen at random, is selected with the probability given by (8). This process is repeated m_4 times.

In this model, the parameters (probabilities) have to satisfy $0 < q < 1$, $0 \leq p, r, s, u < 1$ and $p + q + r + s + u = 1$.

The schematic diagram of the MLW model is illustrated by Fig. 4.

In Fig. 4: (a) The original network has $m = 3$ local-worlds (marked by capital A, B, and C), and there are $m_0 = 3$ nodes (represented by the black circles) and $e_0 = 3$ links in each local-world. (b) A new local-world D is created, depending on the probability p. This new local-world consists of $m_0 = 3$ nodes and $e_0 = 3$ links. (c)-(d) A new node j joins the network. First, it selects the local-world C where it will locate, and then connects an existing node ($m_1 = 1$) in this local-world with preferential attachment with a probability given by (8). (e) Local-world C is chosen, also at random, and then $m_2 = 1$ links are added to this local-world. One end of a link is selected randomly, and the other end of the link is chosen with a probability given by (8). (f) A local-world is randomly chosen, and then $m_3 = 1$ link is deleted within this chosen local-world D. An end of a link is selected at random, and then the other end of the link is chosen with a probability given by (9). (g) Depending on the probability u, $m_4 = 2$ link is added between two nodes located in two different local-worlds, respectively. Both ends of the link are chosen with preferential attachment according to a probability given by (8). (h) At the next time step, one of the five possible operations listed above will be performed, depending on the corresponding probability of occurrence. Hence, a new node may be added to the network.

3.2 Degree Distribution of the MLW Model

Using the mean-field theory [8][11], one can obtain the degree distribution of a node i in the Ωth local-world, which can be derived analytically as follows.

(i) With probability p, create a new local-world:
In this case, the degree of a node i in an existing local-world Ω does not change over time since the original nodes in the newly created local-world have no links with any other nodes in the existing local-worlds. Thus,

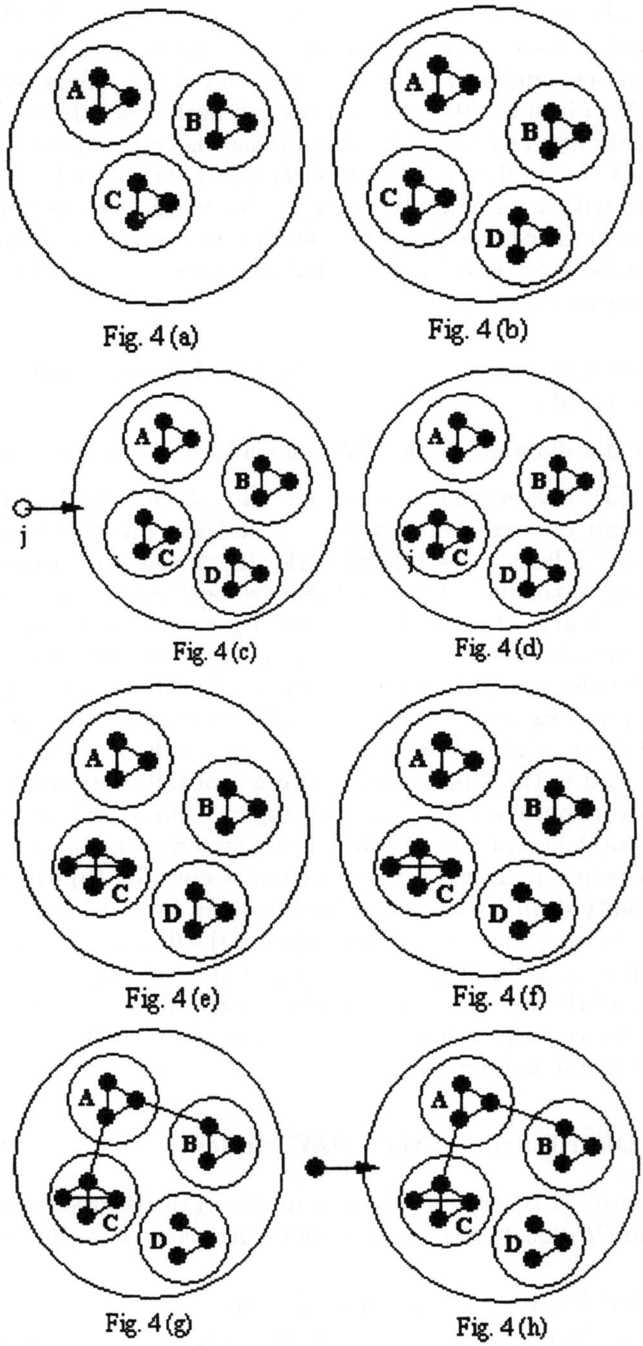

Fig. 4: Schematic illustration of the MLW model

$$\left(\frac{\partial k_i}{\partial t}\right)_{(i)} = 0 \tag{10}$$

(ii) With probability q, add a new node to join the local-world Ω:

$$\left(\frac{\partial k_i}{\partial t}\right)_{(ii)} = \frac{m_1 q}{m+tp}\frac{k_i+\alpha}{\sum\limits_{j\in\Omega}(k_j+\alpha)} \tag{11}$$

The term on the right-hand side of the above corresponds to the random selection of a local-world, and a node selection with preferential attachment according to the probability given by (8). Since there are m_1 links between the new node and the existing nodes, the coefficient is equal to m_1.

(iii) With probability r, add m_2 links to the local-world Ω:

$$\left(\frac{\partial k_i}{\partial t}\right)_{(iii)} = \frac{rm_2}{m+tp}\left[\frac{1}{N_\Omega(t)} + \left(1-\frac{1}{N_\Omega(t)}\right)\frac{k_i+\alpha}{\sum\limits_{j\in\Omega}(k_j+\alpha)}\right] \tag{12}$$

The first term above means the random selection of a node i within a local-world, which is also chosen randomly; the second term represents the preferential selection within the same local-world.

(iv) With probability s, delete m_3 links within a randomly chosen local-world Ω:

$$\begin{aligned}\left(\frac{\partial k_i}{\partial t}\right)_{(iv)} = -\frac{sm_3}{m+tp}&\left[\frac{1}{N_\Omega(t)}\right.\\ &\left.+\left(1-\frac{1}{N_\Omega(t)}\right)\frac{1}{N_\Omega(t)-1}\left(1-\frac{k_i+\alpha}{\sum\limits_{j\in\Omega}(k_j+\alpha)}\right)\right]\end{aligned} \tag{13}$$

The term of the right-hand side implies that the decrease of degree of node i in the local-world Ω comes from two sources: one is that it acts as a randomly chosen end of a deleted link, the other is that it is the end of a deleted link selected with a probability given by (9).

(v) With probability u, add m_4 links between two local-worlds in the network:

$$\left(\frac{\partial k_i}{\partial t}\right)_{(v)} = um_4\left[\frac{2}{m+tp}\frac{k_i+\alpha}{\sum\limits_{j\in\Omega}(k_j+\alpha)} - \frac{1}{m+tp}\frac{1}{m+tp}\frac{k_i+\alpha}{\sum\limits_{j\in\Omega}(k_j+\alpha)}\right] \tag{14}$$

At time step t the total degree of any local-world Ω in the network, on average, is

$$\sum_{j\in\Omega} k_j = 2t(pe_0+qm_1+rm_2-sm_3+um_4)/(m+tp) \tag{15}$$

And the number of nodes in the local-world Ω, on average, is

$$N_\Omega(t) = m_0 + qt/(m+tp) \tag{16}$$

Next, let

$$c = 2(pe_0 + qm_1 + rm_2 - sm_3 + um_4) + q\alpha$$

By combining Eqs. (10)–(16) together, one has

$$\begin{aligned}
\frac{\partial k_i}{\partial t} &= \frac{qm_1}{c}\frac{k_i}{t} + \frac{qm_1\alpha}{c}\frac{1}{t} + \frac{rm_2(q+m_0p-p)}{(q+m_0p)c}\frac{k_i}{t} \\
&+ \left(\frac{rm_2}{(q+m_0p)} + \frac{rm_2(q++m_0p-p)\alpha}{(q+m_0p)c}\right)\frac{1}{t} \\
&- \frac{rm_2m}{(q+m_0p)c}\frac{(k_i+\alpha)}{t^2} + \frac{sm_3p}{(q+m_0p)c}\frac{k_i}{t} \\
&+ \left(\frac{sm_3p\alpha}{(q+m_0p)c} - \frac{2sm_3}{(q+m_0p)}\right)\frac{1}{t} \\
&+ \frac{sm_3m}{(q+m_0p)c}\frac{(k_i+\alpha)}{t^2} + \frac{2um_4}{c}\frac{k_i}{t} \\
&+ \frac{2um_4\alpha}{c}\frac{1}{t} - \frac{um_4}{c}\frac{(k_i+\alpha)}{t(m+tp)} \\
&= \left(\frac{qm_1}{c} + \frac{rm_2(q+m_0p-p)}{(q+m_0p)c} + \frac{sm_3p}{(q+m_0p)c} + \frac{2um_4}{c}\right)\frac{k_i}{t} \\
&+ \left(\frac{q\alpha m_1}{c} + \frac{rm_2}{(q+m_0p)} + \frac{rm_2(q+m_0p-p)\alpha}{(q+m_0p)c}\right)\frac{1}{t} \\
&+ \left(\frac{sm_3p\alpha}{(q+m_0p)c} - \frac{2sm_3}{(q+m_0p)} + \frac{2um_4\alpha}{c}\right)\frac{1}{t} \\
&\qquad \text{(for large } t\text{):}
\end{aligned}$$

Define

$$\begin{aligned}
a &= \frac{qm_1}{c} + \frac{rm_2(q+m_0p-p)}{(q+m_0p)c} + \frac{sm_3p}{(q+m_0p)c} + \frac{2um_4}{c}, \\
b &= \frac{q\alpha m_1}{c} + \frac{rm_2}{(q+m_0p)} + \frac{rm_2(q+m_0p-p)\alpha}{(q+m_0p)c} \\
&+ \frac{sm_3p\alpha}{(q+m_0p)c} - \frac{2sm_3}{(q+m_0p)} + \frac{2um_4\alpha}{c}
\end{aligned}$$

Therefore,

$$\frac{\partial k_i}{\partial t} = a\frac{k_i}{t} + b\frac{1}{t} \tag{17}$$

Since $a \neq 0$, using the initial condition $k_i(t_i) = m_1$, one obtains

$$k_i(t) = -\frac{b}{a} + \left(m_1 + \frac{b}{a}\right)\left(\frac{t}{t_i}\right)^a \tag{18}$$

Define the unit of time in the MLW model as (one local-world creation)/(one node increment)/(one link deletion)/(one new link within a local-world)/(one new link between two local-worlds). Then, the probability density of t_i is

$$P_i(t_i) = 1/(3m + t(1 + 2p))$$

so that

$$P(k_i(t) < k) = P\left(t_i > \left(\frac{m_1 + b/a}{k + b/a}\right)^{1/a} t\right)$$

$$= 1 - \frac{1}{(3m + t(1 + 2p))}\left(\frac{m_1 + b/a}{k + b/a}\right)^{1/a} t \tag{19}$$

Using $P(k) = \frac{\partial(P(k_i(t)<k)}{\partial k}$, one has

$$P(k) = \frac{t}{a(3m + t(1 + 2p))}(m_1 + b/a)^{1/a}(k + b/a)^{-\gamma} \tag{20}$$

where $\gamma = 1 + 1/a$.

To predict real networks, whose power-law exponents are typically between 2 and 3 using the MLW model, the following conditions have to be satisfied:

$$\begin{cases} (m_1 + b/a) > 0 \\ a < 1 \end{cases} \tag{21}$$

Obviously, if one takes $rm_2 \geq 2sm_3$, then condition (21) can be guaranteed. Note that the power-law exponent increases with the increasing of attractiveness of nodes in the MLW model, which indicates that the attractiveness of nodes in a network may play an important role although the underlying mechanism is somewhat ambiguous.

3.3 Some Special Cases of the MLW Model

Now, consider the following special cases of the MLW model.

Case A: When $m = 1$, $q = 1$, and $p = r = s = u = 0$, the network has only one local-world, and the power-law exponent is $\gamma = 3 + \alpha/m_1$. This reduces to the original BA model [11].

Case B: If a network consists of only one local-world, and the evolution of the network only includes the events of addition of node and links, namely, if $m = 1$, $p = 0$, $s = u = 0$, then the exponent $\gamma = 3 + \alpha q/((m_1 - m_2)q + m_2)$. This indicates that the event of addition of links between two existing nodes in the network also has a significant impact on the scale-free feature of this evolving network. However, the exponent keeps unchanged if one takes $\alpha = 0$, which may indicate that the attractiveness of nodes plays a more important role than the addition of links between two existing nodes in the development of the network.

Case C: If a network consists of a fixed number of local-worlds, and the events of addition and deletion of links between two nodes in the same local-world do not occur, then $p = 0$, $r = 0$, $s = 0$. In this case, the power-law exponent is $\gamma = 2 + (m_1 + \alpha)q/((m_1 - 2m_4) + 2m_4)$.

Case D: If $rm_2 = 2\,sm_3$ in the MLW model, then $b = \alpha a$, so that $P(k) \propto (k + \alpha)^{-\gamma}$, different from $P(k) \propto k^{-\gamma}$ in the original BA model. This clearly indicates that the attractiveness of nodes is very important to the evolvement of the network.

To visualize a generated graph by using the MLW model, a simulation is demonstrated here. Randomly assign a sub-region to every local-world in the plane, which the entire network occupies. Within the assigned sub-region, the nodes belonging to the corresponding local-world are distributed randomly. Figure 5 shows an example of the graph generated by using the MLW model, where the network consists of 150 nodes and includes 4 local-worlds.

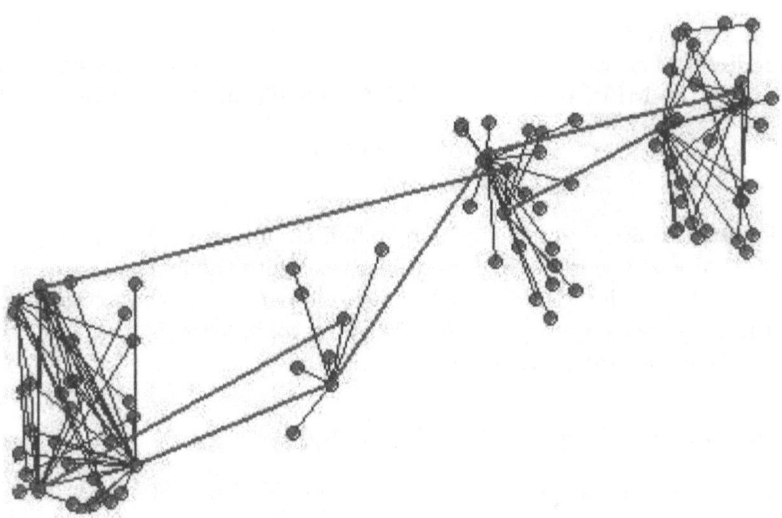

Fig. 5: An example of the graph generated by using the MLW model

In order to model the Internet topology using the MLW model, some real Internet AS graph data shown in Table I, collected by 8 steadily maintained Oregon servers from November 1998 to November 2000 [26], will be used.

Table I. Internet AS map collected by 8 steady Oregon servers

	November 1998	November 2000
AS number	4292	9536
Link number	7298	17,291

During that period of time, there were many ASs and many links between ASs joining or leaving the Internet. Table II shows such "born" and "dead" numbers of ASs and links.

Table II. Birth and death of ASs and links

	Birth number	Death number
ASs	6696	1,452
Links	20,253	10,260

In that real Internet, when a new AS joined the network, it had only one link to the existing nodes in most cases. Also, the nodes with degree 1 were removed from the network with higher probability than the others. Thus, one may set $m_1 = 1$ in the MLW model. On the other hand, the ratio between the number of born links and dead links during November 1998 and November 2000, is about 2. Hence, in the MLW model, one can obtain approximately:

$$\frac{rm_2 + um_4}{sm_3} = 2$$

If one chooses $p = 0$, $\alpha = 0$, $s = 2/11$, and $m_1 = m_2 = m_3 = m_4 = 1$, then by considering the fact that the power-law exponent of the Internet was about 2.2 during that period of time, one obtains $q = 5/11$, $r = 4/33$, $u = 8/33$.

To compare the MLW model with the original BA (OBA) model [11] and the extended BA (EBA) model [8], let $m_0 = 2$, $m = 2$ for the OBA model, and $m_0 = 2$, $m = 1$, $p = 0$, $q = 0.6$ for the EBA model.

Table III. Statistical properties of the Internet in 1997–1999

Year	1997	1998	1999
N	3112	3834	5287
E	5450	6990	10,100
$\bar{k}$	3.5(1)	3.6(1)	3.8(1)
$\bar{C}$	0.18(3)	0.21(3)	0.24(3)
$\bar{d}$	3.8(1)	3.8(1)	3.7(1)

Statistical properties of the Internet topology such as average degree, average clustering coefficient, and average path length are used to evaluate the three different models, i.e., the OBA, EBA, and MLW models. The collected topology data during 1997 and 1999 [21] are shown in Table III, where N represents the number of nodes, E is the number of connections, $\bar{k}$ means the average degree, $\bar{C}$ stands for the average clustering coefficient, and $\bar{d}$ represents the average path length.

A comparison of the results obtained from the three models are shown in Table IV against the actual Internet topology in November 1998.

Table IV. Comparison results for the three models studied

	OBA Model	EBA Model	MLW Model	Real map in Nov. 1998
N	3834	3834	3834	3834
$\bar{k}$	2.41	5.13	4.75	3.6
$\bar{C}$	0.09	0.11	0.24	0.21
$\bar{d}$	6.43	1.68	5.01	3.8
γ	3	2.2	2.2	2.2

One can clearly see from Table IV that the OBA model cannot be used to describe the Internet topology, since it can only predict the real networks with power-law exponent being equal to 3. The EBA model can capture the power-law characteristic of the Internet, but it does not satisfy basic statistical properties. For example, for the average clustering coefficient, the EBA model gives 0.11 while the MLW model gives 0.24, which has a smaller error in comparison against the actual Internet. For the MLW model, as expected, it is better than both the OBA and the EBA models in modelling the Internet AS-level topology, since it can capture the localization property and can reflect the impact of the event of links deletion on the evolvement of the Internet.

4 Conclusions

In this chapter, a multi-local-world (MLW) model has been proposed for the Internet AS-level topology, which is developed based on some careful studies of several intrinsic mechanisms that may be responsible for the emergence of the scale-free characteristic of the Internet. Using the real data collected by NLANR in November 1998, three different models (the original BA model, the extended BA model, and the new MLW model) have been compared, for some basic properties such as the average degree, average clustering coefficient, and average path length. The comparison has clearly demonstrated that the MLW model is the best model for describing the Internet topology.

However, it should be noted that only the inter-domain data of the Internet topology collected in November 1998 was used in this investigation, although it has led to the conclusion that the MLW model is superior to both the original BA model and extended BA model. The main reason that the MLW model can better fit the real Internet data than the other two models is that it can capture the localization property and the impact of the links deletion on the evolving Internet. In order to reach a more convincing conclusion about the nature of the Internet topology, a more complete Internet map is needed and a large database, if available, will be more desirable.

Acknowledgements

This research was supported by the Hong Kong Research Grants Council under the CERG Grants CityU 1004/02E and 1115/03E, the CityU Strategic Research Grant 7001593, and the National Natural Science Foundation of P. R. China under Grant 90412004.

References

1. V. Paxson, S. Floyd: IEEE Simulation Conference Proceedings 1037 (1997)
2. L. A. Adamic, R. M. Lukose, A. R. Puniyani, B. A. Huberman: Phys. Rev. E **64**, 046135 (2001)
3. R. Pastor-Satorras, A. Vespignani: Phys. Rev. Lett **86**, 3200 (2001)
4. M. Doar, I. Leslie: IEEE INFOCOM '93, Proceedings of the Twelfth Annual Joint Conference of the IEEE Computer and Communications Societies: Networking Foundation for the Future **1** 82 (1993)
5. D. Mitzel, S. Shenker: Proceedings of ACM SIGCOMM'94 226 (1994)
6. B. M. Waxman: IEEE J. Selected Areas in Commu. **6**, 1617 (1988)
7. K. I. Calvert, M. B. Doar, E. W. Zegura: IEEE Trans. Commun. **35**, 160 (1997)
8. R. Albert, A. L. Barabasi: Phys. Rev. Lett **85**, 5234 (2000)
9. X. Li, Y. Jin, G. Chen: Physica A **328**, 287 (2003)
10. X. Li, G. Chen: Physica A **328**, 274 (2003)
11. A. L. Barabasi, R. Albert, H. Jeong: Physica. A **272**, 173 (1999)
12. J. J. Pansiot, D. Grad: CM SIGCOMM Comp. Commu.Rev. **28**, 41 (1998)
13. R. Govindan, H. Tangmunarunkit: IEEE INFOCOM 2000, Proceedings of IEEE 19th Annual Joint Conference of Comput. And commu. Societies **3**, 1371 (2000)
14. G. Phillips, H. Tangmunarunkit, S. Shenker: In Proceedings of ACM SIGCOMM'99, Boston, MA, (1999)
15. C. Faloutsos, M. Faloutsos, P. Faloutsos: In Proceedings of ACM SIGCOMM'99, Boston, MA, (1999)
16. A. Vazquez, R. P. Satorras, A. Vespignan: ArXiv: cond-mat/0206084, (2002)
17. A. Danesh, L. Trajkovic, S. H. Rubin, M. H. Smith: IFSA World Congress and 20th NAFIPS Int. Conference **2**, 687 (2001)
18. H. Burch, B. Cheswick: IEEE Computer **32**, 97 (1999)
19. N. Spring, R. Mahajan, D. Wetherall, T. Anderson: IEEE/ACM Trans. Networking **12**, 2 (2004)
20. The National Laboratory for Applied Network Research (NLANR): http://www.moat.nlanr.net/
21. A. Vazquez, R. P. Satorras, A. Vespignani: Phys. Rev. E **65**, 066130 (2002)
22. M. Faloutsos, P. Faloutsos, C. Faloutsos: ACM SIGCOMM'99 Cambridge, MA, (1999)
23. A. Medina, I. Matta, J. Byers: ACM Comput. Commu. Rev. **30**, 18 (2000)
24. C. Jin, Q. Chen, S. Jamin: Dept. EECS, Univ. Michigan, Tech. Rep. CSE-TR-433-00, (2000)
25. Traceroute.org (http://www.traceroute.org/)

26. Q. Chen, H. Chang, R. Govindan, S. Jamin: Proceedings of the Twenty-first Annual Joint Conference of the IEEE Computer and Communications Societies **2**, 608 (2002)
27. G. Siganos, M. Faloutsos, P. Faloutsos, C. Faloutsos: IEEE/ACM Trans. Networking **11**, 514 (2003)
28. C. Labovitz, R. Malan, F. Jahanian: Proceedings of ACM SIGCOMM'97 (1997)
29. B. Huffaker, D. Plummer, D. Moore, K. Claffy: http://www.caida.org/outreach/papers/2002/SkitterOverview/
30. S. Yook, H. Jeong, A. Barabasi: PNAS **99**, 13382 (2002)
31. Q. Chen, D. Shi: Physica. A **335**, 240 (2004)

Evolution of the Internet Topology and Traffic Dynamics of Data Packets

Kwang-Il Goh, Byungnam Kahng, and Doochul Kim

School of Physics and Center for Theoretical Physics, Seoul National University NS50, Seoul 151-747, Korea kahng@phya.snu.ac.kr.

1 Introduction

In recent years, the Internet has become one of the most influential media in our daily life, going beyond in its role as the basic infrastructure in the technological world. Explosive growth in the number of users and hence the amount of traffic poses a number of problems which are not only important in practice for, e.g., maintaining it free from any undesired congestion and malfunctioning, but also of theoretical interests as an interdisciplinary topic [1]. Such interests, also stimulated by other disciplines like biology, sociology, and statistical physics, have blossomed into a broader framework of network science [2, 3, 4]. The Internet is a primary example of complex networks. It consists of a large number of very heterogeneous units interconnected with various connection bandwidths, however, it is neither regular nor completely random. In their landmark paper, Faloutsos et al. [5] showed that the Internet at the autonomous systems (ASes) level is a scale-free (SF) network [6], meaning that the node degree k, the number of connections a node has, follows a power-law distribution,

$$p_d(k) \sim k^{-\gamma}, \tag{1}$$

in node degree k, the number of connections a node has. The degree exponent γ is subsequently measured and confirmed in a number of studies to be $\gamma \approx 2.1$ (Fig. 1). The power-law degree distribution implies presence of a few nodes having a hugh number of connections, called hubs, while most other nodes have a few number of connections.

Emergence of such a heterogeneous degree distribution calls for explanation and understanding of the basic mechanism underlying the growth of the Internet. Once revealed, it can be used to predict what the Internet will be like in the future, as well as how it has evolved into the present shape. In this manuscript, we will address this issue, showing that it can be described by a simple toy model based on the multiplicative stochastic process. By extracting relevant parameters for the stochastic process from the time history of the AS

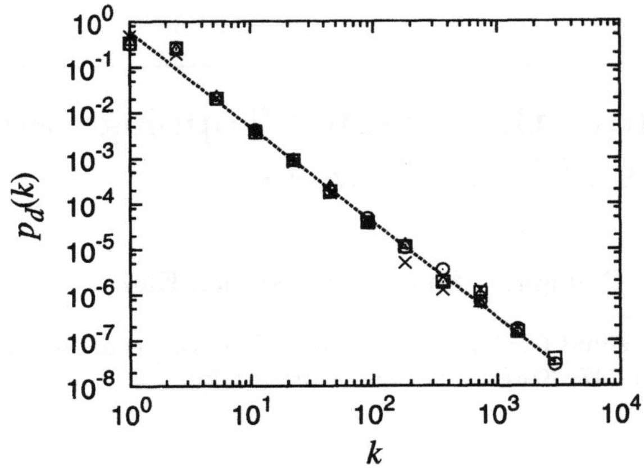

Fig. 1: The degree distribution of the Internet at the autonomous systems level. The data are sampled in January each year, 1998 (×), 2000 (△), 2002 (□), and 2004 (○), from the Oregon route views project. The data are logarithmically binned, and the dashed straight line has the slope −2.1, drawn for the eye.

map archived in the Oregon route views project, we can predict the degree exponent of the Internet accurately.

While the power-law degree distribution is intriguing, it gives only a global snapshot of the Internet structure. The degree-degree correlation function is a quantity which gives more detailed local information of the Internet structure. In this manuscript, we also study the mean degree of the nearest neighbors of nodes with degree k, denoted as $\langle k_{\rm nn}\rangle(k)$, associated with the degree-degree correlation. We show that the behavior of the mean degree function of the Internet is nontrivial. The behavior of the mean-degree function cannot be explained by the toy model with the multiplicative stochastic process only. We improve the toy model by introducing the rule of adaptive rewirings of links. The adaptation model can also generate the clustering coefficient as large as the one obtained from the real-world Internet. However, the adaptation model is not complete yet to reproduce the joint probability $P(k, k')$ that two nodes with degree k and k' are connected.

The Internet is not a quiet object. Data packets are sent and received over it constantly, causing momentary local congestion from time to time. To avoid such undesired congestion, the capacity, or the bandwidth, of the routers should be as large as it can handle the traffic. We will introduce a rough measure of such capacity, called the load and denoted as ℓ. The distribution of the load reflects the high level of heterogeneity of the Internet: It also follows a power law,

$$p_l(\ell) \sim \ell^{-\delta}, \qquad (2)$$

with the load exponent $\delta \approx 2.0$. We will discuss the implication of the power-law behavior of the load distribution.

2 Internet Evolution as a Multiplicative Stochastic Process

The mechanism of the emergence of SF network is mostly captured by the Barabási-Albert (BA) model [7] which assumes the linear growth in numbers of nodes and links in time and the preferential attachment (PA) in establishing links from a new node to other previously existing ones. The PA means that the more links a node already has, the more likely it is to get another from a new node. Precisely, the probability $\Pi_i(t)$ that a node i will receive a link from the new node created at time t is linearly proportional to its present degree $k_i(t)$, i.e.,

$$\Pi_i(t) = \frac{k_i(t)}{\sum_j k_j(t)}. \tag{3}$$

The degree exponent $\gamma = 3$ for the BA model. In the generalized version which can tune γ continuously, k_i in Eq.(3) is replaced by $k_i - a$, with a a constant related to γ by $a = \langle k\rangle(3-\gamma)/2$, $\langle k\rangle$ being the mean degree. The empirical evidence of the PA in the Internet has been reported [8, 9]. As we will see, however, the assumption that the numbers of nodes and links increase linearly in time does not apply to the real situation of the Internet. Rather, the numbers of nodes and links increase exponentially but with different rates. Furthermore, the interconnections between nodes are being updated continually in the Internet, which was not incorporated in the original BA model.

Huberman and Adamic (HA) [10] proposed another scenario for SF networks. They argued that the fluctuation effect arising in the process of connecting and disconnecting links between nodes is an essential feature to describe the dynamics of the Internet topology correctly. In the HA model, the total number of nodes $N(t)$ increases exponentially with time as

$$N(t) = N(0)\exp(\alpha t). \tag{4}$$

Next, they assumed that the degree k_i at a node i evolves through the multiplicative stochastic process,

$$k_i(t+1) = k_i(t)[1 + \zeta_i(t+1)], \tag{5}$$

where $\zeta_i(t)$ is the growth rate of the degree k_i at time t, which fluctuates from time to time. Thus, one may divide the growth rate $\zeta_i(t)$ into two parts,

$$\zeta_i(t) = g_{0,i} + \xi_i(t), \tag{6}$$

where $g_{0,i}$ is the mean value over time, and $\xi_i(t)$ the rest part, representing fluctuations over time. $\xi_i(t)$ is assumed to be a white noise satisfying $\langle \xi_i(t)\rangle = 0$ and $\langle \xi_i(t)\xi_j(t')\rangle = \sigma_{0,i}^2 \delta_{t,t'}\delta_{i,j}$, where $\sigma_{0,i}^2$ is the variance. Here $\langle\cdots\rangle$ is the sample average and $\delta_{a,b}$ is the Kronecker delta symbol. For later convenience, we denote the logarithm of the growth factor as $G_i(t) \equiv \ln[1+\zeta_i(t)]$.

Then a simple application of the central limit theorem ensures that the probability distribution of $k_i(t)/k_i(t_0)$, t_0 being a reference time, follows the log-normal distribution for sufficiently large t. To get the degree distribution, one needs to collect all contributions from different ages τ_i, growth rates $g_{0,i}$, standard deviations $\sigma_{0,i}$ and initial degree $k_i(t_0)$. HA further assumed that ζ_i are identically distributed so that $g_{0,i} = g_0$ and $\sigma_{0,i} = \sigma_0$ for all i. Then the conditional probability for degree, $P_d(k, \tau | \, k_0)$, that $k_i(t_0 + \tau) = k$, given $k_i(t_0) = k_0$ is given by

$$P_d(k, \tau | \, k_0) = \frac{1}{k\sqrt{2\pi\sigma_{\rm eff}^2 \tau}} \exp\left\{ -\frac{(\ln{(k/k_0)} - g_{\rm eff}\tau)^2}{2\sigma_{\rm eff}^2 \tau} \right\}, \tag{7}$$

where $g_{\rm eff} \equiv \langle G_i(t) \rangle$ and $\sigma_{\rm eff}^2 \equiv \langle (G_i(t) - \langle G_i(t) \rangle)^2 \rangle$. $g_{\rm eff}$ and $\sigma_{\rm eff}^2$ are related to g_0 and σ_0^2 as $g_{\rm eff} \approx g_0 - \sigma_0^2/2$ and $\sigma_{\rm eff}^2 \approx \sigma_0^2$, respectively [11]. Since the density of nodes with age τ is proportional to $\rho(\tau) \sim \exp(-\alpha\tau)$, the degree distribution collected over all ages becomes

$$p_d(k) = \int d\tau \rho(\tau) P_d(k, \tau | \, k_0). \tag{8}$$

The integration is evaluated by the steepest descent method in the limit $k/k_0 \to \infty$ to give $p_d(k) \sim k^{-\gamma}$ with

$$\gamma = 1 - \frac{g_{\rm eff}}{\sigma_{\rm eff}^2} + \frac{\sqrt{g_{\rm eff}^2 + 2\alpha\sigma_{\rm eff}^2}}{\sigma_{\rm eff}^2}. \tag{9}$$

Therefore, it is instructive to know the effective values of $g_{\rm eff}$ and $\sigma_{\rm eff}$ to determine the degree exponent. In the next section, we will measure such parameters from the real evolutionary history of the Internet AS map and check if the HA scenario holds.

3 Growth Dynamics of the Internet

A number of projects exists aiming to map the world-wide topology of the Internet. One such is the Route Views project initiated at the University of Oregon [12], the data of which were also archived at the National Laboratory of Applied Network Research (NLANR) [13]. Among the daily data from November 1997 to January 2000, we sample one AS map a month, with the total period of 26 months, and analyze them for various quantities. First we measure the growth rate of the number of ASes α. We also measure directly the growth rate of the number of links β, which can be crosschecked for consistency later.

In Fig. 2(a) and (b), we show the total number of ASes $N(t)$ and the total number of links $L(t)$ as a function of time t. The straight line in log-linear plot means $N(t)$ and $L(t)$ indeed grow exponentially. The growth rates are

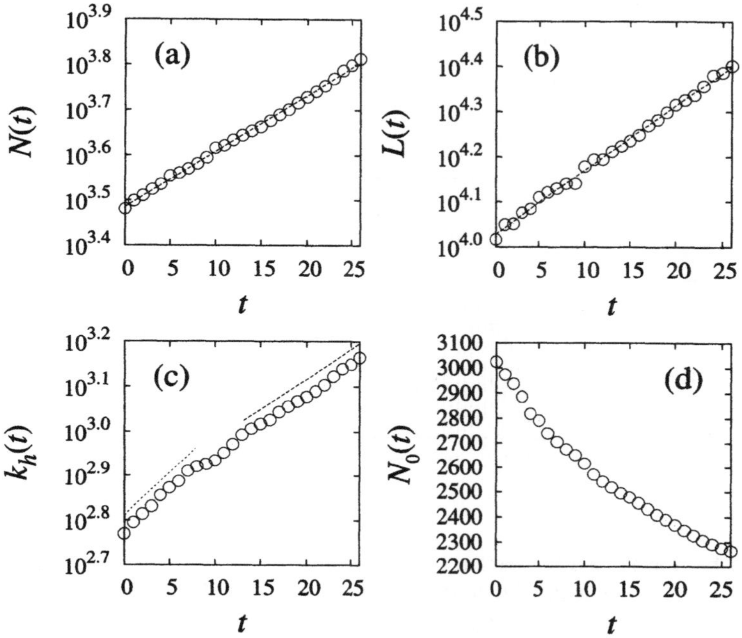

Fig. 2: The time evolution of the number of ASes $N(t)$ (a), the number of links between ASes $L(t)$ (b), the degree of the hub $k_h(t)$ (c), and the number of ASes among those existed at time $t = 0$ that have survived until t, $N_0(t)$ (d). Note that the ordinates in (a), (b) and (c) are in logarithmic scale, indicating the exponential increase of corresponding quantities. The fitted line has a slope 0.029 in (a), 0.034 in (b), and 0.043 (0.030) for the dotted (dashed) one in (c).

determined to be $\alpha \approx 0.029$ and $\beta \approx 0.034$. We also find that the newly appeared AS would connect to only one or two existing ASes so that the average number of links the new AS establishes is $k_{\text{new}} \approx 1.35$. Fig. 2(c) shows the growth of the degree of the hub, the node with the largest degree. It shows a change of growth rate around $t \approx 14$. $N_0(t)$, the number of ASes among those existed at time $t = 0$ that have survived until t, is also shown in Fig. 2(d).

The measurement of g_0 and σ_0 is nontrivial due to the presence of large fluctuations. To this end, we measure the degree growth rate of a node i, $G_i(t)$, defined earlier as $G_i(t) \equiv \ln[1 + \zeta_i(t)] = \ln[k_i(t)/k_i(t-1)]$. To keep $G_i(t)$ well-defined for all t, we consider only the nodes existing for the entire time range $0 \leq t \leq 26$, the set composed of which is denoted by S hereafter. By the existence of a node we mean that its degree is nonzero, since one cannot identify an AS with no connection. For each i ($i \in S$), let $g_i = \langle G_i(t) \rangle_t$ and $\sigma_i^2 = \left\langle (G_i(t) - g_i)^2 \right\rangle_t$, where $\langle \cdots \rangle_t$ means the temporal average over the period $16 < t \leq 26$ ($T = 10$). If the HA scenario holds, the histogram of $\{g_i\}$

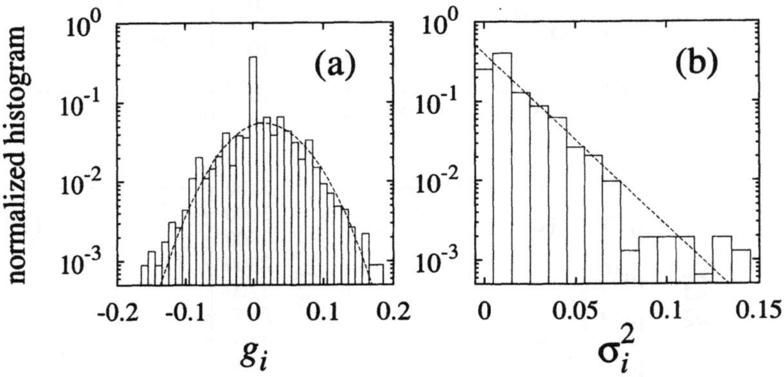

Fig. 3: The normalized histogram of g_i (a) and σ_i^2 (b). In (a), the data is fitted with a Gaussian with the mean 0.016 and the standard deviation 0.04. In (b), the data is fitted with an exponential decay $\exp(-x/x_c)$ with the characteristic scale $x_c \approx 0.02$. The measured value of the average is $\overline{\sigma^2} \approx 0.017$.

for all nodes would follow the Gaussian distribution with the mean g_{eff} and the variance σ_{eff}^2/T. We show such histogram in Fig. 3, the fit of which to the Gaussian gives the mean $\overline{g}$ as 0.016 and the standard deviation σ_g as 0.04. The measured values of $\{\sigma_i^2\}$ give the mean value $\overline{\sigma^2} \approx 0.017$.

It is most likely that $\overline{g}$ and $\overline{\sigma^2}$ would have a distribution over nodes. As HA assumed, however, we try to approximate the growth process by a single process whose effective mean growth rate and standard deviation are g_{eff} and σ_{eff}, respectively. Then Eq. (7) should hold for all i and all t with a suitable choice of those parameters. For this purpose, we consider the distribution $P[k_i(t)/k_i(t_0)]$ in terms of the scaled variables x and y defined as

$$x \equiv \frac{\ln[k_i(t)/k_i(t_0)] - g_d(t-t_0)}{\sqrt{2\sigma_d^2(t-t_0)}}, \tag{10}$$

and

$$y \equiv P[k_i(t)/k_i(t_0)][k_i(t)/k_i(t_0)]\sqrt{2\pi\sigma_d^2(t-t_0)}, \tag{11}$$

where we set $t_0 = 0$ and g_d and σ_d are parameters to be chosen. From Eq. (7), with suitably chosen parameters g_d and σ_d, the distribution for different time t would collapse onto a single curve, $\ln y = -x^2$. We show such data in Fig. 4. While the data collapse breaks down for large x, the best collapse can be accomplished by choosing the parameters $g_d = 0.016$ and

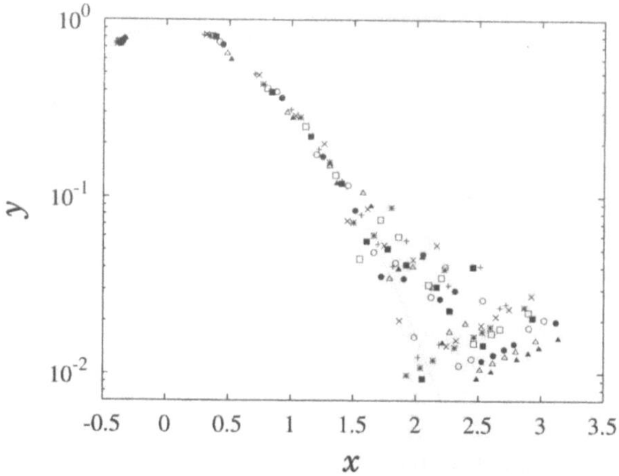

Fig. 4: Plot of $P[k_i(t)/k_i(t_0)]$ versus $k_i(t)/k_i(t_0)$ for different times $t > 16$ in terms of the scaled variables y and x defined in Eqs. (10) and (11). Larger deviations for large x indicate the drawback of the HA approach.

$\sigma_d = 0.14$, which should be identified with g_{eff} and σ_{eff}, respectively. The effective growth parameters found in this way are in accordance with the ones estimated before as $\overline{g} = 0.016$ and $\overline{\sigma^2} = 0.017$. As noted earlier, the consistency of estimated parameters can be checked as, for example, it should satisfy $\beta = \max(\alpha, g_{\text{eff}} + \sigma_{\text{eff}}^2)$, for which we have $\beta = 0.034$ and $g_{\text{eff}} + \sigma_{\text{eff}}^2 = 0.035$, being reasonably consistent with each other. Thus we conclude the parameters $g_{\text{eff}} = 0.016$ and $\sigma_{\text{eff}} = 0.14$ can be regarded as the effective parameters of the degree growth dynamics of the Internet AS map as a single process. Applying those values together with $\alpha = 0.029$ found earlier into Eq. (10), we found $\gamma \approx 2.1$, which is in excellent agreement with the directly measured one shown in Fig. 1.

4 Degree Correlation and Modular Structure

It is known that the degrees of the two nodes located at the ends of a link are correlated each other. As the first step, the degree-degree correlation can be quantified in terms of the mean degree of the neighbors of a given node with degree k as a function of k, denoted by $\langle k_{\text{nn}}\rangle(k)$ [8]. Then the mean degree function $\langle k_{\text{nn}}\rangle(k)$ can be written in terms of the conditional probability $P(k'|k)$ that given a node with degree k is present, another node with degree k' is connected to it. That is $\langle k_{\text{nn}}\rangle(k) = \sum_{k'} k'P(k'|k)$. The conditional probability is then expressed in terms of the joint probability, $P(k',k)$, as $P(k'|k) = P(k',k)/kp_d(k)$. Meanwhile, for the BA model, the joint probability that nodes of degree k' (ancestor) and k (descendent) are connected scales as [14]

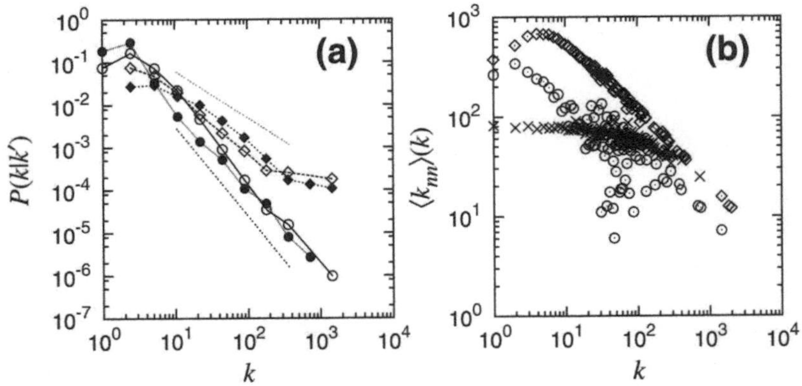

Fig. 5: **(a)** The conditional probability $P(k|k')$ for the dangling nodes with $k' = 1$ (diamonds) and the hub (circles). Data points with filled (open) symbols are from the Internet AS as of January, 2000 (the simulation of the adaptation model). The dashed (dotted) straight line has a slope -2.1 (-1.1), drawn for a guide to the eye. **(b)** The average degree of nearest neighbors of a node whose degree is k, $\langle k_{\mathrm{nn}}\rangle(k) = \sum_{k'} k'P(k'|k)$, as a function of k from the adaptation model (diamond), HA model (cross) and the Internet (circle).

$$P(k', k) \sim k'^{-(\gamma-1)}k^{-2}. \tag{12}$$

Then the function $\langle k_{\mathrm{nn}}\rangle(k)$ is written as [14, 3]

$$\begin{aligned}\langle k_{\mathrm{nn}}\rangle(k) &= \sum_{k'} k'P(k'|k) \\ &= \sum_{k'} \frac{k'P(k', k)}{kp_d(k)} \\ &\sim \sum_{k'} \frac{k'k'^{-(\gamma-1)}k^{-2}}{k^{1-\gamma}} \\ &\sim k^{-3+\gamma}.\end{aligned} \tag{13}$$

More generally, the behavior of the average neighbor degree function is expressed by another power law as

$$\langle k_{\mathrm{nn}}\rangle(k) \sim k^{-\nu}. \tag{14}$$

For the Internet, if Eq.(12) holds, $\langle k_{\mathrm{nn}}\rangle(k)$ would decay as $\sim k^{-0.9}$, i.e., $\nu \approx 0.9$. However, it decays with $\nu \approx 0.5$ for the real-world Internet [8], suggesting that the joint probability does not behave in the way of the BA model does.

On the other hand, the degree-degree correlation can also be described in terms of the assortativity coefficient introduced by Newman [15], which is defined as

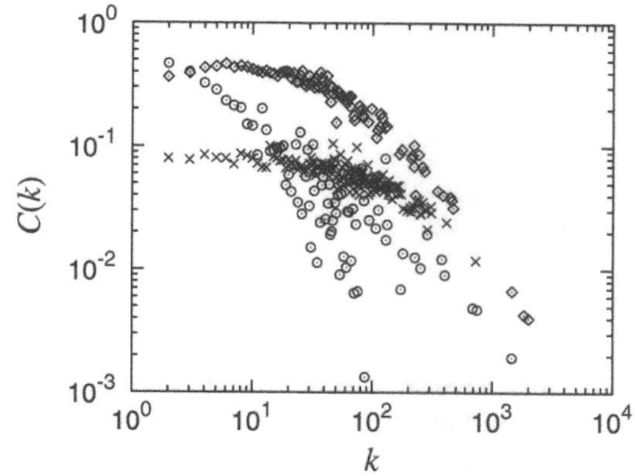

Fig. 6: The clustering function $C(k)$ for the Internet AS as of January, 2000 ($\circ$), HA model ($\times$), and the adaptation model ($\diamond$).

$$r = \frac{\langle k_1 k_2 \rangle - \langle (k_1 + k_2)/2 \rangle^2}{\langle (k_1^2 + k_2^2)/2 \rangle - \langle (k_1 + k_2)/2 \rangle^2}, \tag{15}$$

where k_1 and k_2 are the degree of two end nodes, respectively, of a link, and $\langle \cdots \rangle$ denotes the average over all links. It is nothing but the Pearson correlation coefficient for the degrees of two end nodes over all links, normalized -1 to 1. r is negative when the function $\langle k_{\mathrm{nn}} \rangle(k)$ exhibits decreasing behavior like the case of the Internet, while it is positive when the function increases with k as occurring in social networks. In fact, the assortativity coefficient r was introduced to characterize social networks. We find the assortativity coefficient of the Internet to be $r = -0.18$ as of Jan. 2001.

The Internet has modules within it. Such modular structures arise due to regional control systems, and often form in a hierarchical way [16]. Recently, it was argued that such modular and hierarchical structures can be described in terms of the clustering coefficient. Let C_i be the local clustering coefficient of a node i, defined as $C_i = 2e_i/k_i(k_i - 1)$, where e_i is the number of links present among the neighbors of node i, out of its maximum possible number $k_i(k_i - 1)/2$. The clustering coefficient of a network, C, is the average of C_i over all nodes. $C(k)$ means the clustering function of a node with degree k, i.e., C_i averaged over nodes with degree k. When a network is modular and hierarchical, the clustering function follows a power law,

$$C(k) \sim k^{-\beta}, \tag{16}$$

for large k, and C is independent of system size N, while C depends on N as $C \sim N^{-1}$ for the BA model [17, 18].

For the Internet, the clustering coefficient is measured to be $C_{\mathrm{AS}} \approx 0.25$, which is different from the value obtained from the HA model, $C_{\mathrm{HA}} \approx 0.01(1)$. The clustering function $C_{\mathrm{AS}}(k)$ also behaves differently from what the HA model predicts as shown in Fig. 6.

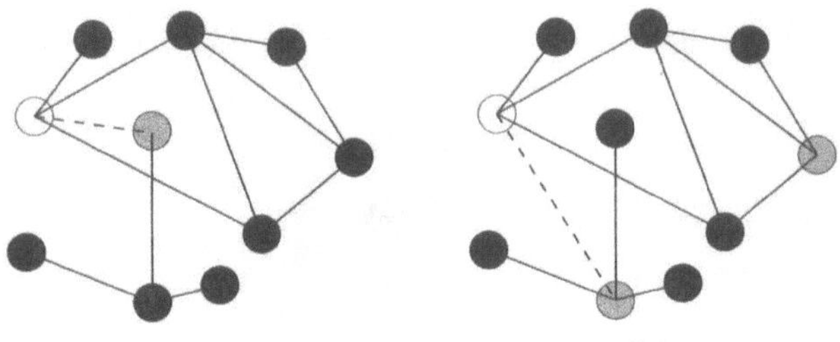

Fig. 7: Shown is the adaptive rewiring rule. A node (white) detaches one of its links from a node (gray) in (a), and attaches it to one of the nodes (gray) with degree 3, larger than 2 of the detached node, in (b).

5 Adaptation Model

Using the measured values, α, $g_{\rm eff}$, and $\sigma_{\rm eff}$, and modifying the HA idea, we construct a stochastic model evolving through the following four rules:

- (i) **Geometrical growth**: At time t, geometrically increased number of new nodes, $\alpha N(t-1)$, are introduced in the system, and following the fact $\langle k_{\rm new}\rangle_t \approx 1.34$, each of them connects to one or two existing nodes according to the PA rule.
- (ii) **Accelerated growth**: Each existing node increases its degree by the factor $g_0 \approx g_{\rm eff} + \sigma_{\rm eff}^2/2$. These new internal links are also connected following the PA rule.
- (iii) **Fluctuations**: Each node disconnects existing links randomly (resp. connects new links following the PA rule) when the noise ($\xi_i(t)$ in Eq.(4)) is chosen to be negative (resp. positive). The variance of the noise is given as $\sigma_0^2 \approx \sigma_{\rm eff}^2$.
- (iv) **Adaptation**: When connecting in step (iii), the PA rule is applied only within the subset of the existing nodes consisting of those having more degree than the one previously disconnected. This last constraint accounts for the adaptation process in our model, which is the distinct feature from the HA model. The adaptive rewiring rule is depicted in Fig. 7.

Through this adaptation process, the model can reproduce generic features of the Internet topologies which are as follows. First, the degree exponent is measured to be $\gamma_{\rm model} \approx 2.2$, close to the empirical result $\gamma_{\rm AS} \approx 2.2(1)$. Second, the clustering coefficient is measured to be $C_{\rm model} \approx 0.15(7)$, comparable to the empirical value $C_{\rm AS} \approx 0.25$. Note that without the adaptation rule, we only get $C \approx 0.01(1)$. The clustering function $C(k)$ also behaves in a similar pattern to that of the real-world Internet shown in Fig. 6, that is, decaying in a similar fashion for large k, but the overall curve shifts upward and the constant behavior for small k appears. Third, the mean degree function $\langle k_{\rm nn}\rangle(k)$ also behaves similarly to that of the real-world networks as shown in Fig. 5,

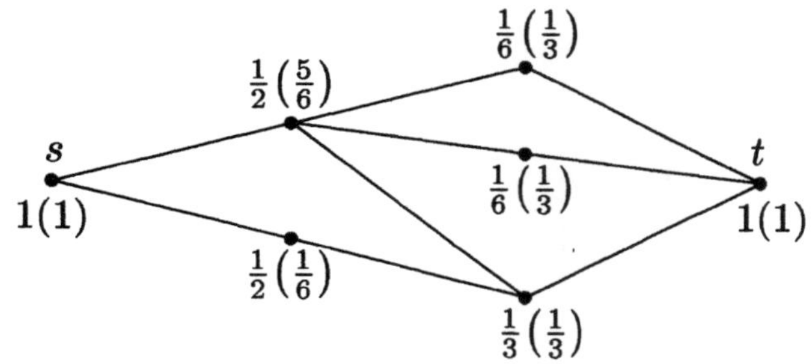

Fig. 8: The load at each node due to a unit packet transfer from the node s to the node t, $\ell_i^{s\to t}$. In this diagram, only the nodes along the shortest paths between (s,t) are shown. The quantity in parentheses is the corresponding value of the load due to the packet from t to s, $\ell_i^{t\to s}$.

but it also shifts upward overall. In short, the behaviors of $C(k)$ and $\langle k_{\mathrm{nn}}\rangle(k)$ of the adaptation model are much closer to those of the real Internet AS map than those of the HA model. However, the discrepancies are still significant.

6 Load Distribution of the Internet

To a large extent, the Internet is the medium of communication. The continuous communication between hosts generates certain amount of data traffic. To make the best use of it, we have to avoid congestions, from which we suffer possible delays and the loss of information. What's worse, one doesn't know when and how much a host will generate the traffic. Absent is the central regulation in the Internet, hence each node should do its best for its own ends. To give a measure of such activity, we define the load ℓ_i as the amount of capacity or bandwidth that a node i can handle in unit time [19]. Not knowing the level of traffic, one assumes that every node sends a unit packet to everyone else in unit time. One further assumes that the packets are transfered from the source to the target only along the shortest paths between them, and divided evenly upon encountering any branching point. To be precise, let $\ell_i^{s\to t}$ be the amount of packet sent from s (source) to t (target) that passes through the node i (see Fig. 8). Then the load of a node i, ℓ_i, is the accumulated sum of $\ell_i^{s\to t}$ for all s and t, $\ell_i = \sum_{s\neq t} \ell_i^{s\to t}$. In other words, the load of a node i gives us the information how much the capacity of the node should be in order to maintain the whole network in a free-flow state. However, due to local fluctuation effect of the concentration of data packets, the traffic could be congested even for the capacity of each node being taken as its load. To calculate load on each node, we use the modified version of the breath-first search algorithm introduced by Newman [20] and independently by Brandes [21], which can evaluate $\{\ell_i\}$ in time of order $\mathcal{O}(N^2)$ for sparse binary graphs. The algorithm works as follows:

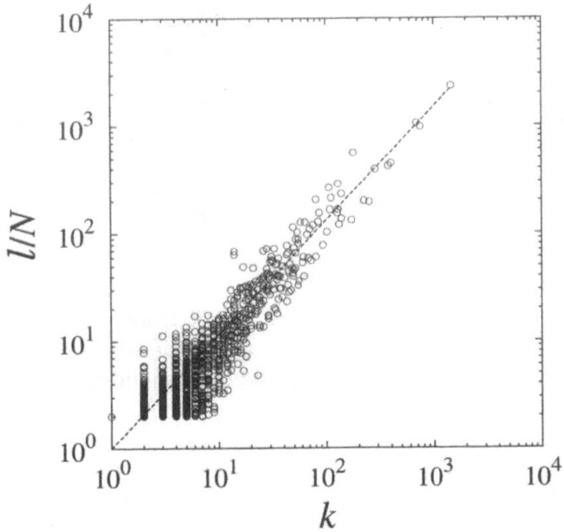

Fig. 9: The scatter plot of the load versus the degree of each node for the AS map as of January 2000. The slope of the dashed line is 1.06, drawn for the eye.

(1) The shortest paths to a node i from every other node are identified by using the breadth-first search and each node j is labeled by the distance d_{ij} from the source node i.
(2) A variable f_j preset to the initial value 1 except for the source node i is assigned to each node j. This will account for the contribution to load due to a unit packet fired from the node i to j. For the source node i, f_i is initially preset to 0 since we do not consider a packet transfer to itself.
(3) Going through the nodes j in descending order of their distance from i, the value of f_j is added to the corresponding variable on the predecessor node of j. Here predecessors of a node j are the nodes connected to j, being closer to the source node i than j. If j has more than one predecessor, then f_j is divided equally by the number of predecessors and each fraction is passed to each predecessor.
(4) When we have completed all nodes in this way, the variable f_j represents the number of packets (out of total $N-1$) that run through the node j. In our convention, the value f_j includes the contributions of the end points of each path. To get the load for all paths, f_j's are added to the load ℓ_j and the entire calculation is repeated for each of the N possible source node i.

For a number of SF networks in nature and society, the load distribution is also found to follow a power law, Eq. (2) [22]. The Internet AS map is of no exception and the load exponent δ of the power law is estimated to be approximately $\delta \approx 2.0$ [22, 23]. The power-law load distribution means that

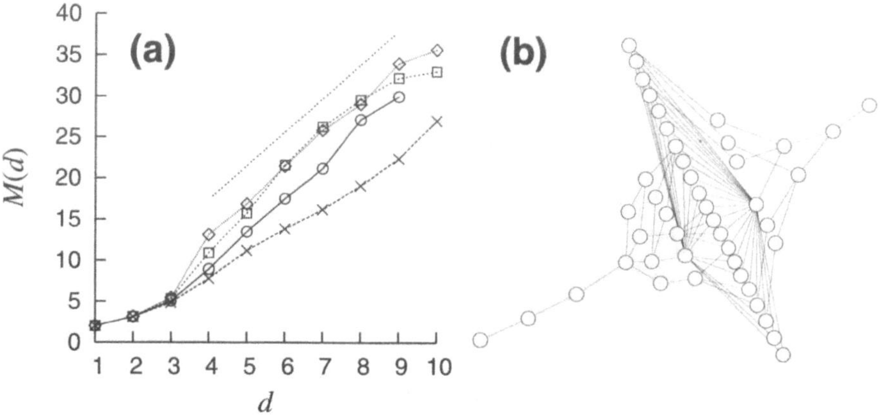

Fig. 10: The mass-distance relation $\mathcal{M}(d)$ for the Internet in the years 1998 ($\times$), 2000 ($\circ$), 2002 ($\Box$), and 2004 ($\diamond$) (a) and the snapshot of the shortest pathways of length 10 as of January 2000 (b)

a few ASes should handle an extraordinarily large amount of load while most others may do only a little.

The load of a node is highly correlated with its degree. The Pearson correlation coefficient between the two quantities is as high as 0.98. This suggests a scaling relation between the load and the degree of a node as

$$\ell \sim k^{\eta} \tag{17}$$

and the scaling exponent η is estimated as $\eta = 1.06 \pm 0.03$ for January 2000 AS map (Fig. 9). In fact, if one assumes that the ranks of each node for the degree and the load are the same, then one can show that the exponent η depends on γ and δ as $\eta = (\gamma - 1)/(\delta - 1)$ with $\gamma \approx 2.1$ and $\delta \approx 2.0$, and we have $\eta \approx 1, 1$, which is consistent with the direct measurement.

Since data packets are normally transferred along the shortest pathway between two nodes in the Internet, it would be interesting to investigate the generic topological feature of the shortest pathways between them. Along the shortest pathways, we count the total number of nodes $\mathcal{M}(d)$ lying on the shortest pathways separated by the distance d, averaged over all pairs of nodes separated by the same distance d. Adopting from the fractal theory, $\mathcal{M}(d)$ is called the "mass-distance" relation. We find that the mass, the total number of nodes lying on the pathways, increases linearly with increasing d, but the increasing rate is roughly 4 larger than 1. If the shortest pathway is singly connected, then the proportionality coefficient would be 1. The fact that it increases linearly in d means that the shortest pathway topology is effectively tree, but that its coefficient is significantly larger than 1. That means that there are many alternative pathways as shown in Fig. 10(b).

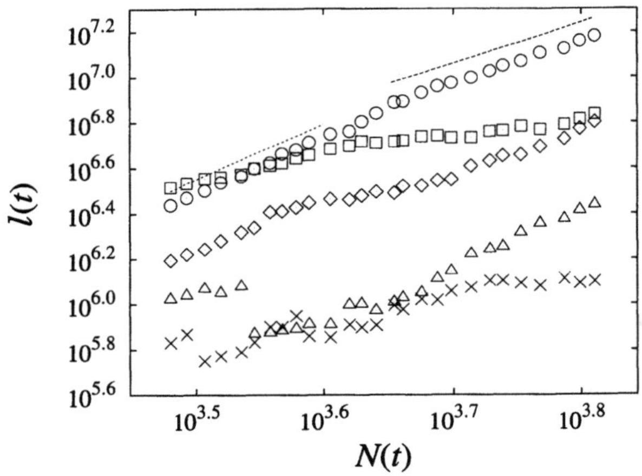

Fig. 11: Time evolution of the load versus $N(t)$ at the ASes of degree-rank 1(○), 2 (□), 3 (◇), 4 (△), and 5 (×). The dashed line for larger N has slope 1.8, drawn for the eye.

The time evolution of the load at each AS is also of interest. Practically, how the load scales with the total number of AS (the size of the AS map) is an important information for the network management. In Fig. 11, we show $\ell_i(t)$ versus $N(t)$ for 5 ASes with the highest rank in degree, i.e., 5 ASes that have largest degrees at $t = 0$. The data of $\{\ell_i(t)\}$ shows large fluctuations in time. Interestingly, the fluctuation is moderate for the hub, implying that the connections of the hub is rather stable. The load at the hub is found to scale with $N(t)$ as $\ell_h(t) \sim N(t)^\mu$, but the scaling shows a crossover from $\mu \approx 2.4$ to $\mu \approx 1.8$ around $t \approx 14$, as it did for the degree.

7 Summary

We have studied the temporal evolution of the Internet AS map and showed that it can be described in the framework of multiplicative stochastic process for the degree growth dynamics. We measured the values of relevant parameters from the history of the AS map monitored by the Oregon Route Views project. With those values, the AS number growth rate $\alpha = 0.029$, the effective degree growth rate $g_{\text{eff}} = 0.019$, and its effective standard deviation $\sigma_{\text{eff}} = 0.14$ are obtained. Using these values, we were able to predict the degree exponent γ as $\gamma \approx 2.1$, which is in excellent agreement with the previously reported empirical values $\gamma_{\text{measured}} = 2.1 \sim 2.2$. We also studied the degree-degree correlation of the Internet in terms of the quantity $\langle k_{\text{nn}}\rangle(k)$, the mean degree of the neighbors of nodes with degree k and the assortativity coefficient. The modularity of the Internet was also examined by measuring the

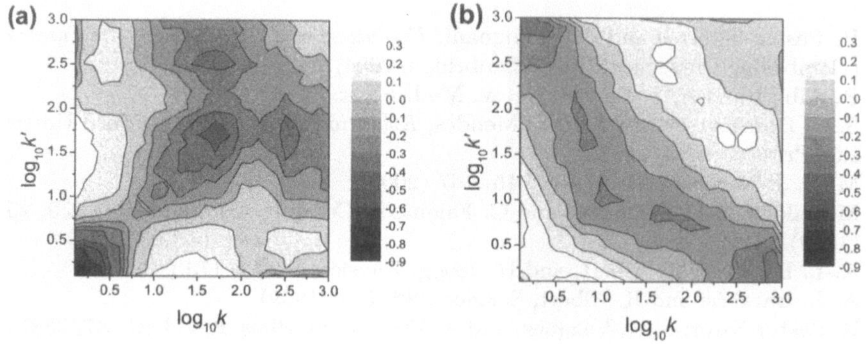

Fig. 12: The plot of the joint probabilities $P(k', k)$ for the Internet as of January, 2000 (a) and the adaptation model (b). The concentration denotes the value of $\log_{10}[P(k, k')/P_{\text{random}}(k, k')]$, where $P_{\text{random}}(k, k')$ is the joint probability of the randomized networks generated for the purpose of reference. They are generated by the switching method [24] conserving the degree sequence.

clustering function $C(k)$. We found that the degree-degree correlation and the modularity cannot be explained by the HA model only with the multiplicative process. We modified the HA model by introducing the adaptive rewiring of links in the process of attaching links. While the adaptation model seems to be successful for reproducing $C(k)$ and $\langle k_{\text{nn}}\rangle(k)$ of the Internet AS roughly, it is not satisfactory yet to describe the real world Internet. For example, the joint probability $P(k, k')$ that two nodes with degrees k and k' are connected via an edge cannot be reproduced by the adaptation model as shown in Fig. 12. The concentration profile of the joint probability for the adaptation model shows different pattern compared to that of the real Internet. Thus, further refinement of the adaptation model reflecting the evolution of the Internet map is needed.

In the second part of the paper, we introduced a quantity called the load as the accumulated sum of the fraction of packets that passes through a given node when a unit packet is transmitted between each pair of nodes along the shortest paths between the two. It can be thought of as the amount of traffic that a node should handle to keep the whole network away from the unwelcome congestion and maintain free-flow state, giving a measure of desired capacity of the nodes. The load distribution also follows a power law. The load and the degree of an AS are highly correlated with each other. The analysis of the temporal change of the load reveals that the load at the hub scales with the system size as $N^{1.8}$. Finally, we note that the contents of this article partly overlaps with our previous studies published in [19, 22, 25].

This work is supported by the KOSEF grants No. R14-2002-059-01000-0 in the ABRL program.

References

1. R. Pastor-Satorras and A. Vespignani, *Evolution and Structure of the Internet* (Cambridge University Press, Cambridge, 2004).
2. R. Albert and A.-L. Barabási, Rev. Mod. Phys. **74**, 47 (2002).
3. S. N. Dorogovtsev and J. F. F. Mendes, *Evolution of Networks* (Oxford University Press, Oxford, 2003).
4. M. E. J. Newman, SIAM Rev. **45**, 167 (2003).
5. M. Faloutsos, P. Faloutsos, and C. Faloutsos, Comput. Commun. Rev. **29**, 251 (1999).
6. A.-L. Barabási, R. Albert, and H. Jeong, Physica A **272**, 173 (1999).
7. A.-L. Barabási and R. Albert, Science **286**, 509 (1999).
8. R. Pastor-Satorras, A. Vázquez, and A. Vespignani, Phys. Rev. Lett. **87**, 258701 (2001).
9. H. Jeong, Z. Néda, and A.-L. Barabási, Europhys. Lett. **61**, 567 (2003).
10. B. A. Huberman and L. A. Adamic, e-print (http://arxiv.org/abs/cond-mat/9901071) (1999).
11. C. W. Gardiner, *Handbook of Stochastic Methods* (Springer-Verlag, Berlin, 1983).
12. D. Meyer, *University of Oregon Route Views Archive Project* (http://archive.routeviews.org).
13. The NLANR project sponsored by the National Science Foundation (http://moat.nlanr.net).
14. P. Krapivsky and S. Redner, Phys. Rev. E **63**, 066123 (2001).
15. M. E. J. Newman, Phys. Rev. Lett. **89**, 208701 (2002).
16. K. A. Eriksen, I. Simonsen, S. Maslov, and K. Sneppen, Phys. Rev. Lett. **90**, 148701 (2003).
17. E. Ravasz, A. L. Somera, D. A. Mongru, Z. N. Oltvai, and A.-L. Barabási, Science **297**, 1551 (2002).
18. E. Ravasz and A.-L. Barabási, Phys. Rev. E **67**, 026112 (2003).
19. K.-I. Goh, B. Kahng, and D. Kim, Phys. Rev. Lett. **87**, 278701 (2001).
20. M. E. J. Newman, Phys. Rev. E **64**, 016132 (2001).
21. U. Brandes, J. Math. Sociol. **25**, 163 (2001).
22. K.-I. Goh, E. Oh, H. Jeong, B. Kahng, and D. Kim, Proc. Natl. Acad. Sci. USA **99**, 12583 (2002).
23. A. Vázquez, R. Pastor-Satorras, and A. Vespignani, Phys. Rev. E **65**, 066130 (2002).
24. S. Maslov, K. Sneppen, and A. Zaliznyak, e-print (http://arxiv.org/abs/cond-mat/0205379).
25. K.-I. Goh, B. Kahng, and D. Kim, Phys. Rev. Lett. **88**, 108701 (2002).

Modeling TCP/RED: a Dynamical Approach

Hui Zhang, Mingjian Liu, Vladimir Vukadinović, and Ljiljana Trajković

Simon Fraser University, Burnaby, BC, V5A 1S6 Canada {hzhange, jliu1, vladimir, ljilja}@cs.sfu.ca

1 Introduction

Today's Internet applications, such as World Wide Web, file transfer, Usenet news, and remote login, are delivered via Transmission Control Protocol (TCP). With an increasing number and variety of Internet applications, congestion control becomes a key issue. Active Queue Management (AQM) interacts with TCP congestion control mechanisms and plays an important role in meeting today's increasing demand for performance of Internet applications. Random Early Detection (RED), a widely deployed AQM algorithm, is a gateway-based congestion control mechanism. An accurate model of TCP with RED may help understand and predict the dynamical behavior of the network. In addition, the model may help analyze the stability margins of the system and provide design guidelines for selecting network parameters. The design guidelines are important to network designers who aim to improve network robustness. Therefore, modeling TCP with RED is an important step toward improving the service provided to Internet users and the network efficiency.

Modeling TCP performance has gained increased attention during the last few years, due to the benefits that TCP models offer to the networking community. Analytical TCP models enable researchers to closely examine the existing congestion control algorithms, address their shortcomings, and propose methods for improvement. They may also be used to compare various TCP flavors and implementations, and to determine their performance under various operating conditions. Moreover, these models help examine the interactions between TCP and the queuing algorithms implemented in network routers. Hence, they help improve the existing and design better algorithms, such as AQM techniques. Finally, such models offer the possibility of defining TCP-friendly behavior in terms of throughput for non-TCP flows that coexist with TCP connections in the same network.

The goal of TCP modeling is to investigate the nonlinear phenomena in a TCP/RED system. We use an iterative map to model the system. We derive a second-order discrete-time model to capture the interactions of TCP congestion control algorithm with the RED mechanism. We use the concepts proposed in [31] and construct a nonlinear dynamical model of TCP/RED that employs two state variables: the window size and the average queue size. The change of window size reflects the dynamics of TCP congestion control, while the average queue size captures the queue dynamics in RED gateway. The novelty of the proposed model is in capturing detailed dynamical behavior of TCP/RED. The proposed model considers slow start phase and takes into account timeout events common in TCP.

This article is organized as follows: In Section 2, we briefly describe TCP congestion control and the RED algorithm. In Section 3 we survey related work. A nonlinear second-order discrete-time model named S-model is introduced in Section 4. In Section 5, we compare its performance to the ns-2 simulation results and to an existing model. Conclusions are given in Section 6.

2 TCP and RED Algorithms

In this section, we describe TCP congestion control mechanisms and the RED algorithm.

2.1 TCP Congestion Control Algorithms

To adjust the window size, the TCP congestion control mechanism employs four algorithms: slow start, congestion avoidance, fast retransmit, and fast recovery, as shown in Fig. 1. They were introduced by Jacobson [18], [19] and are described in RFCs 1122 [4] and 2581 [1]. We briefly describe here their basic characteristics.

In order to avoid congesting the network with large bursts of data, an established TCP connection first employs the slow start algorithm to detect the available bandwidth in the network. Typically, a TCP sender initializes its congestion window (*cwnd*) to one or two segments, depending on the TCP implementation. Upon receipt of each acknowledgment (ACK) that acknowledges receipt of new data by the receiver, TCP increments *cwnd* by one segment size.

When *cwnd* exceeds a threshold (*ssthresh*), the sender's mechanism leaves the slow start and enters the congestion avoidance phase. During the congestion avoidance, *cwnd* is incremented by one segment size per round trip time (RTT). A timer is set every time a sender sends a packet. A packet loss is detected by the timeout mechanism if the timer expires before the receipt of the packet has been acknowledged. If a packet loss is detected by the timeout mechanism, the TCP sender adjusts its *ssthresh* and switches back to the slow start.

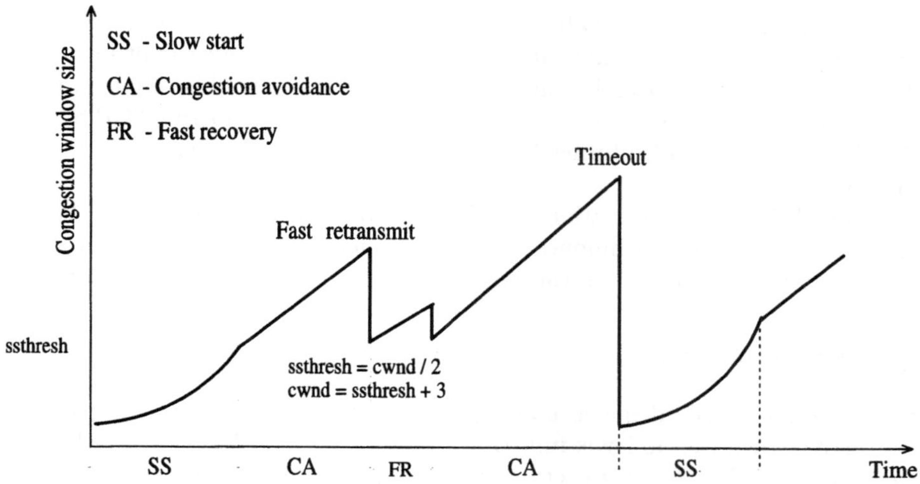

Fig. 1: Evolution of the window size in TCP Reno. It consists of slow start, congestion avoidance, fast retransmit, and fast recovery phase.

The fast retransmit algorithm is used for recovery from losses detected by triple duplicate ACKs. Whenever a TCP receiver receives an out-of-order segment, it immediately sends a duplicate ACK, which informs the sender of the sequence number of the packet that the receiver expects. The receipt of triple duplicate ACKs (four consecutive ACKs acknowledging the same packet) is used as an indication of packet loss. The TCP sender reacts to the packet loss by halving *cwnd* and re-transmitting the lost packet, without waiting for the retransmission timer to expire.

The fast recovery algorithm is used to control data transmission after fast retransmission of the lost packet. During this phase, the TCP sender increases its *cwnd* for each duplicate ACK received. The fast recovery algorithm recognizes each duplicate ACK as an indication that one packet has left the channel and has reached the destination. Since the number of outstanding packets has decreased by one, TCP sender is allowed to increment its *cwnd*. When a non-duplicate ACK is received, TCP switches from the fast recovery to the congestion avoidance phase.

2.2 TCP Implementations

Older TCP implementation, released in the early 1980s, employed a simple window-based congestion control specified in RFC 793 [34]. TCP Tahoe, released in the late 1980s, employed the slow start, congestion avoidance, and fast retransmit algorithms. TCP Reno, introduced in the early 1990s, added the fast recovery algorithm.

Using ns-2 simulations, Fall and Floyd [11] demonstrated that TCP Reno exhibits poor performance in terms of link utilization whenever multiple packets are dropped from a single window of data. To alleviate this problem, they introduced two modifications to TCP Reno: TCP New-Reno and TCP SACK [23]. A large number of Internet Web servers still use TCP Reno and its variants [32].

Other TCP implementations, such as TCP Vegas [5] and TCP Westwood [7], use various techniques to avoid congestion. They adjust the congestion window size based on estimates of the throughput at the bottleneck.

2.3 RED Algorithm

A traditional DropTail queue management mechanism discards the packets that arrive when the buffer is full. However, this method has two drawbacks. First, it may allow few connections to monopolize the queue space so that other flows are starved. Second, DropTail allows queues to be full for a long period of time. During that period, incoming packets are dropped in bursts. This causes severe reduction in throughput of TCP flows. One solution, recommended in RFC 2309 [2], is to deploy active queue management (AQM) algorithms. The purpose of AQM is to react on incipient congestion, before the buffer overflows. Active queue management allows responsive flows, such as TCP flows, to react timely and reduce their sending rates in order to prevent congestion and severe packet losses.

The most popular active queue management algorithm is Random Early Detection (RED), proposed by Floyd and Jacobson [14]. The RED mechanism calculates exponentially weighted moving average of the queue size. Let w_q be the weight factor and q_{k+1} be the current queue size. At every packet arrival, RED gateway updates the average queue size as:

$$\bar{q}_{k+1} = (1 - w_q)\bar{q}_k + w_q \cdot q_{k+1}. \tag{1}$$

The average queue size is compared to two parameters: minimum queue threshold q_{min} and maximum queue threshold q_{max}. If the average queue size is smaller than q_{min}, the packet is admitted to the queue. If it exceeds q_{max}, the packet is marked or dropped. If the average queue size is between q_{min} and q_{max}, the packet is dropped with a drop probability p that is a function of the average queue size:

$$p_{k+1} = \begin{cases} 0 & \text{if } \bar{q}_{k+1} \le q_{min} \\ 1 & \text{if } \bar{q}_{k+1} \ge q_{max} \\ \frac{\bar{q}_{k+1} - q_{min}}{q_{max} - q_{min}} p_{max} & otherwise \end{cases}, \tag{2}$$

where p_{max} is the maximum packet drop probability. The relationship between the drop probability and the average queue size is shown in Fig. 2.

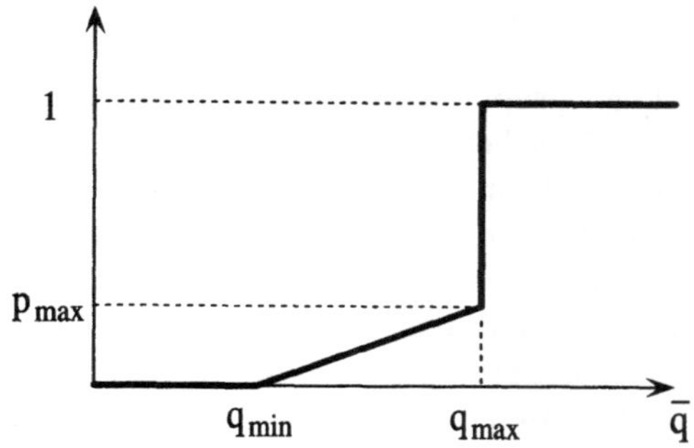

Fig. 2: RED drop probability as a function of the average queue size.

3 Modeling Methodologies

From the viewpoint of flow characteristics, analytical TCP models can be classified in three categories based on the duration of the TCP flows, which determines the dominant TCP congestion control algorithms to be modeled, and the aspects of TCP performance that can be captured by the model [20]. The first category models short-lived flows, where TCP performance is strongly affected by the connection establishment and slow start phases [27]. These models typically approximate average latency or completion time, i.e., the time it takes to transfer a certain amount of data. The second category models long-lived flows that characterize the steady-state performance of bulk TCP transfers during the congestion avoidance phase [26], [29], [31], [37]. These models approximate aspects such as the average throughput and window size evolution. The final category includes models for flows of arbitrary duration, i.e., those that can accommodate both short and long-lived flows [6], [8].

From the control theoretic point of view, the developed models of TCP and TCP/RED [16], [17], [21], [22], [28], [35], [36] can be classified into two types: averaged models and iterative map models. An averaged model is described by a set of continuous differential equations. It neglects the detailed dynamics and only captures the "low frequency characteristics" of the system. It can be used to analyze the steady-state performance and to predict low frequency slow-scale bifurcation behavior, such as Hopf bifurcations. Examples of such models are given in [16], [17], [21]. In contrast, an iterative map model has a discrete-time form and employs a set of difference equations. It provides relatively complete dynamical information. Iterative maps are adequate to explore nonlinear phenomena, such as period-doubling and saddle-node bifurcations, which may appear across a wide spectrum of frequencies and cause the existence of solutions in the high frequency range. Examples of iterative maps are given in [22], [35], [36].

3.1 Survey of Related TCP/RED Models

Several models have been proposed recently in order to analyze and understand the performance of packet networks. A simple steady-state model of TCP Reno, introduced in [31], models the steady-state sending rate as a function of the loss rate and the round trip time (RTT) of a bulk data transfer TCP flow. The model not only captures the essence of TCP's fast retransmit and congestion avoidance, but it also takes into account the effect of timeout mechanism, which is important from a modeling perspective. Measurements demonstrated that the proposed model was adequate over a range of loss rates.

A simplified first-order discrete-time nonlinear dynamical model was developed for TCP with RED control in [22], [35], [36]. An exponentially weighted average queue size has been used as the state variable. The model describes the system dynamics over large parameter variations, and employs sampling the buffer occupancy at certain time instances. This dynamical model was used to investigate the stability, bifurcation, and routes to chaos in a network for various system parameters. Based on the developed model, the authors demonstrated that nonlinear phenomena, such as bifurcation and chaos, might occur if the system parameters were not properly selected. However, this discrete-time model neglects the dynamics of TCP. The derived map is:

$$q_{k+1}^{ave} = \begin{cases} (1-w)q_k^{ave} & \text{if } q_k^{ave} \geq q_u^{ave} \\ (1-w)q_k^{ave} + w \cdot B & \text{if } q_k^{ave} \leq q_l^{ave} \\ (1-w)q_k^{ave} + w(\frac{N \cdot K}{\sqrt{p_k}} - \frac{C \cdot d}{M}) & \text{otherwise} \end{cases}, \quad (3)$$

where

$q_{k+1}^{ave} \doteq$ average queue size in round $k+1$
$q_k^{ave} \doteq$ average queue size in round k
$q_u^{ave} \doteq$ upper bound of average queue size
$q_l^{ave} \doteq$ lower bound of average queue size
$p_k \doteq$ drop probability at round k
$w \doteq$ queue weight in RED algorithm
$B \doteq$ buffer size
$N \doteq$ number of TCP connections
$K \doteq$ constant $[1, \sqrt{8/3}]$
$C \doteq$ link capacity
$d \doteq$ round trip propagation delay
$M \doteq$ packet size.

In [28], a second-order nonlinear dynamical model was developed to examine the interactions of a set of TCP flows with RED routers. The model employs fluid-flow and stochastic differential equations. Window size and average queue length are used as state variables. From a set of coupled ordinary differential equations, the authors develop a numerical algorithm to obtain the transient behavior of average queue length, round trip time, and throughput of TCP flows. This model is described by nonlinear differential equations that employ the average values of network variables:

$$\begin{aligned} \dot{W}(t) &= \frac{1}{R(t)} - \frac{W(t)W(t-R(t))}{2R(t-R(t))} p(t-R(t)) \\ \dot{q}(t) &= \frac{N(t)}{R(t)} W(t) - C \end{aligned} \quad , \tag{4}$$

where

$W(t) \doteq$ expectation of TCP window size
$q(t) \doteq$ expectation of queue length
$R(t) \doteq$ round trip time
$N(t) \doteq$ load factor (number of TCP sessions)
$p(t) \doteq$ probability of packet mark/drop
$C \doteq$ link capacity.

A third-order dynamical model that describes the interaction of TCP flows with an RED-controlled queue was developed in [21]. The state variables of the model are average queue length, instantaneous queue length, and throughput. TCP sources are idealized to operate only in congestion avoidance phase where congestion window follows the rule of linear increase and multiplicative decrease. This dynamical model is used to explore various parameter settings and observe transient and equilibrium behavior of the system. The validity of the model is verified by comparison with simulation results. The interaction between TCP and RED is modeled as:

$$\begin{aligned} &\tfrac{d}{dt}\bar{s}(t) = \bar{\lambda}(t-R/2)\beta(\bar{q}(t)-\bar{s}(t)) \\ &\tfrac{d}{dt}\bar{q}(t) = \bar{\lambda}(t-R/2)(1-\pi_K(\bar{q}(t)))(1-p(\bar{s}(t))) - \mu(1-\pi_0(\bar{q}(t))) \\ &\tfrac{d}{dt}\bar{\lambda}(t) = -\frac{P_L(t-R/2)}{2m}\bar{\lambda}(t)\bar{\lambda}(t-R) + (1-P_L(t-R/2))\frac{m}{R^2}\frac{\bar{\lambda}(t-R)}{\bar{\lambda}(t)} \\ &P_L(t) = p(\bar{s}(t)) + \pi_K(\bar{q}(t)) - p(\bar{s}(t))\pi_K(\bar{q}(t)) \end{aligned} \quad , \tag{5}$$

where

$\bar{s}(t) \doteq$ expectation of the exponentially averaged queue length
$\bar{q}(t) \doteq$ expectation instantaneous queue length
$\bar{\lambda}(t) \doteq$ expectation of TCP sending rate
$P_L(t) \doteq$ loss probability in the queue at time t
$\pi_K(\bar{q}(t)) \doteq$ steady-state probability for the queue to be full
$\pi_0(\bar{q}(t)) \doteq$ steady-state probability for the queue to be empty
$p \doteq$ drop probability
$R \doteq$ round trip time
$\beta \doteq$ queue weight in RED algorithm
$m \doteq$ number of identical TCP sources.

In [16], [17], the authors obtained a second-order linear model for TCP and RED by linearizing a fluid-based nonlinear TCP model. Window size and average queue length are used as state variables of the system. The authors performed analysis of TCP interactions with RED from a control theoretic viewpoint. They presented design guidelines for choosing RED parameters that lead to local stability of the system. In addition, they proposed two alternative controllers to improve the transient behavior and stability of the system: a proportional (P) controller that possesses good transient response but suffers from steady-state errors in queue regulation, and a proportional-integral (PI) controller that exhibits zero steady-state regulation error and acceptable transient behavior. An important contribution of this paper is a good example how to use classical control theory to solve problems in complex communication systems. The model linearized around the operating point is described as:

$$\begin{aligned}\delta\dot{W}(t) &= -\tfrac{N}{R_0^2\cdot C}(\delta W(t)+\delta W(t-R_0)) - \tfrac{1}{R_0^2\cdot C}(\delta q(t)-\delta q(t-R_0)) \\ &\quad -\tfrac{R_0\cdot C^2}{2N^2}\delta p(t-C) \\ \delta\dot{q}(t) &= \tfrac{N}{R_0}\delta W(t) - \tfrac{1}{R_0}\delta q(t)\end{aligned} \quad , \qquad (6)$$

where

$\delta \mathrm{W} \doteq W - W_0$
$\delta \mathrm{q} \doteq q - q_0$
$\delta \mathrm{p} \doteq p - p_0$
$W_0,\ q_0,\ p_0 \doteq$ the set of operating points
$W \doteq$ expectation of TCP window size
$q \doteq$ expectation of queue length
$R_0 \doteq$ round trip time
$C \doteq$ link capacity
$T_p \doteq$ propagation delay
$\mathrm{N} \doteq$ load factor (number of TCP sessions)
$p \doteq$ probability of packet mark/drop.

A multi-link multi-source model [25] was used to study the stability of a general TCP/AQM system. A local stability condition were derived for the case of a single link with heterogeneous sources and the stability region of TCP/RED. The state variables of this model are window size, instantaneous queue length, and average queue length. large link capacities. Finally, they devised a new distributed congestion control algorithm that maintains local stability for arbitrary delay, capacity, and traffic load. They provided preliminary simulation results to illustrate the model's behavior.

4 Discrete-time Dynamical Model of TCP/RED

The basic idea behind RED is to sense impending congestion before it occurs and to try to provide feedback to senders by either dropping or marking packets. Hence, from the control theoretic point of view, the network may be considered as a complex feedback control system. TCP adjusts its sending rate depending on whether or not it has detected a packet drop in the previous RTT interval. The drop probability of RED can be considered as a control law of the network system. Its discontinuity is the main reason for oscillations and chaos in the system. Hence, it is natural to model the network system as a discrete-time model. In this article, we model the TCP/RED system using a "stroboscopic map", which is the most widely used type of discrete-time maps for modeling power converters [3], [9], [10], [39]. This map is obtained by periodically sampling the system state. In our study, the sampling period is one RTT. Since window size and queue size behave as step functions of RTT, one RTT is the sampling period that captures their changes [13]. Higher sampling rate would not significantly improve the accuracy of the model. On the other hand, lower sampling rate would ignore the changes and affect the model accuracy.

State variables employed in the proposed S-model are window size and average queue size. These state variables capture the detailed behavior of TCP/RED. Variations of the window size reflect the dynamics of TCP congestion control. The window size increases exponentially and linearly in slow start and congestion avoidance phases, respectively. It multiplicatively decreases when loss occurs. The average queue size captures the queue dynamics in RED because it is updated upon every packet arrival. In our study, we do not consider instantaneous queue size as an independent state variable.

4.1 TCP/RED Model

We consider a simple network shown in Fig. 3. It consists of one TCP source, two routers, and a destination. RED is employed in Router 1. A TCP Reno connection is established between the source and the destination. Data packets are sent from the source to the destination, while traffic in the opposite direction consists of ACK packets only.

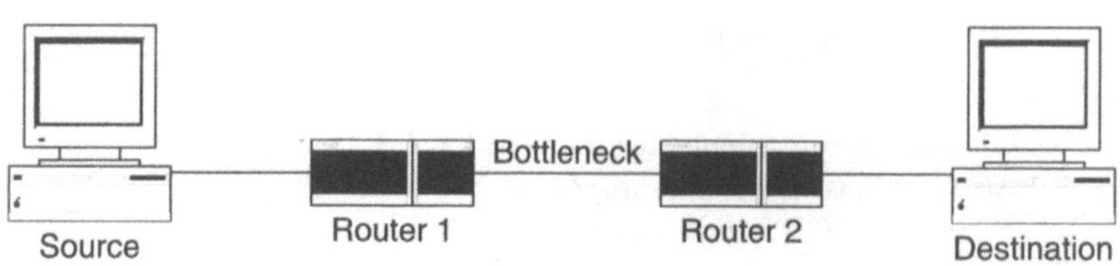

Fig. 3: Topology of the modeled and simulated network.

We made several assumptions in order to construct an approximate model. We assume that ACK packets are never lost. The connection is long-lived and the source always has sufficient data to send. Round trip propagation delay d between the source and destination and the data packet size M are kept constant. The link that connects the two routers is the only bottleneck in the network. We also assume that the timeout is caused only by packet loss and that the duration of the timeout period is 5 RTTs [13]. The state variables of the system are sampled at the end of every RTT period. We assume that the queue size is constant during each sampling period. The model includes three cases, depending on the number of packets lost in the previous RTT period: no loss, single loss, and multiple packet losses.

4.2 Case 1: No Loss

Let W_k, q_k, and $\bar{q}$ be the window size, queue size, and average queue size at the end of the sampling period k. If no packet is dropped during the last RTT period, TCP Reno increases its window size. The window size is increased exponentially in the slow start phase and linearly in the congestion avoidance phase:

$$W_{k+1} = \begin{cases} \min(2W_k, ssthresh) & \text{if } W_k < ssthresh \\ \min(W_k + 1, rwnd) & \text{if } W_k \geq ssthresh \end{cases}, \tag{7}$$

where *rwnd* is the receiver's advertised window size, i.e., the largest window size that the receiver could accept in one round. Usually, *rwnd* is greater than window size. In this case, *rwnd* does not affect the variations of window size. In case when window size increases linearly and reaches the value *rwnd*, window size is kept at *rwnd* until loss occurs in the network.

In order to calculate the average queue size given by Eq. (1), we need to find the queue size at the sampling period $k+1$. This queue size depends on the queue size in the previous period, the current window size, and the number of packets that have left the queue during the previous sampling period. Therefore, the current queue size is:

$$\begin{aligned} q_{k+1} &= q_k + W_{k+1} - \tfrac{C \cdot RTT_{k+1}}{M} \\ q_{k+1} &= q_k + W_{k+1} - \tfrac{C}{M}(d + \tfrac{q_k \cdot M}{C}), \\ q_{k+1} &= W_{k+1} - \tfrac{C \cdot d}{M} \end{aligned} \tag{8}$$

where

$q_{k+1} \doteq$ instantaneous queue size in round $k+1$
$q_k \doteq$ instantaneous queue size in round k
$W_{k+1} \doteq$ current TCP window size in round $k+1$
$RTT_{k+1} \doteq$ round trip time in round $k+1$
$C \doteq$ link capacity
$M \doteq$ packet size
$d \doteq$ round trip propagation delay.

Substituting q_{k+1} in (1) gives the average queue size:

$$\bar{q}_{k+1} = (1 - w_q)\bar{q}_k + w_q \cdot \max(W_{k+1} - \frac{C \cdot d}{M}, 0). \tag{9}$$

RED updates the average queue size at every packet arrival. Hence, $\bar{q}$ is updated W_{k+1} times during the current sampling period. We assume that the queue size is constant during each period and that $\bar{q}$ is given as:

$$\bar{q}_{k+1} = (1 - w_q)^{W_{k+1}} \bar{q}_k + (1 - (1 - w_q)^{W_{k+1}}) \cdot \max(W_{k+1} - \frac{C \cdot d}{M}, 0). \quad (10)$$

Finally, if $p_k W_k < 0.5$, which implies that no packet loss occurred in the previous sampling period, the state variables of the model are:

$$\begin{aligned} W_{k+1} &= \begin{cases} \min(2W_k, ssthresh) & \text{if } W_k < ssthresh \\ \min(W_k + 1, rwnd) & \text{if } W_k \geq ssthresh \end{cases} \\ \bar{q}_{k+1} &= (1 - w_q)^{W_{k+1}} \bar{q}_k + (1 - (1 - w_q)^{W_{k+1}}) \cdot \max(W_{k+1} - \tfrac{C \cdot d}{M}, 0). \end{aligned} \quad (11)$$

4.3 Case 2: One Packet Loss

If $0.5 \leq p_k W_k < 1.5$, which implies that one packet loss occurred in the previous RTT period, the congestion control mechanism of TCP Reno halves the window size in the current sampling period:

$$W_{k+1} = \frac{1}{2} W_k. \quad (12)$$

The average queue size is updated in a manner similar to Case 1:

$$\begin{aligned} W_{k+1} &= \tfrac{1}{2} W_k \\ \bar{q}_{k+1} &= (1 - w_q)^{W_{k+1}} \bar{q}_k + (1 - (1 - w_q)^{W_{k+1}}) \cdot \max(W_{k+1} - \tfrac{C \cdot d}{M}, 0) \end{aligned} \quad (13)$$

4.4 Case 3: At Least Two Packet Losses

In this case $p_k W_k \geq 1.5$, which implies that at least two packets are lost in the previous RTT period. When multiple packets are lost from the same window, TCP Reno may not be able to send a sufficient number of new packets in order to receive three duplicate ACKs for each packet lost. TCP source will often have to wait for the timeout before retransmitting the lost packet [11]. During the timeout period, the source does not send packets into the network. In S-model, window size is equivalent to the number of packets that are sent by the source during one RTT period. Hence, we assume that the window size is zero during the timeout period.

RED mechanism updates the average queue size for each packet arrival. However, during timeout period there are no packet arrivals. Average queue size is not updated and has the same value as in the previous RTT period. RED takes this "idle time" period into account when it updates the average queue size upon the next packet arrival. However, S-model does not take into account the "idle time". TCP/RED system during the timeout period is modeled as:

$$\begin{aligned} W_{k+1} &= 0 \\ \bar{q}_{k+1} &= \bar{q}_k \end{aligned} . \tag{14}$$

The pseudo code describing the S-model is shown in Algorithm 1.

Algorithm 1 *S-model*

```
Initialization:
q̄_0 ← 0
q_0 ← 0
p_0 ← 0
for every round
calculate the product of p_k W_k
if p_k W_k < 0.5 then
   compare W_k with ssthresh
   if W_k < ssthresh then
      W_{k+1} ← min(2W_k, ssthresh)
      q̄_{k+1} ← (1 − w_q)^{W_{k+1}} q̄_k + (1 − (1 − w_q)^{W_{k+1}}) max(W_{k+1} − Cd/M, 0)
   else
      W_{k+1} ← min(W_k + 1, rwnd)
      q̄_{k+1} ← (1 − w_q)^{W_{k+1}} q̄_k + (1 − (1 − w_q)^{W_{k+1}}) max(W_{k+1} − Cd/M, 0)
   end if
   calculate the drop probability using Eq. (2)
else if 0.5 ≤ p_k W_k < 1.5 then
   W_{k+1} ← (1/2) W_k
   q̄_{k+1} ← (1 − w_q)^{W_{k+1}} q̄_k + (1 − (1 − w_q)^{W_{k+1}}) max(W_{k+1} − Cd/M, 0)
   calculate the drop probability using Eq. (2)
else
   W_{k+1} ← 0
   q̄_{k+1} ← q̄_k
end if
end
```

4.5 Properties of the S-model

The proposed second-order discrete model captures interactions between TCP Reno congestion control algorithm and RED mechanism. It models the dynamics of the TCP/RED system. Unlike past models [21], [22], [31], [35], [36], the S-model includes the slow start phase. It also takes into account timeout, common in TCP [31], which models [21], [22], [35], [36] ignore. However, the S-model does not capture all details of the fast recovery phase. Congestion window size in the S-model is not increased for each duplicate ACK received after retransmitting the lost packet. Instead of "inflating" the window we assume that TCP sender switches to the congestion avoidance phase without performing slow start. Evolution of the window size in the S-model is shown in Fig. 4.

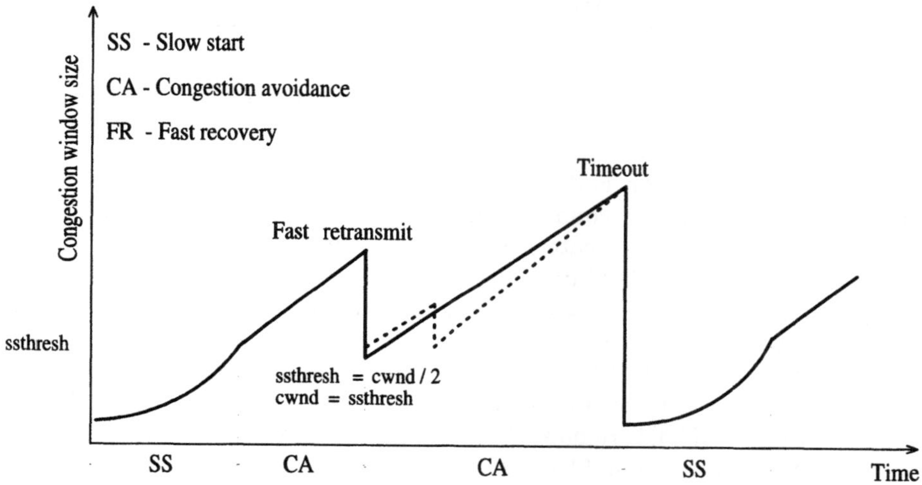

Fig. 4: Evolution of the window size in the proposed model. The fast recovery phase has been simplified.

The S-model captures the most important characteristics of the RED algorithm. The average queue size is updated after each packet arrival, i.e., W_{k+1} times in the sampling period k+1. In contrast, models presented in [22], [35], [36] update the average queue size only once in every RTT period. However, in deriving the new model, we have also made simplifications in the RED algorithm: we ignored the counter that counts the number of packet arrivals since the last packet drop. RED uses this counter to modify drop probability (2). We have also ignored the "idle time" period, since it has no significant impact on the dynamics of the system.

5 S-Model Validation and Comparison

In order to verify the accuracy of the S-model , we compare its performance with the ns-2 simulation results. The topology of the simulated network is shown in Fig. 5. It consists of one source, one sink, and two routers: R1 and R2. RED is employed in router R1. The link between R1 and R2 is the only bottleneck in the network. Its capacity is 1.54 Mbps and propagation delay is 10 ms. The capacity of the links between the source and R1 and between

R2 and the sink is 100 Mbps. This is sufficient to guarantee no congestion in these links. Their propagation delay is 0 ms. We compared window size and average queue size in the S-model and the ns-2 simulation results.

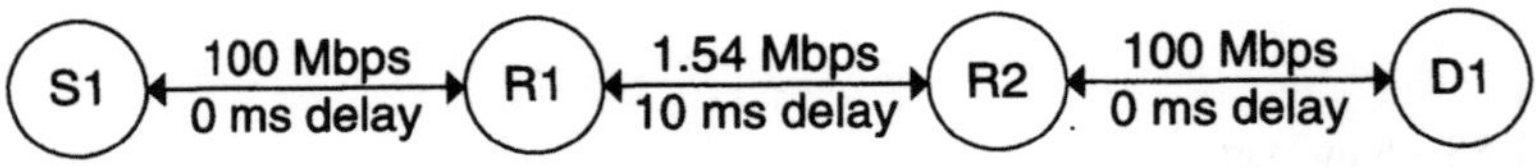

Fig. 5: Topology of the simulated network.

The model validation is divided into four stages. First, we used default ns-2 RED parameters. Second, we choose various queue weights w_q while keeping other system parameters constant. In the third scenario, we varied the maximum drop probability p_{max}. Finally, we varied the minimum and maximum queue thresholds q_{min} and q_{max} simultaneously, while keeping their ratio q_{max}/q_{min}= 3. In each simulation scenario, we observed system behavior and measured average RTT, sending rate, and drop probability.

We also compared the proposed S-model with a discrete-time nonlinear dynamical model of TCP Reno with RED gateway proposed in [22], [35], [36], named here M-model. The M-model is a first-order discrete-time dynamical model with the average queue size as the state variable.

The default RED parameters are shown in Table 1. Other system parameters for the proposed S-model and the M-model are shown in Table 2.

Table 1: Default ns-2 RED parameters.

Packet size M (bytes)	500
Maximum drop probability (p_{max})	0.1
Minimum queue threshold (q_{min}) (packets)	5
Maximum queue threshold (q_{max}) (packets)	15
Queue weight (w_q)	0.002

5.1 Default RED Parameters

In order to verify that the S-model can capture the detailed information of the system behavior, we evaluated the waveforms of the two state variables: window size and average queue size.

The waveforms of the window size for various time scales are shown in Fig. 6. The S-model and ns-2 simulation results are quite similar, especially during the steady-state response.

The waveforms of the average queue size for various time scales are shown in Fig. 7. The average queue size using the S-model is approximately one

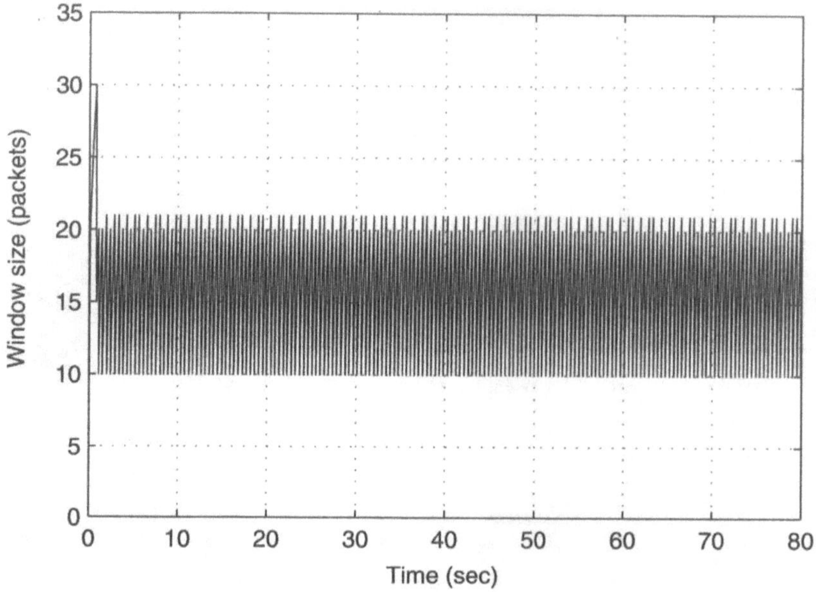

(a)

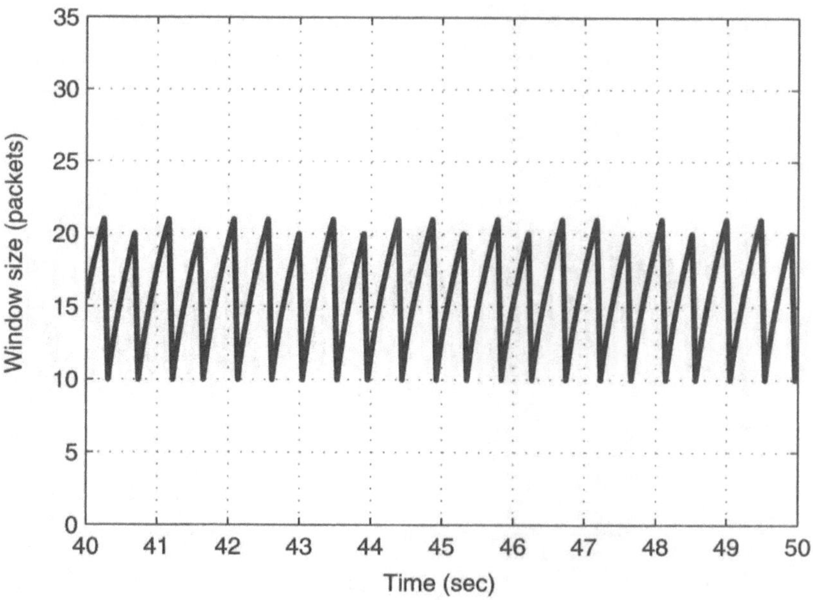

(b)

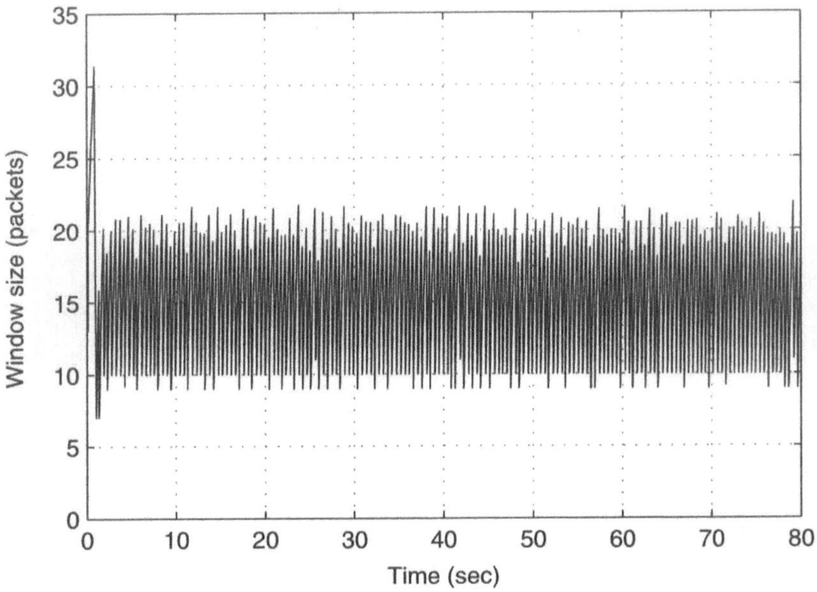

(c)

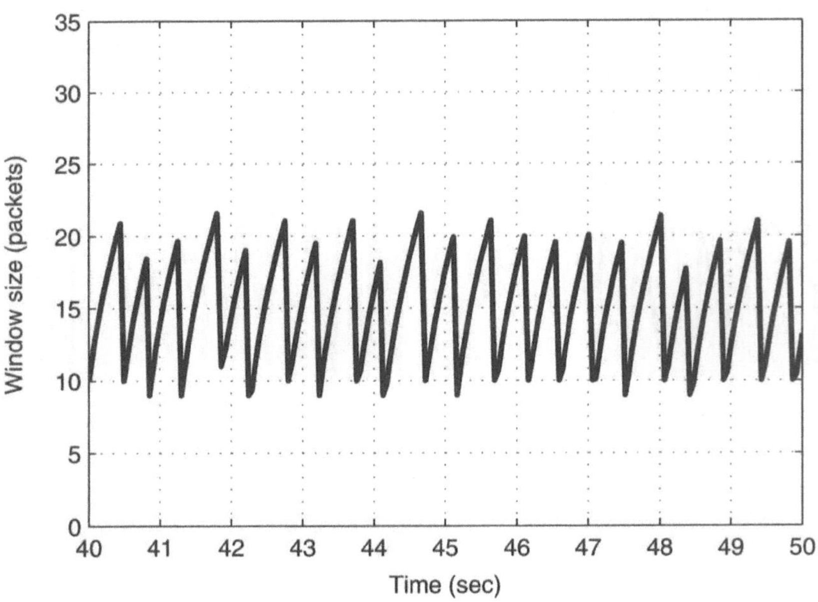

(d)

Fig. 6: Evolution of the window size with default RED parameters: (a) S-model and (b) zoom-in, (c) ns-2 simulation results and (d) zoom-in. Waveforms show good match between the S-model and the ns-2 simulation results.

Table 2: System parameters.

	S-model	M-model
Link capacity C (bit/s)	1.54e+6	1.54e+6
Packet size M (bytes)	500	500
Round trip propagation delay d (ms)	22.8	22.8
Buffer size B (packets)	-	100
Slow start threshold size ssthresh (packets)	20	-
Number of TCP connection N	1	1
Constant K	-	$\sqrt{3/2}$

packet size larger than the average queue size obtained by the ns-2 simulations. This difference is due to introduced simplifications. The new model employs packet-marking/drop probability calculated by Eq. (2), while RED algorithm adopts a smooth packet-drop probability p_a that increases slowly as the *count* increases:

$$p_a \leftarrow p_b/(1 - count \cdot p_b), \tag{15}$$

where p_b is the drop probability given by Eq. (2) and *count* measures the number of packets that have arrived since the last dropped packet. In the S-model, p_b is used as the final drop probability and the counter is ignored. Since $p_b < p_a$, the average queue size of the S-model is larger than that obtained via ns-2 simulations.

The statistics for the two state variables (window size and average queue size) and comparison with ns-2 simulation results are shown in Table 3.

The M-model matches closely the ns-2 results shown in Figs. 7(c) and (d). The time waveform of the average queue size for the M-model is shown in Fig. 8. After the transient state, the average queue size reaches a constant value (5.71 packets), while the average queue size in ns-2 simulations varies around its fixed point. To the contrary, the proposed S-model captures the dynamical characteristic of the average queue size.

Table 3: State variables: S-model and ns-2 simulation results.

	Window size (packets)			Average queue size (packets)		
	S-model	ns-2	Δ (%)	S-model	ns-2	Δ (%)
Average	15.35	15.01	2.25	7.24	5.95	21.63
Max	30	31.38	-1.20	7.69	8.14	-5.60
Min	2	2	0	0.12	0.05	77.45
Max (steady-state)	21	21.86	-3.93	7.69	6.51	18.16
Min (steady-state)	10	9	11.11	7.02	5.41	29.79

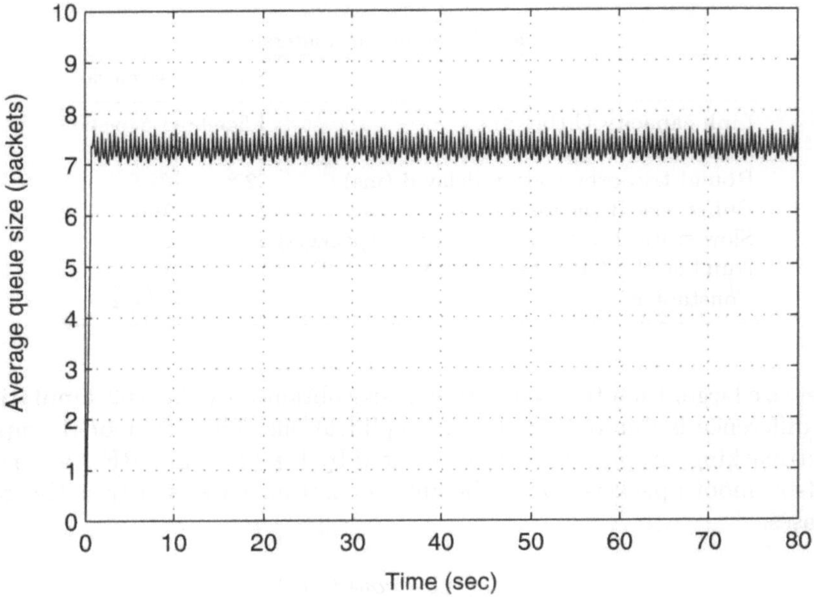
Average queue size (packets)
Time (sec)
0
1
2
3
4
5
6
7
8
9
10
0
10
20
30
40
50
60
70
80

(a)

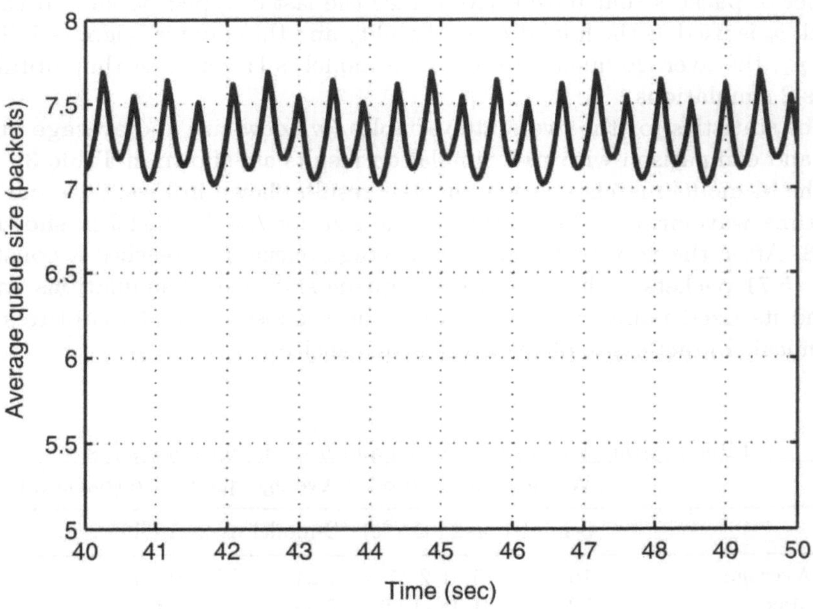
Average queue size (packets)
Time (sec)
5
5.5
6
6.5
7
7.5
8
40
41
42
43
44
45
46
47
48
49
50

(b)

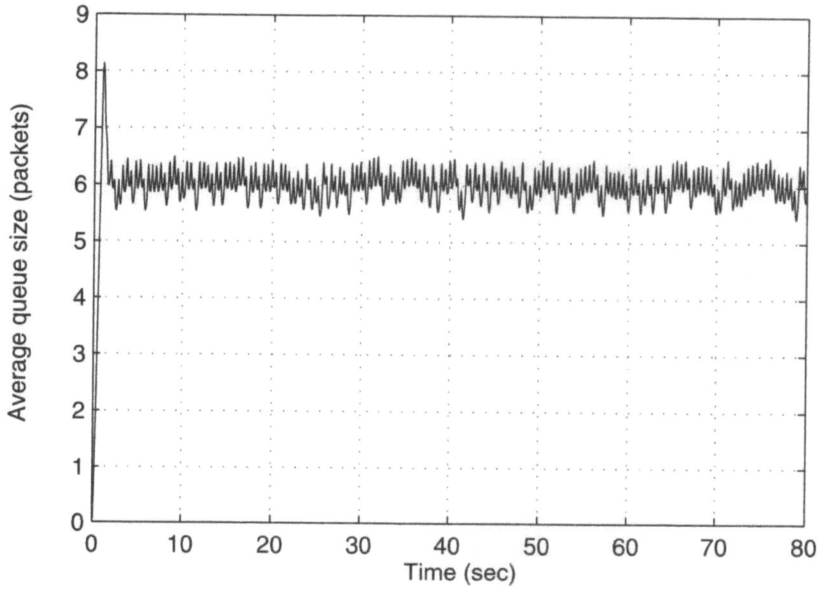

(c)

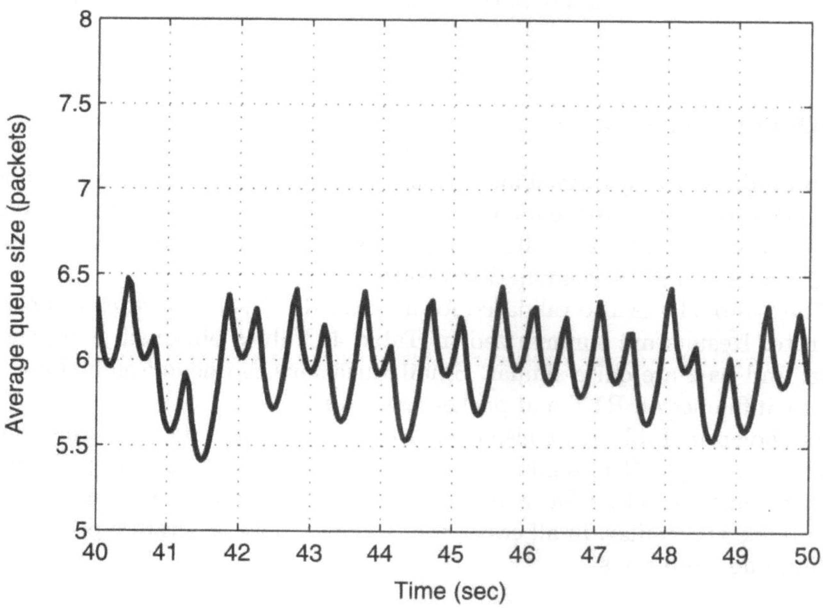

(d)

Fig. 7: Evolution of the average queue size with default RED parameters: (a) S-model and (b) zoom-in, (c) ns-2 simulation result and (d) zoom-in. The average queue size obtained using the S-model is higher than the average queue size obtained using ns-2 simulations.

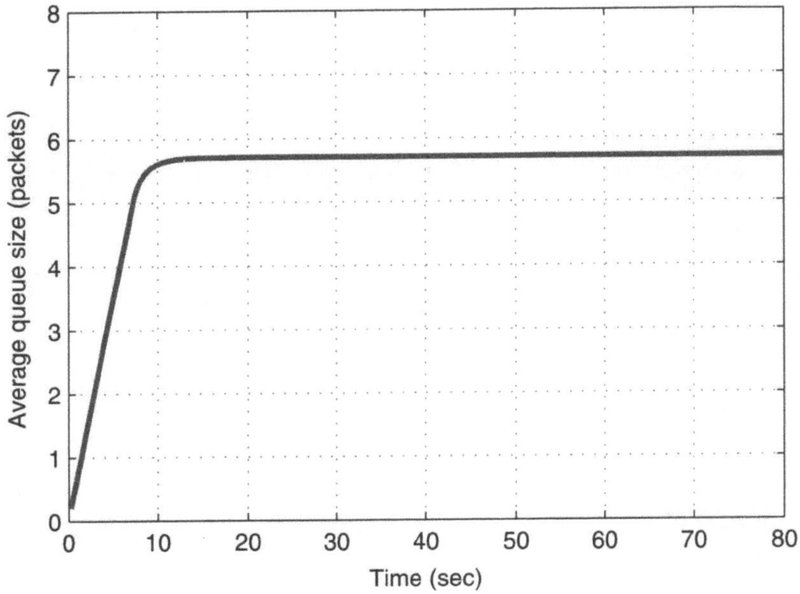

Fig. 8: Average queue size in the M-model with default RED parameters.

5.2 Queue Weight w_q

The S-model was also verified for various queue weight w_q. The average queue size during the steady-state for various values of w_q is shown in Fig. 9. When the queue weight increases, the average queue size slightly decreases in both the S-model and in ns-2 simulation results.

The S-model was also validated for average RTT, sending rate, and packet loss rate. Results are summarized in Table 4. Values obtained using the S-model and ns-2 are quite similar. Small variations in queue weight have significant influence on RTT and packet loss rate.

As shown in Table 4, we also evaluated the M-model for the same system variables: average RTT, sending rate, and packet drop rate. Except for $w_q = 0.001$, when S-model performs better, the discrepancy in predicting RTT for two models are similar. In all cases, the S-model more accurately predicts the sending and drop rates.

5.3 Drop Probability p_{max}

We also evaluated the S-model for various p_{max}. When the maximum drop probability is set to a very small value, RED algorithm has small influence on the system behavior. In this case, the system behaves as TCP with DropTail,

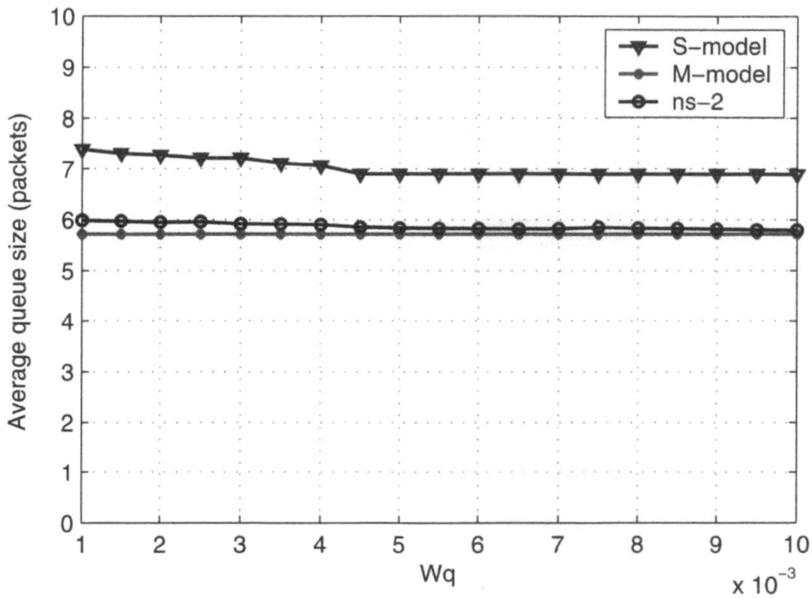

Fig. 9: Comparison of the average queue size for various w_q.

which leads to bursty packet losses and longer queuing delays. However, if the value of p_{max} is close to one, high drop rate will cause the system to become unstable. Simulation results are shown in Fig. 9. The average queue size decreases as the maximum drop rate increases. Results obtained from the S-model and ns-2 simulation results show the same trend.

Validation results for the average RTT, sending rate, and drop rate are listed in Table 5. They show that system variables in S-model and in ns-2 simulations change in a similar manner. As expected, when the maximum drop probability increases, the actual drop rate increases. At the same time, the average RTT decreases indicating a lower queuing delay.

The average RTT, sending rate, and drop rate for the M-model are also summarized in Table 5. Under various drop probabilities p_{max}, the S-model better estimates the average RTT, the sending rate, and the drop rate.

5.4 Thresholds q_{min} and q_{max}

The S-model is also evaluated for various queue thresholds. Values of q_{min} and q_{max} are varied simultaneously, while maintaining the ratio q_{max}/q_{min}=3, as recommended in [2]. Simulation results for the average queue size are shown in Fig. 9.

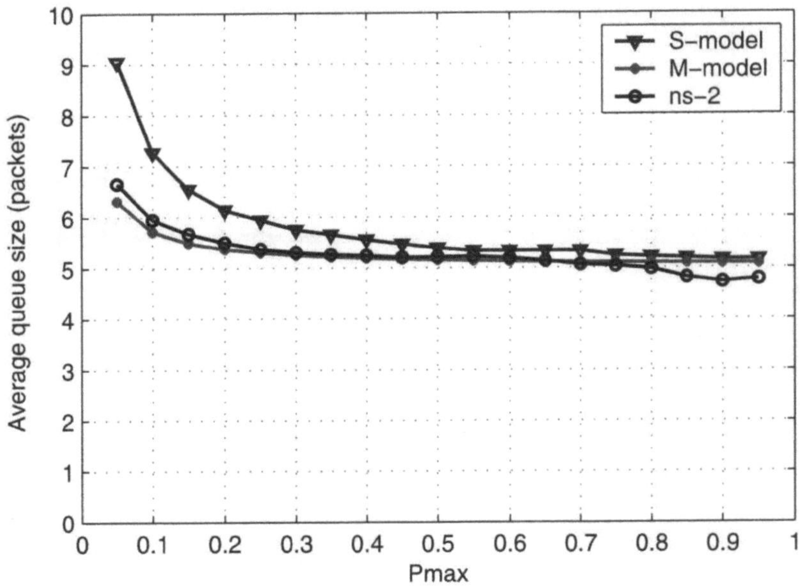

Fig. 10: Comparison of the average queue size for various p_{max}.

The average queue size increases when thresholds q_{max} and q_{min} increase. At the same time, the average RTT increases and the drop rate decreases, as shown in Table 6.

A comparison with the M-model suggests that the proposed S-model is more accurate in predicting average RTT, sending rate, and drop rate.

5.5 Validation Summary

The S-model was evaluated by comparing the waveforms of its two state variables (window size and average queue size) to the ns-2 simulation results. While the window sizes match well, the steady-state values of the average queue size differ. Nevertheless, the average queue size of the S-model and ns-2 results have similar trend as the system parameters w_q, p_{max}, q_{min}, and q_{max} vary.

The difference in average queue size between the S-model and ns-2 is due to simplifications to the RED's packet discarding algorithm: S-model employs probability p_b (Eq. 2) as the final drop probability, while RED in ns-2 uses p_a (Eq. 15). If a modified drop probability $p_a = \alpha p_b$ is used, the window size and the average queue size would evolve as shown in Figs. 12(a) and (b), respectively. Comparison shows that the average queue size matches well the ns-2 simulation results for modified drop probabilities with $\alpha = 1.8$.

Table 4: System variables for various w_q.

Parameters	Average RTT (ms)				
weight (w_q)	S-model	Δ (%)	M-model	Δ (%)	ns-2
0.001	40.3	11.63	45.8	26.87	36.1
0.002	39.9	10.83	41.3	14.72	36.0
0.004	39.4	8.80	39.4	8.84	36.2
0.006	39.0	8.93	38.8	8.38	35.8
0.008	39.0	8.90	38.5	7.54	35.8
0.01	38.9	8.96	38.3	7.28	35.7
	Sending rate (packets/s)				
weight (w_q)	S-model	Δ (%)	M-model	Δ (%)	ns-2
0.001	384.99	0.07	325.89	-15.29	384.71
0.002	384.98	0.06	355.26	-7.67	384.77
0.004	385.11	0.08	370.04	-3.83	384.79
0.006	385.08	0.09	374.90	-2.55	384.73
0.008	385.10	0.11	377.25	-1.93	384.68
0.01	385.02	0.08	378.37	-1.65	384.70
	Drop rate (%)				
weight (w_q)	S-model	Δ (%)	M-model	Δ (%)	ns-2
0.001	0.55	1.29	0.67	23.39	0.54
0.002	0.56	2.56	0.70	28.21	0.55
0.004	0.59	6.12	0.71	27.70	0.56
0.006	0.60	7.91	0.71	27.70	0.56
0.008	0.61	11.11	0.71	29.33	0.55
0.01	0.61	11.72	0.71	30.04	0.55

6 Conclusions

TCP/RED system can be viewed as a complex feedback control system where TCP adjusts its sending rate depending on the packet loss probability determined by RED. In this article, we have introduced a second-order discrete model for interaction between TCP Reno and RED algorithms. We used an iterative map to construct the discrete-time model of the system. The S-model captures the dynamical behavior of TCP/RED and may be used to study its nonlinear behavior. Unlike other models, it takes into account the TCP slow start and timeout events. We evaluated the model by comparing its performance to the ns-2 simulation results and an existing TCP/RED model. Validation of the proposed model illustrates the performance of the model for various RED parameters.

Table 5: System variables for various p_{max}.

Parameters	Average RTT (ms)				
p_{max}	S-model	Δ (%)	M-model	Δ (%)	ns-2
0.05	44.3	16.27	43.4	13.91	38.1
0.1	39.9	10.83	41.3	14.72	36.0
0.25	36.5	5.80	39.9	15.65	34.5
0.5	35.3	3.80	39.4	15.88	34.0
0.75	34.8	-0.85	39.2	11.68	35.1
	Sending rate (packets/s)				
p_{max}	S-model	Δ (%)	M-model	Δ (%)	ns-2
0.05	385.13	0.11	354.23	-7.92	384.70
0.1	384.98	0.06	355.26	-7.67	384.77
0.25	384.93	0.05	356.02	-7.46	384.73
0.5	384.98	1.48	356.27	-6.09	379.37
0.75	384.63	7.60	356.33	-0.34	357.55
	Drop rate (%)				
p_{max}	S-model	Δ (%)	M-model	Δ (%)	ns-2
0.05	0.45	-11.76	0.63	23.53	0.51
0.1	0.56	2.56	0.70	28.21	0.55
0.25	0.65	11.28	0.74	26.50	0.59
0.5	0.73	19.09	0.76	23.98	0.61
0.75	0.74	14.37	0.77	19.01	0.65

7 Acknowledgment

The authors thank I. Khalifa, W. G. Zeng, N. Cackov, B. Vujičič, S. Vujičić, Q. Shao, and J. Chen for comments, discussions, and suggestions that helped improve the content of the article.

References

1. M. Allman, V. Paxson, and W. Steven, "TCP Congestion Control," *Request for Comment (RFC) 2581*, Apr. 1999.
2. B. Barden et al., "Recommendations on queue management and congestion avoidance in the Internet," *Request for Comments (RFC) 2309*, Apr. 1998.
3. M. di Bernardo and C. K. Tse, "Chaos in power electronics: an overview," *Chaos in Circuits and Systems*, New York: World Scientific, pp. 317–340, 2002.
4. R. Braden, "Requirements for Internet hosts–communication layers," *Request for Comment (RFC) 1122*, Oct. 1989.

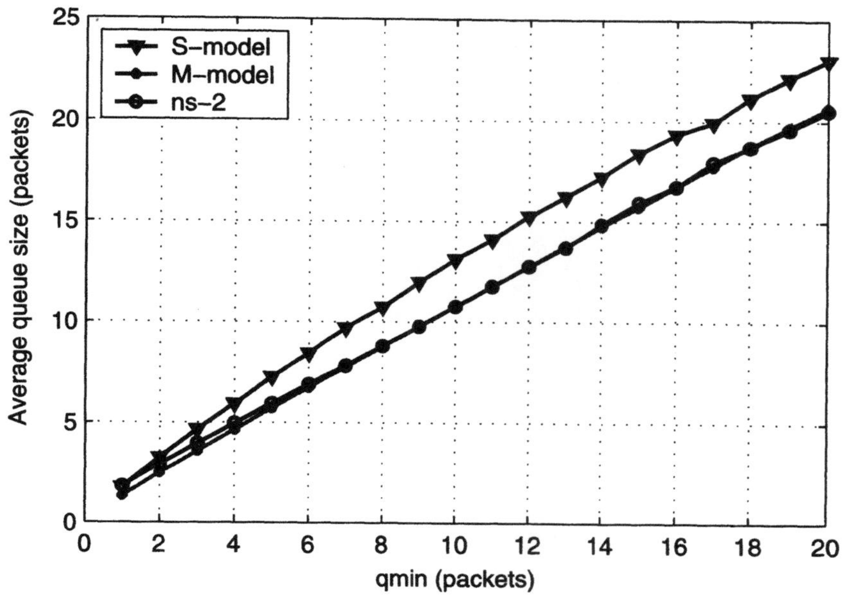

Fig. 11: Comparison of the average queue size for various q_{min} and q_{max}.

5. L. Brakmo and L. Peterson, "TCP Vegas: end to end congestion avoidance on a global Internet," *IEEE Journal on Selected Areas in Communication*, vol. 13, no. 8, pp. 1465–1480, Oct. 1995.
6. N. Cardwell, S. Savage, and T. Anderson, "Modeling TCP latency," in *Proc. IEEE INFOCOM 2000*, Tel Aviv, Israel, Mar. 2000, vol. 3, pp. 1742–1751.
7. C. Casetti, M. Gerla, S. Lee, S. Mascolo, and M. Sanadidi, "TCP with faster recovery," in *Proc. MILCOM* 2000, Los Angeles, CA, USA, Oct. 2000, vol. 1, pp. 320–324.
8. C. Casetti and M. Meo, "A new approach to model the stationary behavior of TCP connections," in *Proc. INFOCOM 2000*, Tel Aviv, Israel, Mar. 2000, vol. 1, pp. 367–375.
9. W. C. Y. Chan and C. K. Tse, "Study of bifurcation in current-programmed DC/DC boost converters: from quasi-periodicity to period-doubling," *IEEE Trans. Circuit and Systems I*, vol. 44, no. 12, pp. 1129–1142, Dec. 1997.
10. K. W. E. Cheng, M. Liu, and J. Wu, "Chaos study and parameter-space analysis of the DC-DC buck-boost converter," *IEE Proceedings-Electric Power Applications*, vol. 150, pp. 126–138, Mar. 2003.
11. K. Fall and S. Floyd, "Simulation-based comparison of Tahoe, Reno, and SACK TCP," *ACM Communication Review*, vol. 26, no. 3, pp. 5–21, July 1996.
12. K. Fall and K. Varadhan, "The ns Manual," UC Berkeley, LBL, USC/ISI, and Xerox PARC, June 2003: http://www.isi.edu/nsnam/ns/doc/ns_doc.pdf.
13. V. Firoiu and M. Borden, "A study of active queue management for congestion control," in *Proc. IEEE INFOCOM 2000*, Tel-Aviv, Israel, Mar. 2000, vol. 3, pp. 1435–1444.

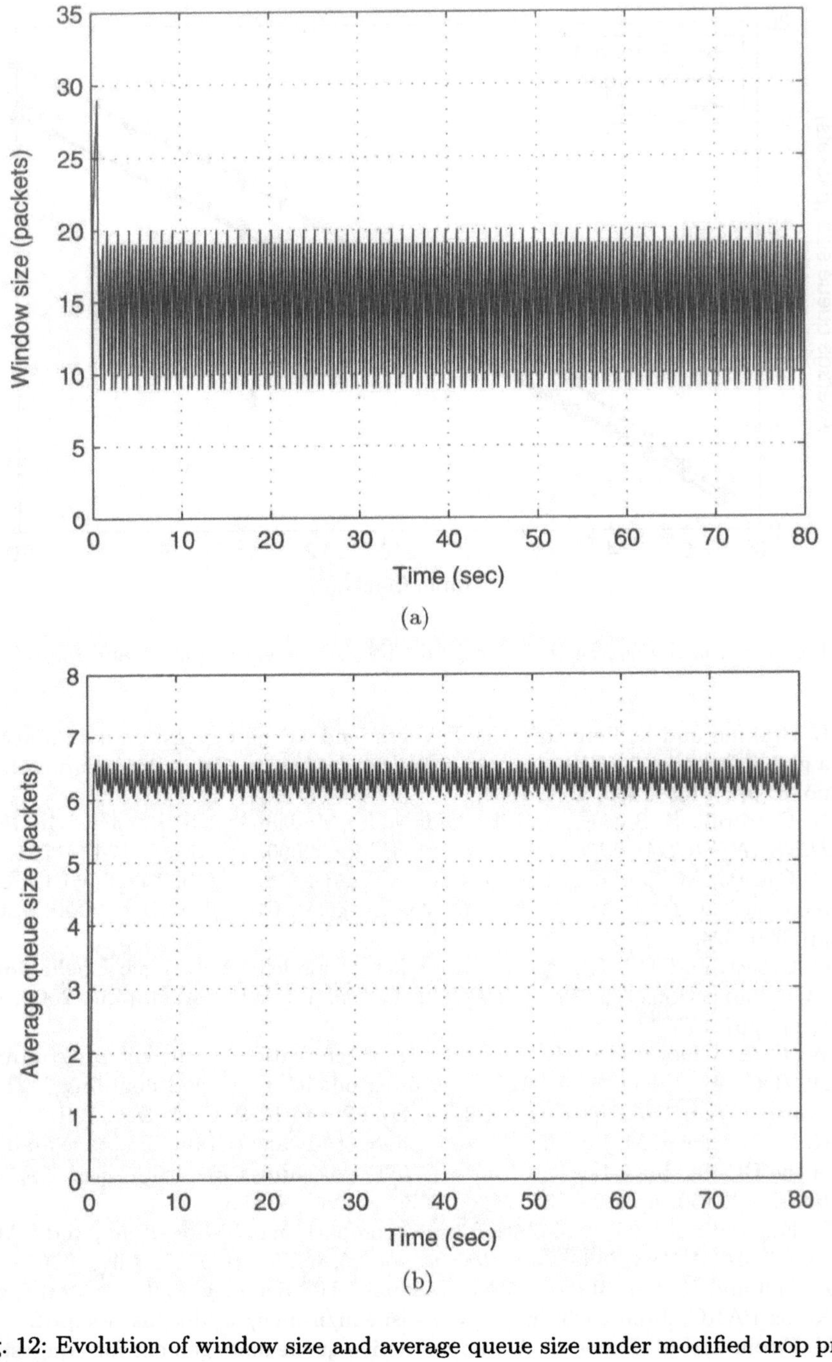

Fig. 12: Evolution of window size and average queue size under modified drop probability.

Table 6: System variables for various q_{min} and q_{max}.

Parameters	Average RTT (ms)				
q_{min} (packets)	S-model	Δ (%)	M-model	Δ (%)	ns-2
3	33.4	7.40	34.1	9.65	31.1
5	39.9	10.83	41.3	14.72	36.0
10	54.7	13.72	60.8	26.40	48.1
15	67.7	12.27	83.1	37.81	60.3
20	79.1	8.36	109.1	49.45	73.0
	Sending rate (packets/s)				
q_{min} (packets)	S-model	Δ (%)	M-model	Δ (%)	ns-2
3	383.22	0.20	366.13	-4.26	382.44
5	384.98	0.06	355.26	-7.76	384.77
10	385.10	0.06	330.94	-14.01	384.85
15	385.06	0.06	311.01	-19.19	384.83
20	385.30	0.09	296.27	-23.04	384.95
	Drop rate (%)				
q_{min} (packets)	S-model	Δ (%)	M-model	Δ (%)	ns-2
3	0.78	10.01	0.96	35.40	0.71
5	0.56	2.56	0.70	28.21	0.55
10	0.31	-6.34	0.37	11.78	0.33
15	0.20	-10.71	0.22	-1.79	0.22
20	0.15	-5.66	0.14	-11.95	0.16

14. S. Floyd and V. Jacobson, "Random early detection gateways for congestion avoidance," *IEEE/ACM Trans. Networking*, vol. 1, no. 4, pp. 397–413, Aug. 1993.
15. S. Floyd, "RED: discussions of setting parameters," Nov. 1997: http://www.icir.org/floyd/REDparameters.txt.
16. C. V. Hollot, V. Misra, D. Towsley, and W. B. Gong, "A control theoretic analysis of RED," in *Proc. IEEE INFOCOM 2001*, Anchorage, AK, Apr. 2001, vol. 3, pp. 1510–1519.
17. C. V. Hollot, V. Misra, D. Towsley, and W. B. Gong, "Analysis and design of controllers for AQM routers supporting TCP flows," *IEEE Trans. on Automatic Control*, vol. 47, no. 6, pp. 945–959, June 2002.
18. V. Jacobson, "Congestion avoidance and control," *ACM Computer Communication Review*, vol. 18, no. 4, pp. 314–329, Aug. 1988.
19. V. Jacobson, "Modified TCP congestion avoidance algorithm," ftp://ftp.isi.edu/end2end/end2end-interest-1990.mail, Apr. 1990.
20. I. Khalifa and Lj. Trajković, "An overview and comparison of analytical TCP models," in *Proc. IEEE International Symposium on Circuits and Systems*, Vancouver, BC, Canada, May 2004, vol. V, pp. 469–472.

21. P. Kuusela, P. Lassila, J. Virtamo, and P. Key, "Modeling RED with idealized TCP sources," *9th IFIP Conference on Performance Modeling and Evaluation of ATM and IP Networks 2001*, Budapest, Hungary, June 2001, pp. 155–166.
22. R. J. La, P. Ranjan, and E. H. Abed, "Nonlinearity of TCP and instability with RED," in *Proc. SPIE ITCom*, Boston, MA, USA, July 2002, pp. 283–294.
23. Y. Lai and C. Yao, "TCP congestion control algorithms and a performance comparison," in *Proc. 10th International Conference on Computer Communications and Networks*, Scottsdale, AZ, USA., Oct. 2001, pp. 523–526.
24. S. H. Low and D. E. Lapsley, "Optimization flow control—I: basic algorithm and convergence," *IEEE/ACM Trans. Networking*, vol. 7, no. 6, pp. 861–874, Dec. 1999.
25. S. H. Low, F. Paganini, J. Wang, S. Adlakha, and J. C. Doyle, "Dynamics of TCP/RED and a scalable control," *Proc. IEEE INFOCOM 2002*, New York, NY, USA., June 2002, vol. 1, pp. 239–248.
26. M. Mathis, J. Semke, J. Mahdavi, and T. Ott, "The macroscopic behavior of the TCP congestion avoidance algorithm," *ACM Computer Communication Review*, vol. 27, no. 3, pp. 67–82, July 1997.
27. M. Mellia, I. Stoica, and H. Zhang, "TCP model for short lived flows," *IEEE Communications Letters*, vol. 6, no. 2, pp. 85–87, Feb. 2002.
28. V. Misra, W. B. Gong, and D. Towsley, "Fluid-based analysis of a network of AQM routers supporting TCP flows with an application to RED," in *Proc. ACM SIGCOMM 2000*, Stockholm, Sweden, Aug. 2000, pp. 151–160.
29. A. Misra, J. Baras, and T. Ott, "Generalized TCP congestion avoidance and its effect on bandwidth sharing and variability," in *Proc. IEEE GLOBECOM*, San Francisco, CA, USA., Nov. 2000, vol. 1, pp. 329–337.
30. ns-2 Network Simulator: http://www.isi.edu/nsnam/ns.
31. J. Padhye, V. Firoiu, and D. F. Towsley, "Modeling TCP Reno performance: a simple model and its empirical validation," *IEEE/ACM Trans. Networking*, vol. 8, no. 2, pp. 133–145, Apr. 2000.
32. J. Padhye and S. Floyd, "On inferring TCP behavior", in *Proc. ACM SIGCOMM 2001*, San Diego, CA, USA, Aug. 2001, pp. 287–298.
33. V. Paxson and M. Allman, "Computing TCP's Retransmission Timer," *Request for Comment (RFC) 2988*, Nov. 2000.
34. J. Postel, "Transmission control protocol," *Request for Comment (RFC) 793*, Sept. 1981.
35. P. Ranjan, E. H. Abed, and R. J. La, "Nonlinear instabilities in TCP-RED," in *Proc. IEEE INFOCOM 2002*, New York, NY, USA, June 2002, vol. 1, pp. 249–258.
36. P. Ranjan, R. J. La, and E. H. Abed, "Bifurcations of TCP and UDP traffic under RED," in *Proc. 10th Mediterranean Conference on Control and Automation (MED) 2002*, Lisbon, Portugal, July 2002.
37. R. Roy, R. C. Mudumbai, and S. S. Panwar, "Analysis of TCP congestion control using a fluid model," in *Proc. IEEE ICC 2001*, Helsinki, Finland, June 2001, vol. 8, pp. 2396–2403.
38. W. R. Stevens, *TCP/IP Illustrated, Volume 1: The protocols.* New York: Addison-Wesley, 1994.
39. C. K. Tse and M. D. Bernardo, "Complex behavior in switching power converters," *Proceedings of the IEEE*, vol. 90, no. 5, pp. 768–781, May 2002.

Nonlinear Instabilities in TCP-RED

Priya Ranjan, Eyad H. Abed, and Richard J. La

Institute for Systems Research,
University of Maryland, College Park, MD-20742
{priya,abed,hyongla}@isr.umd.edu

1 Introduction

With the growing size and popularity of the Internet, congestion control has emerged as an important problem. Poor management of congestion can render a network partly or fully inaccessible and significantly degrade the performance of networking applications. Researchers have proposed various approaches for addressing this issue. One approach is to keep the network simple and place most of the required intelligence at the end hosts by implementing a more sophisticated end-user rate control allocation [7]. Another approach is to control the congestion level at each router through Active Queue Management (AQM) mechanisms, *e.g.,* Random Early Detection (RED) [11], Random Early Marking (REM) [3], and Virtual Queue (VQ) [12]. A common goal of these AQM mechanisms is to detect early signs of congestion and provide feedback to the adaptive sources so that congestion can be avoided without causing a significant degradation in network performance.

The RED mechanism, proposed by Floyd and Jacobson [11], attempts to control the congestion level at a bottleneck by monitoring and updating the average queue size. The basic idea of RED is to sense impending congestion before it happens and provide feedback to the sources by either dropping or marking their packets. The packet marking/dropping probability is the control administered by the RED gateways when they detect queue build-up beyond a certain threshold. Although the RED mechanism is conceptually very simple and easy to understand, its interaction with Transmission Control Protocol (TCP) connections has been found to be rather complex and is not well understood. Most of the rules for setting the parameters of the RED mechanism are based on limited empirical data and come from networking experience. These rules have been evolving, as our understanding of the effects of controller parameters increases. There are reports that discourage wide deployment of RED (*e.g.,* [24]), arguing that there is insufficient consensus on how to select controller parameter values, and that RED does not provide a drastic improvement in performance.

As noted above, the behavior of a network with TCP at the end nodes and RED at the routers is not well understood, and indeed has been found to at times exhibit erratic behavior. To improve the understanding of TCP-RED network dynamics in congested regimes, we follow a *nonlinear* modeling and analysis framework. In congested regimes, when the number of connections is large the stochastic nature of incoming traffic is of less importance, and the network can be approximated as a deterministic system. This allows us to use nonlinear analysis and detailed simulations to explore network behavior over a wide range of parameter values. The major difference between the present work and earlier deterministic studies of TCP-RED is that here we are able to model the generic nonlinear effects of TCP beyond linear operating regimes.

We use a deterministic nonlinear dynamical model of a simple network with TCP connections and a RED gateway. The basic model that we consider was originally proposed by Firoiu and Borden [8]. We modify their model with a simpler TCP throughput function [14, 23] to facilitate the analysis while keeping the dominant nonlinearity. It is also shown that the model proposed in [8] can be rewritten as a first-order discrete-time nonlinear dynamical model. This modeling framework is very much in the spirit of self-clocked models proposed by Jacobson [17]. Our work goes beyond a simple linear stability analysis, and studies regions where nonlinear instabilities occur due to nonlinearity of the throughput function combined with buffer space limits becoming active. The effect of these nonlinearities and the dynamics of network protocols in the large have not been explored thoroughly. We show that the model exhibits a rich variety of bifurcation behavior, leading to irregular network operation. As parameters are varied, the system dynamics are shown to transition between a stable fixed point and oscillatory or chaotic behavior via a period doubling bifurcation.

Motivated by the observation of period doubling bifurcation as basic instability initiation mechanism, we also illustrate a simple delayed feedback control algorithm to control instabilities [1, 29]. The basic idea behind this control is to modulate one of the control parameters by feeding back a function of the difference between the state and the desired fixed point [27]. Our control only enhances the stability of operating point without actually changing it. This can be contrasted with other approaches like adaptive RED (ARED) [10] where control itself oscillates in time using additive increase and multiplicative decrease (AIMD) algorithm.

The rest of the chapter is organized as follows. Section 3 presents the nonlinear first-order discrete-time model that is used in the analysis. In Section 4, the fixed point of the model is determined and an associated period doubling bifurcation is analyzed. The border collision bifurcation from the period doubled orbit is studied in Section 5. Section 6 contains numerical examples illustrating bifurcations and nonlinear instabilities in the model. An analytical study of sufficient conditions for chaotic behavior is given in Section 7. We describe our control algorithm in Section 8.

2 Background on TCP and RED

In this section we first briefly explain the main protocols, the interaction of which is the main subject of this chapter.

2.1 Transmission Control Protocol

Transmission Control Protocol (TCP) is the most popular form of congestion control protocol adopted by responsive end user applications. The transmission rate of a TCP connection is controlled by its congestion window size that determines the maximum number of outstanding packets that have not been acknowledged. TCP operates in two different modes. When a TCP connection is first initiated, it starts in Slow Start (SS) mode. In SS the connection increases its congestion window size by one for each acknowledged packet until it receives the first congestion notification, *e.g.*, packet drop or marking, at which time it switches to Congestion Avoidance (CA) mode. In CA the congestion window size is increased by one during the course of a round-trip time (RTT) if the connection does not receive any congestion notification. When a connection receives a congestion notification, the congestion window size is reduced to half of the current value. This is often referred to as the Additive-Increase-Multiplicative-Decrease (AIMD) mechanism.

Due to its popularity and important role in proper management of congestion inside the network, the behavior of TCP has been much studied and it is well understood in the context of single flow. It has been shown that given the RTT R and packet loss probability p, the stationary throughput of a TCP Reno connection can be approximated by

$$T(p,R) = \frac{MK}{\sqrt{p}R} + o\left(\frac{1}{\sqrt{p}}\right) , \tag{1}$$

where K is some constant in $\left[1, \sqrt{\frac{8}{3}}\right]$ (see [14, 23, 28, 31]), when packet losses are detected through triple-duplicate acknowledgements. We will use this formula for TCP throughput in our analysis, assuming TCP flows are in congestion avoidance (CA) mode unless mentioned otherwise.

2.2 Random Early Detection

Random Early Detection (RED) is one of the first active queue management (AQM) mechanisms proposed by Floyd and Jacobson [11]. A RED gateway estimates the congestion level by monitoring and updating its average queue size. In order to maintain a relatively small (average) queue size, rather than waiting until the buffer overflows it drops a packet with a certain probability to provide an early signal of impending congestion when the average queue size exceeds a threshold. This packet dropping probability p is a function of the average queue size q^{ave} of the following form [11]:[1]

$$p(q^{ave}) = \begin{cases} 0 & \text{if } q^{ave} < q_{min} \\ 1 & \text{if } q^{ave} > q_{max} \\ \frac{q^{ave} - q_{min}}{q_{max} - q_{min}} p_{max} & \text{otherwise} \end{cases} , \tag{2}$$

where q_{min} and q_{max} are the lower and upper threshold values, and p_{max} is the selected drop probability when $q^{ave} = q_{max}$. The average queue size is updated at the time of packet arrival according to the exponential averaging

$$q^{ave}_{new} = (1 - w) q^{ave}_{old} + w \cdot q_{curr} , \tag{3}$$

where q_{curr} is the current queue size, *i.e.*, the queue size at the time of arrival, and w is the exponential averaging weight, which determines the time constant of the averaging mechanism and how fast the RED mechanism can react to a time-varying load. On one hand, the averaging weight w should be selected small enough so that transient, temporary congestion does not result in an oscillation of the packet drop probability. On the other hand, the averaging weight should be set large enough so that the RED mechanism can react to changes in load in a timely manner. These are two conflicting goals, and the selection of the parameters will affect the interaction of the RED mechanism with adaptive sources, such as TCP. In this chapter, however, we show that the averaging weight cannot be set arbitrarily large without causing an oscillatory behavior at the bottlenecks, which affects TCP performance.

3 Discrete-Time Feedback Model for TCP-RED

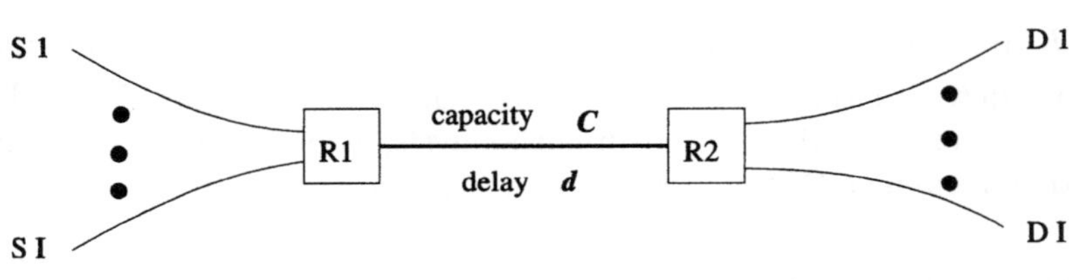

Fig. 1: Topology of the network.

We consider a simple network, where a single bottleneck link is shared by many connections. This is shown in Fig. 1. This can be viewed as a network, where there exists a dominant bottleneck shared by connections, *e.g.*, a

[1]In practice a RED gateway drops a packet with a modified probability in order to lead to a more uniform drop pattern [11].

bottleneck intercontinental Internet link. Let $\mathcal{I}, \mathcal{I} = \{1, \cdots, N\}$, denote the set of connections. The set of connections is assumed to remain fixed for the time period of interest. The capacity of the shared link is denoted by C. We assume that the RED queue management mechanism is implemented at each node in order to control the average queue size at the router. If an Explicit Congestion Notification (ECN) mechanism is implemented, the RED gateway marks the packet by setting the ECN bit in the IP header of the packet if the transport layer is ECN capable. This is indicated in the packet through an ECN Capable Transport (ECT) bit in the IP header. If the source is not ECN capable, the RED gateway drops the packet [9]. The goal of the controller is to keep the aggregate transmission rate of the connections close to the link capacity, while maintaining a reasonably small average queue size between q_{min} and q_{max}.

All connections are assumed to be a long-lived TCP Reno connections. We assume that the connections are uniform and have the same round-trip propagation delay (without any queueing delay), which is denoted by d. However, rather than interpreting this assumption as a requirement that the connections must have the same propagation delay, one should consider the delay d as the effective delay that represents the overall propagation delay of the connections, or this could describe a case where the bottleneck link has a large propagation delay that dominates the round-trip delays of the connections, *e.g.*, an intercontinental Internet link. We denote the rate or throughput of a connection by x, and the packet size by M.[2]

A network with an AQM mechanism can be modeled as a feedback system, where sources adjust their transmission rates based on feedback from the AQM mechanism in the form of marked or dropped packets [9, 11]. We use a dynamic discrete-time feedback system model, first introduced by Firoiu and Borden [8], to analyze the interaction of a RED gateway with TCP connections.

The control system is defined as follows. At period $k, k = 1, 2, \ldots,$ the RED controller at the router provides the feedback signal p_k in the form of a packet drop probability. This feedback signal is a function of the average queue size q_k^{ave} evaluated at period k. Due to a feedback delay introduced by the RTT, the packet drop probability p_k at period k, $k \geq 1$, determines the throughput of the connections and the queue size q_{k+1} at period $k+1$, based on system constraints (such as capacity constraints). The queue size q_{k+1} at period $k+1$ is used to compute the average queue size q_{k+1}^{ave} at period $k+1$ according to the exponential averaging rule in (3). Then, the average queue size q_{k+1}^{ave} is used to calculate the packet drop probability p_{k+1} at period $k+1$, which is the control variable of the AQM mechanism. This can be expressed mathematically as

[2]For the simplicity of analysis we assume that all connections use the same packet size. Again, this should be interpreted as the average packet size of the connections rather than a strict requirement.

$$\text{plant function: } q_{k+1} = G(p_k) \tag{4}$$
$$\text{averaging function: } q_{k+1}^{ave} = A(q_k^{ave}, q_{k+1}) \tag{5}$$
$$\text{control function: } p_{k+1} = H(q_{k+1}^{ave}) \tag{6}$$

where $A(q_k^{ave}, q_{k+1})$ is the averaging function

$$A(q_k^{ave}, q_{k+1}) = (1 - w)q_k^{ave} + w \cdot q_{k+1} \tag{7}$$

as given in (3), and the RED control function $H(q_{k+1}^{ave}) = p(q_{k+1}^{ave})$ as given in (2). The exact form of the plant function $G(\cdot)$ depends on system parameters such as the number N of connections, the nature of the connections, round-trip delays d, etc. We describe the plant function subsection 3.1.

Let us first motivate our discrete-time model and explain the relationship with some of previously proposed models. Since the queue size and average queue size are updated upon packet arrivals, the queue dynamics at a RED gateway evolve at a faster time scale than RTT of connections. However, because the reaction times of TCP connections are fundamentally limited by their RTTs, the average queue size should not change much over the course of one RTT in order to allow the connections enough time to react to the current level of feedback signal and filter out oscillations due to transient congestion in order for RED mechanism to work properly as mentioned earlier. Therefore, the detailed dynamics of the interaction over one round-trip time will be averaged out by the RED averaging mechanism and will not play a significant role.

This observation has been verified in [34] using a discrete-time stochastic model, where a period is assumed to be a RTT of connections. They show that as the number of flows becomes large, both queue and average queue sizes converge to deterministic processes (*i.e.*, macro-scale model) with details of TCP dynamics filtered out. These results suggest that when modeling a large number of TCP connections the detailed dynamics of interaction with the RED can be simplified using a macro-scale model that captures the larger time scale dynamics roughly at the time scale of round-trip times of the connections. Similar results have also been obtained using stochastic differential equations [4]. In addition, our results in [32] suggest that the model used here can be interpreted as the underlying discrete-time model corresponding to the system given by delay-differential equations in [4, 15, 22] which attempt to approximate packet level dynamics using differential equations.

Let us denote the duration of a period in our discrete-time model by T_{period} and the number of packets the link can transmit in a period by n_{period}. Since a period in our model is much larger than typical inter-arrival times of packets as explained above, the exponential averaging weight in (7) is approximately $w \approx 1 - (1 - w_{red})^{n_{period}} \approx n_{period} \cdot w_{red}$ if $w_{red} \ll 1$, where w_{red} is the exponential averaging weight at a RED gateway.

3.1 Plant Function

In this subsection we describe the plant function (4) that will be used for our analysis. In order to compute the plant function we assume the following: Given the packet drop probability at period k the aggregate throughput of the connections is given by the stationary throughput formula in (1). In this chapter we follow [28] in taking $K = \sqrt{\frac{3}{2}}$. The exact value of K is not crucial to our analysis.

We use this simple approximation for TCP throughput to facilitate our analysis. However, our qualitative results do not depend on this particular form of TCP throughput approximation, and are consequences of the nonlinear dependence of TCP throughput on drop probability p, of which (1) is one instance. Similar results to those obtained here hold for more detailed TCP throughput function models.

Since the aggregate throughput of connections cannot be larger than the link capacity, this determines the queue size at the next period $k+1$ as follows. First, we can compute the steady-state packet drop probability p_u such that the bandwidth capacity constraint is satisfied,

$$\sum_{i \in \mathcal{I}} T(p_u, d) = N \cdot \frac{MK}{\sqrt{p_u}d} = C . \tag{8}$$

This is the smallest probability that results in a queue size of zero at the next period, and for all $p_k > p_u$, the queue size is zero at the next period. Hence, if $p_k \geq p_u$, we know that the throughput of the TCP connections is given by $\frac{MK}{\sqrt{p_k}d}$ and the queue size at period $k+1$ is zero, *i.e.*, $q_{k+1} = 0$. From (8) we can derive that

$$p_u = \left(\frac{NMK}{dC}\right)^2 , \tag{9}$$

and the corresponding average queue size q_u^{ave} such that for any $q_k^{ave} \geq q_u^{ave}$, q_{k+1} is identically zero is given by

$$q_u^{ave} = \begin{cases} \frac{p_u(q_{max} - q_{min})}{p_{max}} + q_{min} & \text{if } p_{max} \geq p_u \\ q_{max} & \text{otherwise} \end{cases}$$

Suppose first that the buffer size B is infinite. If $p_k < p_u$, the bottleneck link capacity is fully utilized. Thus, if $p_k < p_u$ one can obtain the queue size q_{k+1} at period $k+1$ as the solution of the following equation:

$$\frac{MK}{\sqrt{p_k}(d + \frac{q_{k+1} \cdot M}{C})} = \frac{C}{N} . \tag{10}$$

The interpretation of (10) is as follows. Assuming symmetric TCP connections, the bottleneck link capacity is equally divided among the TCP connections. In this case, the throughput of a TCP connection will be given by

$T(p_k, R(q_{k+1})) = \frac{MK}{\sqrt{p_k}(d+\frac{q_{k+1}M}{C})} = \frac{C}{N}$. Hence, the queue occupancy q_{k+1} is given by

$$q_{k+1} = \frac{C}{M}\left(\frac{MKN}{\sqrt{p_k}C} - d\right) \tag{11}$$

Now let the buffer be of finite size B. From (11) we see that q_{k+1} is a strictly decreasing function of p_k, and hence we can compute the largest p_k such that the queue size q_{k+1} equals the buffer size B. This probability, which we denote by p_l, is given by $\left(\frac{NMK}{dC+BM}\right)^2$. The corresponding average queue size q_l^{ave} is

$$q_l^{ave} = \frac{p_l(q_{max} - q_{min})}{p_{max}} + q_{min} \ .$$

It is obvious that for all $p_k \leq p_l$, *i.e.*, $q_k^{ave} \leq q_l^{ave}$, we have $q_{k+1} = B$. From (8) and (11) we have the full definition of the plant function

$$\begin{aligned} G(p_k) &= \begin{cases} 0 \ , & \text{if } p_k \geq p_u \\ B \ , & \text{if } p_k \leq p_l \\ \frac{NK}{\sqrt{p_k}} - \frac{Cd}{M} \ , & \text{otherwise} \end{cases} \\ &= q_{k+1} \end{aligned} \tag{12}$$

This type of plant function has been verified by ns-2 simulation by Firoiu and Borden [8] using a particular TCP throughput function similar to that used here.

From (4)-(6) and (12), we obtain the mapping

$$\begin{aligned} &q_{k+1}^{ave} \\ &\quad = (1-w)q_k^{ave} + w \cdot A(G(H(q_k^{ave}))) \\ &\quad = \begin{cases} (1-w)q_k^{ave} & \text{if } q_k^{ave} \geq q_u^{ave} \\ (1-w)q_k^{ave} + w \cdot B & \text{if } q_k^{ave} \leq q_l^{ave} \\ (1-w)q_k^{ave} + w \cdot \left(\frac{NK}{\sqrt{p_k}} - \frac{Cd}{M}\right) & \text{otherwise} \end{cases} \\ &\quad := g(q_k^{ave}, \rho) \ , \end{aligned} \tag{13}$$

where ρ summarizes the system parameters, including the exponential averaging weight w, and $p_k = \frac{q_k^{ave} - q_{min}}{q_{max} - q_{min}} p_{max}$ from (2). This mapping gives the dynamical relationship of the average queue size at period $k+1$ to the average queue size at period k as shown in Fig. 2. There are three segments in this map showing either increasing and decreasing behaviors of average queue size in different regimes. Most of the interesting dynamics occur due to the middle segment of the map. There are two types of forces in this segment, one of which arises from averaging and the other arises from the RED control action. Their relative contributions in the queue occupancy in the next period are determined by the averaging parameter w. This interaction of averaging and RED control law is crucial to the kind of instabilities and instability cascades that occur as a system or RED parameter is slowly varied.

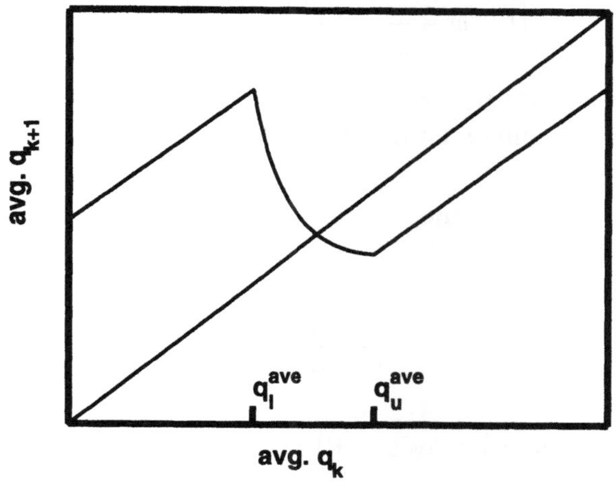

Fig. 2: First Return Map for TCP-RED

4 Fixed Point and Its Bifurcation

A fixed point of the mapping $g(\cdot)$ is an average queue size q^* such that $q^* = g(q^*, \rho)$. If the RED parameters are properly configured, then the average queue size should remain between q_{min} and q_{max}.

Assumption 1 *$p_{max} > p_u$, where p_u is the largest probability that yields the full utilization, defined in (9).*

This is natural from a practical point of view since it disallows a disconnected RED law wherein the drop probability jumps from p_{max} to 1.

Under Assumption 1, solving (13) for a fixed point q^* leads to a third order polynomial, which does not depend on the exponential averaging weight w because neither the "queue law" nor the "feedback control law" is a function of w. The corresponding probability p^* of the fixed point q^* is given as the square of the positive real solution of the polynomial

$$\frac{CM}{\nu} y^3 + (CMq_{min} + dC^2)y - NMKC = 0 , \tag{14}$$

where $\nu = p_{max}/(q_{max} - q_{min})$.

The linear stability of the fixed point q^* can be studied by considering the associated eigenvalue:

$$\begin{aligned} \frac{\partial}{\partial q_k^{ave}} g(q_k^{ave}, \rho)\bigg|_{q^*} &= 1 - w - \frac{wNK}{2\sqrt{\nu}(q^* - q_{min})^{\frac{3}{2}}} \\ &:= \lambda(q^*(\rho), \rho) , \end{aligned} \tag{15}$$

The linear stability condition is $|\lambda(q^*(\rho), \rho)| < 1$, or

$$\left|1 - w - \frac{wNK}{2\sqrt{\nu}(q^* - q_{min})^{\frac{3}{2}}}\right| < 1 . \tag{16}$$

In order to simplify the analysis, we reduce the number of parameters in the model by performing a normalization.

4.1 Normalization scheme

Define the parameter γ as

$$\gamma := \frac{q_{max} - q_{min}}{p_{max}B} = \frac{1}{\nu B} > 0$$

The normalized state variables and RED queue thresholds are defined as

$$q_k^n := \frac{q_k^{ave}}{B}, \quad q_{min}^n := \frac{q_{min}}{B}$$

$$q_l^n := \frac{q_l^{ave}}{B} = \left(\frac{NK}{B + \frac{dC}{M}}\right)^2 \gamma + q_{min}^n$$

$$q_u^n := \frac{q_u^{ave}}{B} = p_u \cdot \gamma + q_{min}^n$$

$$\begin{aligned} q_{k+1}^n &= \begin{cases} (1-w)q_k^n \ , & \text{if } q_k^n > q_u^n \\ (1-w)q_k^n + w \ , & \text{if } q_k^n < q_l^n \\ (1-w)q_k^n + w\left(\frac{NK}{B\sqrt{\frac{q_k^n - q_{min}^n}{\gamma}}} - \frac{dC}{MB}\right) & \\ & \text{otherwise} \end{cases} \\ &:= f(q_k^n, \rho) \end{aligned} \tag{17}$$

Eq. (17) maps the unit interval into itself.

4.2 Bifurcation analysis

Local stability of an one-dimensional map in the neighborhood of a fixed point is determined by the eigenvalue of the linearized map evaluated there. For the normalized map (17), this eigenvalue is

$$\begin{aligned} \left.\frac{\partial f(q_k^n, \rho)}{\partial q_k^n}\right|_{q_n^k = q^{n*}} &= 1 - w - \frac{wNK}{2B(q^{n*} - q_{min}^n)^{\frac{3}{2}}}\sqrt{\gamma} \\ &:= \lambda(\rho) \end{aligned} \tag{18}$$

where q^{n*} is the fixed point of the normalized map. The linear stability criterion ($|\lambda(\rho)| < 1$ [16]) is

$$\left|1 - w - \frac{wNK}{2B(q^{n*} - q^n_{min})^{\frac{3}{2}}}\sqrt{\gamma}\right| < 1 \tag{19}$$

Note that the eigenvalue depends on the fixed point. Of significant interest here are the parameter settings which may lead to loss of stability of the fixed point, giving rise to nonlinear instabilities through a system bifurcation. Numerical simulations of the system show the presence of oscillatory regimes as control and system parameters are varied, and indicate that a period doubling bifurcation occurs from the fixed point with the variation of any of the system or control parameters. Thus, we are led to consider cases in which the eigenvalue given by (18) becomes -1, giving a period doubling bifurcation (PDB) leading to oscillatory behavior in the system. To demonstrate existence of such bifurcations, it is easiest to focus on the exponential averaging parameter w as the distinguished bifurcation parameter. The critical value of w is one for which the eigenvalue given by (18) is -1. The critical value can be expressed in a closed form as follows:

$$w_{crit} = \frac{2}{1 + \frac{nK}{2(q_e^* - q_{min})^{\frac{3}{2}}}\sqrt{\frac{q_{max} - q_{min}}{p_{max}}}} \tag{20}$$

where q_e^* is a fixed point of the system whose corresponding probability is given as a square of the solution from (14).

Period doubling bifurcation can be supercritical or subcritical. In the supercritical case, attracting period two orbits emerge from the fixed point on the unstable side of the fixed point. In the subcritical case, repelling period two orbits emerge on the stable side. The ramifications of these two types of period doubling bifurcation for system behavior are very different, with supercritical bifurcation leading to a steady oscillatory behavior near the original fixed point, and subcritical bifurcation leading to divergent oscillations. It is possible to determine analytically which of these two cases will arise [16]. To do so, we need to compute the second and third derivatives of the normalized map

$$\left.\frac{\partial^2 f}{\partial q_k^{n2}}\right|_{q_n^k = q^{n*}} = \frac{3wNK}{4B(q^{n*} - q^n_{min})^{\frac{5}{2}}}\sqrt{\gamma} \tag{21}$$

$$\left.\frac{\partial^3 f}{\partial q_k^{n3}}\right|_{q_n^k = q^{n*}} = \frac{-15wNK}{8B(q^{n*} - q^n_{min})^{\frac{7}{2}}}\sqrt{\gamma} \tag{22}$$

to analyze the nature of this bifurcation. The quantity

$$S = \left(\frac{1}{2}\left(\frac{\partial^2 f}{\partial q_k^{n2}}\right)^2 + \frac{1}{3}\left(\frac{\partial^3 f}{\partial q_k^{n3}}\right)\right)$$

(evaluated at the fixed point and the selected parameter values) determines the nature of a period doubling bifurcation (see [13], pp.158). A positive S implies that the bifurcation is supercritical, and a negative S implies a subcritical bifurcation. For the system given by (17), S is

$$S = \frac{wNK\sqrt{\gamma}}{B(q^{n*} - q^n_{min})^{\frac{7}{2}}} \left[\frac{9}{32} \frac{wNK\sqrt{\gamma}}{B(q^{n*} - q^n_{min})^{\frac{3}{2}}} - \frac{5}{8} \right] . \tag{23}$$

The expression for S in (23) shows that it may change sign giving rise to a subcritical bifurcation if the parameters are in certain ranges. This should be kept in mind when designing a TCP-RED system to avoid any unexpected oscillations in router queues.

First, suppose that the system and control parameters are fixed, except for the averaging weight w. Then, from (15) we see that the eigenvalue is a linearly decreasing function of w, becoming more negative as w is increased. Now consider the critical averaging weight w^* to be a function of N, and denote it as $w^*(N)$. Then

$$w^*(N) = \frac{2}{1 + \frac{NK}{2\sqrt{\nu}(q^* - q_{min})^{\frac{3}{2}}}} .$$

The next lemma states that the largest value of the averaging weight that can be used without resulting in loss of stability is an increasing function of N.[3]

Lemma 1. *The critical parameter value $w^*(N)$ is an increasing function of N.*

This lemma tells us that when the load is light, the averaging weight must be selected small in order to avoid an oscillatory behavior in the queue size due to a period doubling bifurcation. The importance of the bifurcation point is that the system quickly becomes very unstable in the sense that the queue size oscillates widely, often resulting in an empty queue, reducing the system throughput and increasing the RTT variance of TCP connections. One can show in a similar manner that the initial period doubling bifurcation point $w^*(\cdot)$ is a decreasing function of the round-trip propagation delay d and q_{min} and an increasing function of q_{max} when these parameters are varied in isolation while other parameters are fixed.

Below, some analytical properties of the map (17) are given and proved, in preparation for the study of possible instability routes in the next section.

Assumption 2 *Assume that the left derivative of the normalized map in (17) is negative for $q^n_k = q^n_u$.*

This assumption is not very restricting. It simply asserts that as the (normalized) average queue size q^n_k increases from q^n_l to q^n_u the average queue size

[3]For the rest of the chapter we limit our interests to the region where $q^* \leq q_{max}$.

in the next period computed according to the map in (17) decreases with q_k^n. This assumption will be true if the negative feedback component of the RED is larger than the contribution retained by the averaging mechanism in the middle segment of the map shown in Fig. 2.

Lemma 2. *The map given by (17) is piecewise monotone under Assumptions 1 and 2.*

Next, we analyze the parametric dependence of TCP-RED system and show that it is smooth in parameter with respect to w.

Lemma 3. *The map given by (17) depends smoothly on w.*

We refer the reader to [30] for a proof of Lemmas 2 and 3. Properties of TCP-RED map outlined by these lemmas will be used in the next section to leverage Border Collision bifurcation (BCB) theory to the understand the dynamics in different regions.

5 Border Collision Bifurcation (BCB)

In this section we use the *border collision bifurcation* theory [5, 6] to analyze the bifurcations due to the variation of parameter w. BCBs occur for piecewise smooth maps, and involve a nonsmooth bifurcation occurring when a parameter change results in a fixed point (or other operating condition) crossing a border between two regions of smoothly defined dynamics in state space.

If a fixed point collides with the border(s) with a change in the parameters, there is a discontinuous change in the derivative $\frac{\partial f}{\partial x}$ of map $f(x)$, and the resulting phenomenon is called *border collision bifurcation.* This kind of bifurcation has been reported widely in economics [26], mechanical systems, and power electronic models [5, 6, 26].

Border collision is a local bifurcation and hence it can be studied by characterizing the local properties of a map in the neighborhood of the colliding border. It is shown in [26] that a normal form which is an affine approximation of f in the border neighborhood is sufficient to quantify the possible border collision bifurcations. This normal form is

$$G(x,\mu) = \begin{cases} ax + \psi, & \text{if } x \leq 0 \\ bx + \psi, & \text{if } x \geq 0 \end{cases} \tag{24}$$

where

$$a = \lim_{x \to x_b^-} \frac{\partial}{\partial x} f(x, \psi^*) \ , \quad b = \lim_{x \to x_b^+} \frac{\partial}{\partial x} f(x, \psi^*) \tag{25}$$

and ψ^* is the parameter for which border collision happens. It can be assumed to be 0 without any loss of generality.

There are various types of bifurcation scenarios possible depending on the values of coefficients a and b in the normal form given in (25). For the sake of simplicity, we will discuss only the case relevant to the observed phenomena in our model and provide a numerical proof by computing the one sided coefficients (eigenvalues) for the same.

The following lemma from [5] shows that the border has a crucial role if a certain bifurcation sequence occurs.

Lemma 4. *If a fixed point of the map given by (17) undergoes a smooth (eigenvalue = −1) period doubling bifurcation at* w_1 *and the resulting period two orbit also goes through a smooth period doubling for* $w_2 > w_1$*, then under the piecewise monotonicity condition, the periodic orbit must collide with the border for some* $w \in [w_1\ w_2]$.

We will see this kind of smooth and nonsmooth bifurcations in the next section when we present numerical examples.

For our model, the case of interest in border collision theory is when

$$0 < a < 1 \text{ and } b < -1 \tag{26}$$

This is mentioned as case 8 in [26]. It is shown that in this case a fixed point attractor can bifurcate into a periodic attractor or a chaotic attractor as ψ is varied from negative to positive. This is the exact phenomenon we observe for our model when the bifurcation parameter w is varied and a stable period two orbit transitions to chaos. Essentially, if we take the second iterate of our map, it exhibits a fixed point bifurcating into a chaotic orbit. Existence of chaos can be confirmed by computing the Lyapunov exponents [31]. A numerical example that provides evidence for our claim is given in the next section.

6 Numerical Examples

The behavior of the map in (13) can be explored numerically in parameter space to look for interesting dynamical phenomena. When the eigenvalue exits the unit circle, the fixed point becomes unstable. Depending on the nature of the ensuing bifurcation, there can be new fixed points, higher period orbits, or chaos. There is also a possibility of an orbit (original fixed point or a bifurcated orbit) colliding with either border q_u^{ave} or q_l^{ave}, leading to a rich set of possible bifurcations.

In this section we numerically validate our analysis using bifurcation diagrams. A bifurcation diagram shows the qualitative changes in the nature and the number of fixed points of a dynamical system as parameters are quasistatically varied. The horizontal axis is the parameter that is being varied, and the vertical axis represents a measure of the steady states (fixed points or higher period orbits). For generating the bifurcation diagrams, in each run we randomly select four initial average queue sizes, $q_1^{ave}(0), q_2^{ave}(0), q_3^{ave}(0)$

and $q_4^{ave}(0)$, and these average queue sizes evolve according to the map $g(\cdot)$ in (13), *i.e.*,

$$q_i^{ave}(k) = g(q_i^{ave}(k-1), \rho) \ , \ \text{for } k = 1, \cdots , \ 1{,}000 \\ \text{and } i = 1, \ 2, \ 3, \ \text{and } 4 \ .$$

We plot $q_i^{ave}(k), k = 991, \cdots, 1{,}000$ and $i = 1, 2, 3$, and 4. Hence, if there is a single stable fixed point or attractor q^* of the system at some value of the parameter, all $q_i^{ave}(k)$ will converge to q^* and there will be only one point along the vertical line at the value of the parameter. However, if there are two stable fixed points, $\tilde{q}_1^{ave}$ and $\tilde{q}_2^{ave}$, with a period of two, *i.e.*, $g(\tilde{q}_i^{ave}, \rho) \neq \tilde{q}_i^{ave}$ and $g(g(\tilde{q}_i^{ave}, \rho)) = \tilde{q}_i^{ave}, i = 1, 2$, then there will be two points along the vertical lines and the average queue size will alternate between $\tilde{q}_1^{ave}$ and $\tilde{q}_2^{ave}$.

Next, we study the effects of various system and control parameters on average and queue behavior as each of these parameters is varied while the others are fixed. More specifically, we study how the averaging weight w, lower threshold q_{min}, the number of connections N, and the round-trip propagation delay d affect system stability, queue behavior and their sensitivity to these parameters.

6.1 Effect of Exponential Averaging Weight

We use the following parameters for the numerical examples presented in this subsection:

$$q_{max} = 750, \ q_{min} = 250, \ C = 75 \text{ Mbps}, \\ K = \sqrt{3/2}, \ B = 3{,}750 \text{ packets}, \ M = 4{,}000 \text{ bits}, \\ N = 250, \ d = 0.1 \text{ sec}, \ w = \text{bifurcation parameter}$$

The bifurcation plots in Fig. 3 and 4, show the effect of varying the averaging weight w for different values of p_{max}, namely $p_{max} = 0.1$ and $p_{max} = 0.03$. Fig. 3(a) and 4(a) show the exponentially averaged queue sizes, and Fig. 3(b) and 4(b) plot the actual queue sizes. For small w, these plots have a fixed point, which shows up as a straight line until some critical value of w is reached, at which point the straight line splits into two. The emergence of two stable fixed points of period two is a consequence of a period doubling bifurcation. This is the first indication of oscillatory behavior appearing in the system due to the inherent nonlinearity of the interaction between RED mechanism and TCP, as opposed to a discontinuity in "queue or control law" which has been suggested in the past. This period two oscillation starts batching load at the router as shown in the plots.

Increasing w further results in one of the period two fixed points colliding with the upper border of the map, giving a chaos type phenomenon. This is basically a bifurcation sequence expressed briefly as $1 \rightarrow 2 \rightarrow$ *chaos*. This is a case of border collision bifurcation as shown in the analysis earlier. It can be

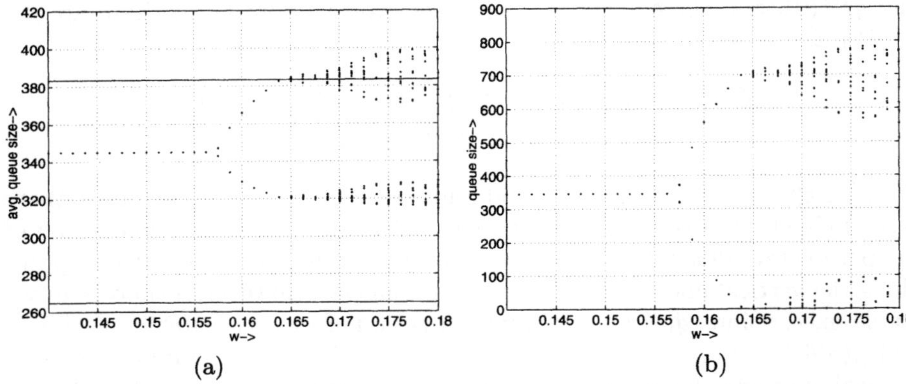

Fig. 3: Bifurcation diagram of average and actual queue length with respect to the averaging weight w ($p_{max} = 0.1$).

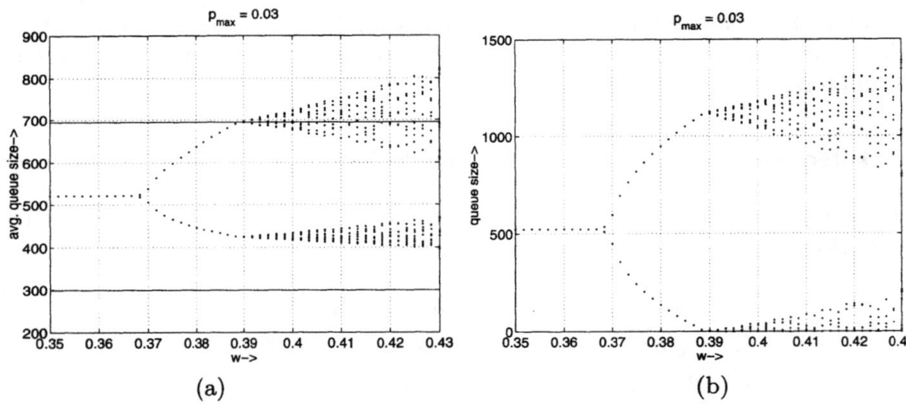

Fig. 4: Bifurcation diagram of average and actual queue length with respect to the averaging weight w ($p_{max} = 0.03$).

seen that when the bifurcation diagram for q_k^{ave} collides with the border q_u^{ave}, the queue empties, underutilizing the bottleneck link capacity. The implication of a relatively small oscillation in the average queue length is rather serious for the queue length since the buffer starts getting empty and overly filled in every alternate cycle. This dynamical phenomenon is common to both plots in Fig. 3 and 4. We note that the distance between the initial period doubling bifurcation point and the border collision bifurcation point is short in both cases. This suggests that an effective way of controlling the instability may be to control the first period doubling bifurcation point.

To illustrate the period doubling bifurcation in the system we compute the eigenvalue for the fixed point as w is varied. It can be seen that this eigenvalue leaves unit circle along negative real line indicating a period doubling bifurca-

Table 1: Eigenvalues computed for different values of parameter w to illustrate PDB, $q_u^{ave} = 0.102222$.

w	q_k^n	$\lambda(q_k^n, w)$	Legend
0.1561	0.092028	-0.978111	Close to PDB
0.1572	0.092028	-0.992051	Closer to PDB
0.1583	0.092028	-1.005990	After PDB
0.1594	0.092028	-1.019929	After PDB

Table 2: Eigenvalues computed for different values of parameter w to illustrate BCB, $q_u^n = 0.102222$.

w	q_k^n	q_{k-1}^n	q_{k-2}^n	q_{k-3}^n
0.1620	0.100108	0.086412	0.100108	0.086412
0.1631	0.085846	0.101330	0.085846	0.101330
0.1642	0.102193	0.085488	*0.102283*	0.085447

w	$\lambda^2(k, k-1)$	$\lambda^2(k-2, k-3)$	Legend
0.1620	0.786415	0.786415	Before BCB
0.1631	0.729421	0.729421	Before BCB
0.1642	0.692157	-1.815238	After BCB

tion. We also track the unstable fixed point and compute the corresponding eigenvalue to show that it indeed crosses the unit circle as shown in Table 1. We also notice that both stable and unstable fixed point ($q^{n*} = 0.092028$) is smaller than $q_u^{ave} = 0.102222$ for the normalized model. Hence, it lies on the same side of the border even after smooth period doubling bifurcation.

To provide evidence for our claim for a BCB, we further compute the eigenvalue of a period two orbit of the map numerically and show that indeed one sided eigenvalues obey the condition given in (26). This computation is done for the set of parameters corresponding to Fig. 3. We define $\lambda^2(i, j) = \lambda(q_i^n) * \lambda(q_j^n)$.

In Table 2, the first and second rows show the four consecutive states (the exponentially averaged queue size at the router) corresponding to the parameter w just before the BCB but after PDB. We note that all the states stay on the same side of the border with eigenvalue corresponding to a period-two orbit being less than unity. This implies the existence of stable period-2 orbit.

The third row depicts the same data just after a border collision bifurcation from a fixed point to chaos for the second iterate. Comparing the states with the border (q_u^n) reveals that q_{k-2}^n and q_{k-3}^n lie on different sides of the border. The eigenvalues corresponding to these two points, *i.e.*, $\lambda^2(k-2, k-3)$, is negative. This eigenvalue $\lambda^2(k-2, k-3)$ can be used to approximate b in

(25) in this case. Similarly the eigenvalue corresponding to q^n_{k-1} and q^n_k, *i.e.*, $\lambda^2(k, k-1)$, can be used to approximate a in (25). Since a lies between 0 and 1 as shown in Table 2, and b is smaller than -1, these values satisfy the condition given by (26). Note the eigenvalues change discontinuously as w is varied. This supports our contention that there is a *border collision* bifurcation in the system through which the system may become chaotic. It also stresses the role played by a border. We also note that there is a possibility of other rich nonlinear instabilities with different periodicity based on different parameter settings.

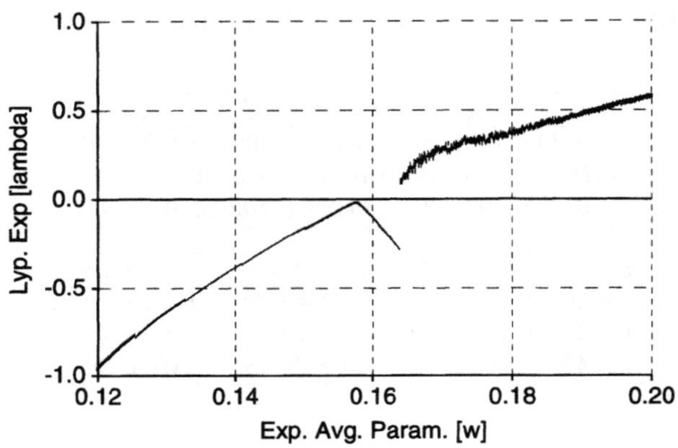

Fig. 5: Lyapunov exponent computed for average queue length with respect to the averaging weight w ($p_{max} = 0.1$).

We also plot the Lyapunov exponents for the bifurcation scenario in Fig. 3 where $p_{max} = 0.1$. This is useful since a positive Lyapunov exponent indicates the presence of chaotic behavior (page 110, [2]). Fig. 5 shows that for small w the exponent is negative, which corresponds to the single stable fixed point. It slowly increases to zero near the period doubling bifurcation, and then becomes negative again due to a stable period two orbit. Finally, it jumps to a positive value when one of the period two fixed points collides with one of the borders.

7 Chaotic Behavior

The purpose of this section is to give an analytical proof of the presence of chaos in the TCP-RED dynamic model. The tool we use is a well known theorem of Sharkovsky [33] which was also proved by Li and Yorke [20] and goes by the name "period three implies chaos." It applies to continuous one-dimensional maps, and thus can be applied to piecewise smooth but continuous systems such as the system studied here.

The main result of Li and Yorke [20] is as follows.

Theorem 1. *Let J be an interval and let $F : J \to J$ be continuous. Assume that there is a point $a' \in J$ for which the points $b' = F(a'), c' = F^2(a')$ and $d' = F^3(a')$, satisfy*

$$d' \leq a' < b' < c' \text{ or } d' \geq a' > b' > c'$$

Then
T1: for every $k = 1, 2, \ldots$ there is a periodic point in J having period k; and, furthermore,
T2: there is an uncountable set $S \subset J$ (containing no periodic points), which satisfies the following conditions:
(A) For every $p, q \in S$ with $p \neq q$,

$$\limsup_{n\to\infty} |F^n(p) - F^n(q)| > 0$$

and

$$\liminf_{n\to\infty} |F^n(p) - F^n(q)| = 0$$

(B) For every $p \in S$ and periodic point $q \in J$,

$$\limsup_{n\to\infty} |F^n(p) - F^n(q)| > 0$$

In our case $J = [0\ B]$, and F is given by the function $g(\cdot,\cdot)$ which defines the TCP-RED map in (13). It is also clear that the TCP-RED map is continuous by construction as long as Assumption 1 is in force. Also, note that the existence of a period three orbit, *i.e.*, $d' = a' > b' > c'$ or $d' = a' < b' < c'$, is a special case of the hypotheses of the theorem and proves the existence of chaos.

We have proved earlier [30] that the map (13) is strictly increasing for $0 \leq q^{ave} \leq q_l^{ave}$ and for $q_u^{ave} \leq q^{ave} \leq B$ but it can be strictly decreasing in the segment where $q_l^{ave} \leq q^{ave} \leq q_u^{ave}$ under certain conditions.

To apply the theorem, we need to choose a starting point a' and iterate on it using the map g three times and then apply the conditions stated in the theorem. We select $a' = \frac{q_u^{ave}}{(1-w)}$. This choice is made based on earlier numerical studies (Matlab) which showed a strong tendency toward bifurcation and chaos when the system state q_k^{ave} nears q_u^{ave}. With this choice for a', we find that $b' = g(a') = (1-w)a' = q_u^{ave}$ and $c' = g^2(a') = g(q_u^{ave}) = (1-w)q_u^{ave}$. Looking at Fig. 2, it is clear that there are two possible cases for the location of c': either $q_l^{ave} < c'$ (Case I) or $q_l^{ave} \geq c'$ (Case II). If w is small, then $c' = (1-w)q_u^{ave}$ will be close to q_u^{ave} and therefore Case I will hold. However, conditions for Theorem 1 to apply will be found below for both Case I and Case II.

Case I: Let $q_l^{ave} < (1-w)q_u^{ave}$ (this corresponds to $(1-w)q_u^{ave}$ lying in the interval $[q_l^{ave}, q_u^{ave}]$. Then

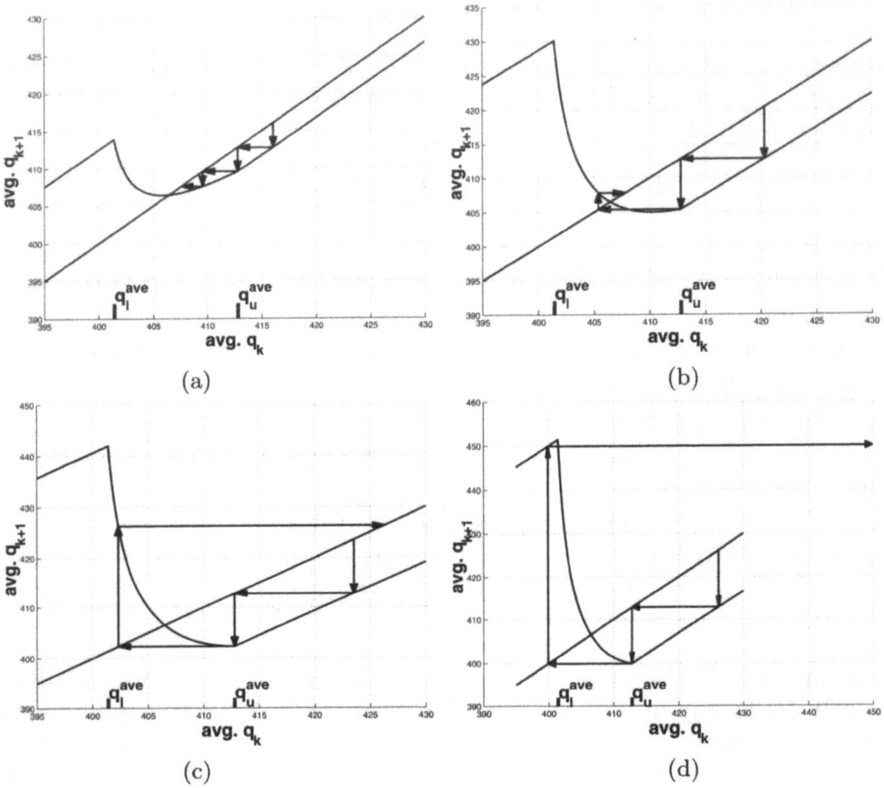

Fig. 6: a. First return map and period three condition for $w = 2^{-7}$, b. for $w = 2^{-5.8}$, c.for $w = 2^{-5.3}$ and d. for $w = 2^{-5}$

$$g^3(a') = (1-w)^2 q_u^{ave} + w \left(\frac{NK}{\sqrt{\frac{p_{max}((1-w)q_u^{ave} - q_{min})}{(q_{max} - q_{min})}}} - \frac{dC}{M} \right)$$

Hence the criterion $d' \geq a' > b' > c'$ of Theorem 1 ensuring existence of chaos gives

$$(1-w)^2 q_u^{ave} + w \left(\frac{NK}{\sqrt{\frac{p_{max}((1-w)q_u^{ave} - q_{min})}{(q_{max} - q_{min})}}} - \frac{dC}{M} \right) \geq a' \tag{27}$$

Case II: Alternatively, suppose $q_l^{ave} \geq (1-w)q_u^{ave}$. Then $g^3(a') = (1-w)^2 q_u^{ave} + wB$. Now the criterion $d' \geq a' > b' > c'$ of Theorem 1 ensuring existence of chaos in this case gives $g^3(a') - a' \geq 0$, which reduces to following simple condition.

$$
\begin{aligned}
& (1-w)^2 q_u^{ave} + wB \geq \frac{q_u^{ave}}{(1-w)} \\
& \Rightarrow (1-w)^3 q_u^{ave} + w(1-w)B \geq q_u^{ave} \\
& \Rightarrow w(1-w)B - (1-(1-w)^3)q_u^{ave} \geq 0 \\
& \Rightarrow w(1-w)B - (1-1+3w-3w^2+w^3)q_u^{ave} \geq 0 \\
& \Rightarrow q_u^{ave} \leq \frac{(1-w)B}{(3-3w+w^2)}
\end{aligned}
\tag{28}
$$

Summarizing, we have the following lemma. Here, "chaotic in the sense of Li and Yorke" means satisfying the conclusions of Theorem 1.

Theorem 2. *The TCP-RED system given by (13) is chaotic in the sense of Li and Yorke if either (27) and $q_l^{ave} < (1-w)q_u^{ave}$ hold, or (28) and $q_l^{ave} \geq (1-w)q_u^{ave}$ hold.*

The progression of nonlinear instabilities towards period-three and chaos is illustrated in Fig. 6. It is shown that as exponential averaging weight w is increased initially the condition given by Case-I holds, and for larger values of w condition given by Case-II is satisfied.

8 Feedback Control of Instabilities

In this section, we illustrate a simple delayed feedback control algorithm to control instabilities [1, 29]. The basic idea behind this control is to modulate one of control parameters by feeding back a function of the difference between the state and the desired fixed point [27]. This will delay the occurrence of bifurcation. We will also describe a nonlinear control strategy that can be used to achieve stabilization without changing the critical parameter value, and a combination of linear and nonlinear control terms achieve both a delay in parameter space and stabilization of ensuing bifurcations and hence reduction in the amplitude of oscillations.

8.1 Washout Filter Based Control

The washout filter mechanism has been successfully utilized to control a number of bifurcations in nonlinear models with uncertainty [1]. This approach for TCP-RED systems differs considerably from other schemes where the control scheme tries to keep the operating point invariant under significant parametric variations [3, 10]. For example, the adaptive RED (ARED) scheme also modulates a control parameter, namely p_{max}, to adapt to dynamically changing operating conditions, using an additive increase and multiplicative decrease algorithm [10]. However, the adaptation is done based on the difference between the current average queue size and *fixed* target queue size, and hence keeps system operation independent of other parameter variations. An inherent problem with such an approach is that the range over which it is effective may be severely limited in the parameter space [19].

A simple discrete time high-pass filter can be used as an analogue of washout filter in continuous time. Consider the following high-pass filter discussed in [1].

$$G(z) = \frac{1 - z^{-1}}{1 - \tau z^{-1}}$$

This can have the following time domain implementation:

$$\begin{aligned} z_{k+1} &= x_k + (1 - \tau) z_k \\ y_k &= x_k - \tau z_k \end{aligned} \tag{29}$$

where $\{x_k\}$ is the input sequence to the washout filter, $\{y_k\}$ is the output sequence, and the washout filter constant τ should satisfy $0 < \tau < 2$. At steady state, $z_{k+1} = z_k$ and $x_{eq} - \tau \cdot z_{eq} = 0$. Hence, from (29), we have $y_k \equiv 0$ and the output of the washout filter vanishes at the steady state.

Now, we can consider a scalar nonlinear dynamical system with washout filter control:

$$x_{k+1} = f(x_k, u_k) \tag{30}$$

where u_k is a scalar control input. If washout filter is put in the feedback loop with feedback function $h(\cdot)$, we have following modified system:

$$\begin{aligned} x_{k+1} &= f(x_k, u_k) \\ z_{k+1} &= x_k + (1 - \tau) z_k \\ y_k &= x_k - \tau z_k \\ u_k &= h(y_k) \end{aligned}$$

where $h : R \to R$ is any smooth function such that $h(0) = 0$. It can be shown that this type of feedback control does not modify the equilibrium point of the original system under no control, *i.e.*, $u_k = 0$ [1]. However, with a proper choice of feedback function $h(\cdot)$ and washout filter constant, it can enhance the stability of the original equilibrium point without the need for accurate knowledge of the system model or equilibrium value.

8.2 Application to TCP-RED

In this section we look at the stabilization of map in (13) with linear control terms in the neighborhood of fixed point q^*, *i.e.*, $q^* = f(q^*, \rho)$. For this we need to compute the linearization of the map ($x_{n+1} = Ax_n + bu_n$) around the intended fixed point of the system.

$$\left.\frac{\partial f(q_{k+1}^{ave}, \rho)}{\partial q_{k+1}^{ave}}\right|_{q_{k+1}^{ave}=q^*} = 1 - w - \frac{0.5wNK}{(q^* - q_{min})^{\frac{3}{2}}}$$
$$:= \lambda_0(\rho) \tag{31}$$

Depending on the RED parameter to be modulated, $b(p_{max}) = \frac{\partial f}{\partial p_{max}}$ or $b(q_{max}) = \frac{\partial f}{\partial q_{max}}$ can be computed.

$$b(p_{max}) = -\frac{0.5wNK}{\sqrt{\frac{(q_k^{ave} - q_{min})}{(q_{max} - q_{min})}} p_{max}^{1.5}} \tag{32}$$

$$b(q_{max}) = \frac{0.5wNK}{\sqrt{(q_k^{ave} - q_{min})(q_{max} - q_{min})p_{max}}} \tag{33}$$

It is clear from above that $b(\cdot) \neq 0$ for nominal range of parameters. For one dimensional system with nonzero eigenvalue, both left (l) and right (r) eigenvectors are 1.

From above two observations we conclude that $l \cdot b(\cdot) \neq 0$. This has consequences for linear stabilizability due to Popov-Belevitch-Hautus (PBH) eigenvector test for controllability of modes of linear time invariant systems [18], and tells us that linear stabilizing feedback exists in this case. This also means that cubic feedback exists, which we study in Section 8.5.

In the view of PBH test for controllability and wash out filter described above, we can view the averaged queue size of RED as input to the state estimation filter that provides the estimate y_k. This estimate can be used to construct the control depending on the functional form of h. In this section we consider only the linear control law, *i.e.*, $u_k = k_l \cdot y_k$, because in linear analysis all the nonlinear terms vanish when the system is linearized at the fixed point. Throughout this section we assume that we modulate q_{max} unless stated otherwise. In this framework, the TCP-RED system given by (13) when augmented by washout filter, can be rewritten as follows:

$$z_{k+1} = q_k^{ave} + (1 - \tau)z_k \tag{34}$$

$$u_k = h(q_k^{ave} - \tau z_k) \tag{35}$$

$$q_{k+1}^{ave} = \begin{cases} (1-w)q_k^{ave} & \text{if } q_k^{ave} > q_u^{ave} \\ (1-w)q_k^{ave} + wB & \text{if } q_k^{ave} < q_l^{ave} \\ (1-w)q_k^{ave} + w\left(\frac{NK}{\sqrt{\frac{(q_k^{ave} - q_{min})p_{max}}{(q_{max}^{wo} - q_{min})}}} - \frac{dC}{M}\right) & \text{otherwise} \end{cases} \tag{36}$$

where $q_{max}^{wo} = \min\{\frac{B}{2}, \max\{\alpha \cdot q_{min}, q_{max} + u_k\}\}$. We upper limit q_{max} to 0.5 B due to the consideration of GENTLE_ mode of RED and lower limit it to $\alpha \cdot q_{min}$, where $1 < \alpha < 2$.

8.3 Stability Analysis with Washout Filter

In this section we analyze the stability of washout enabled TCP-RED given by (36). Clearly, $[\frac{q^*}{\tau}, q^*]$ is the fixed pointof the new system given by (34) -

(36) for $\tau \neq 0$. The Jacobian matrix evaluated at the fixed point $[\frac{q^*}{\tau}, q^*]$ is given by

$$A = \begin{pmatrix} 1-\tau & 1 \\ b\frac{\partial h(q_k^{ave}-\tau z_k)}{\partial z_k} \frac{\partial f(q_k^{ave},\rho)}{\partial q_k^{ave}} + b\frac{\partial h(q_k^{ave}-\tau z_k)}{\partial q_k^{ave}} \end{pmatrix} \tag{37}$$

where $b = b(q_{max})$ given in (33).

If we evaluate (37) at the fixed point $[\frac{q^*}{\tau}, q^*]$ with linear control, *i.e.*, $u_k = k_l(q_k^{ave} - \tau \cdot z_k)$, (37) simplifies to

$$A = \begin{pmatrix} 1-\tau & 1 \\ -\tau b k_l & \lambda_0 + bk_l \end{pmatrix} \tag{38}$$

where $\lambda_0 = \frac{\partial f(q_k^{ave},\rho)}{\partial q_k^{ave}}\Big|_{q_k^{ave}=q^*}$ from (31).

Next we recall Jury's stability test for second order discrete-time systems:

Lemma 5. *(Jury's stability test for second order systems [21]) A necessary and sufficient condition for the zeros of the polynomial*

$$p(\lambda) = a_2\lambda^2 + a_1\lambda + a_0$$

($a_2 > 0$) to lie within unit circle is

$$p(1) > 0,\ p(-1) > 0 \ \ and \ |a_0| < a_2$$

The characteristic equation for matrix in (38) is given by

$$\lambda^2 - \lambda((1-\tau)) + \lambda_0 + bk_l) + (1-\tau)\lambda_0 + bk_l = 0$$

Using Jury's test for stability, the conditions for linear asymptotic stability are given as follows.

$$\tau(1-\lambda_0) > 0 \tag{39}$$

$$2 + 2bk_l + 2\lambda_0 - \tau(1+\lambda_0) > 0$$

$$\Rightarrow k_l > \frac{(\tau-2)(1+\lambda_0)}{2b} \text{ for } b > 0 \tag{40}$$

$$|\lambda_0(1-\tau) + bk_l| < 1$$

$$\Rightarrow \frac{-1-\lambda_0(1-\tau)}{b} < k_l < \frac{1-\lambda_0(1-\tau)}{b} \text{ for } b > 0 \tag{41}$$

Similar inequalities can be formulated for linear stability in the case of $b < 0$, *e.g.*, p_{max} is modulated. As we see here the stability region for pair (τ, k_l) is made up of three straight lines in (τ, k) plane, which are described below:

$$(l_1): k = \frac{(1+\lambda_0)\tau}{2b} - \frac{1+\lambda_0}{b} \tag{42}$$

$$(l_2): k = \frac{\lambda_0\tau}{b} - \frac{(1+\lambda_0)}{b} \tag{43}$$

$$(l_3): k = \frac{\lambda_0\tau}{b} + \frac{(1-\lambda_0)}{b} \tag{44}$$

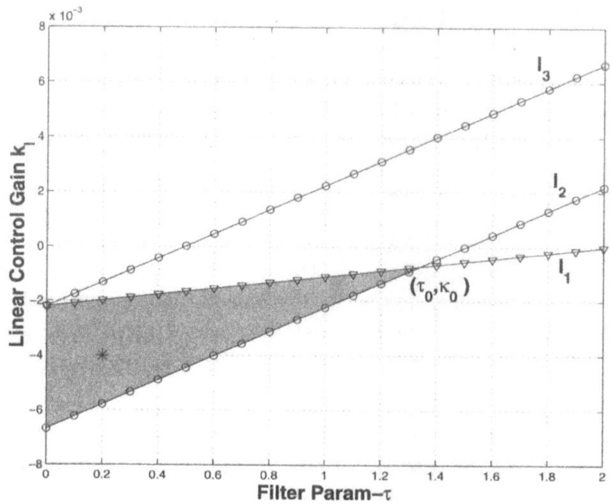

Fig. 7: Control Region for washout augmented TCP-RED

Under the generic assumption of $\lambda_0 < -1$ and $b > 0$, we can see that lines (l_2) and (l_3) are parallel as they have the same slope. Lines (l_1) and (l_3) intersect each other at $(\tau_0, k_0) = (\frac{4}{1-\lambda_0}, \frac{(1+\lambda_0)^2}{(1-\lambda_0)b})$. Similarly, lines (l_1) and (l_2) intersect each other at $(\tau_1, k_1) = (0, \frac{(1+\lambda_0)}{b})$.

Proposition 8.1 *For a (τ, k) pair to be stabilizing, it must lie within the triangle with the vertices* $(0, \frac{(1+\lambda_0)}{b}), (0, \frac{(1-\lambda_0)}{b})$, *and* $(\frac{4}{1-\lambda_0}, \frac{(1+\lambda_0)^2}{(1-\lambda_0)b})$

The parameter that will be modulated for control will determine the value of b, *e.g.*, $b < 0$ for p_{max} and $b > 0$ for q_{max}. Gain k_l needs to be chosen accordingly. Washout filter parameter τ is chosen such that $0 < \tau < 2$, and $\lambda_0 < -1$ in the regime after period doubling bifurcation. This shows that, theoretically, it is possible to control the average queue size of RED locally near critical parameter value, although allowable range for parameters such as p_{max} or q_{max} is limited by physical constraints in the real system. Also, these control gains need to be limited so as to not cross the basin of attraction for the fixed point. Hence, though local stabilization near the critical value of a parameter is possible, it may not be possible to stabilize in an arbitrarily large parameter range. Next, using Jury's test we compute the parameter range where stabilization is possible for a fixed value of k_l as exponential averaging weight w is varied. Similar results can be obtained for round-trip propagation delay d, and the number of active connections N.

8.4 Stabilization with Respect to Exponential Averaging Weight

Stabilization with respect to exponential averaging weight w is simpler to analyze since the fixed point is independent of w and the eigenvalue λ_0 decreases linearly with w from (31). Hence, due to linear stabilizability of the original system, *i.e.*, $l \cdot b \neq 0$, it is possible to stabilize the RED averaged queue by picking appropriate k_l and τ that obey the conditions given by (39) - (41). Here we are interested in investigating the possibility of local linear stabilizing over all values of $0 < w < 1$. It turns out that due to some interesting properties of λ_0 and $b = b(q_{max})$ as given by (33) it is possible to pick a (τ, k_l) pair to stabilize the system for all possible values of $w > w_{crit}$, where w_{crit} is the value of w at which first period doubling bifurcation happens in the uncontrolled system and is given by (20).

From (31) and (33) one can show that $\frac{(1-\lambda_0)}{b(q_{max})}$ is independent of w. This provides important insight into the locus of triangular stability region as given by Proposition 8.1. It shows that one of the vertices $(0, \frac{(1-\lambda_0)}{b})$ is invariant of w. We now need to understand the behavior of λ_0 and $b(q_{max})$ and that of $\frac{\lambda_0}{b(q_{max})}$ as w is varied in unit interval. It is clear from (31) that λ_0 decreases linearly as a function of w. Similarly, $b(q_{max})$ as given by (33) increases linearly with w, and $\frac{\lambda_0}{b(q_{max})}$ is strictly decreasing with w, which can be seen directly by differentiating the expression. This means that all three constraint lines given by (42) - (44) become steeper with increasing w. This leads to decreasing area of stability triangle shown in Fig. 7. Finally, we use the fact that w is bounded by one from above and evaluate the worst case stability region. Clearly, the eigenvalue remains finite for w=1. Evaluating the vertices for $w = 1$ will provide the smallest triangle. Hence, if the stabilizing pair (τ, k_l) lies within this triangle, then it does for all other values of $w > w_{crit}$.

Theorem 3. *TCP-RED system along with washout filter for a given washout control parameter and linear control gain pair (τ, k_l) and all other parameters held fixed, will be stable for $w_{crit} < w < 1$ where w_{crit} is the value of w corresponding to the first period doubling bifurcation, if (τ, k_l) lies within the triangle with vertices $(0, \frac{(1+\lambda_0(w=1))}{b}), (0, \frac{(1-\lambda_0(w=1))}{b}), (\frac{4}{1-\lambda_0(w=1)}, \frac{(1+\lambda_0(w=1))^2}{(1-\lambda_0(w=1))b})$ with $b = b(q_{max})$.*

Now, that we know about the linear stabilizability of the TCP-RED system, it is possible to use explicit nonlinear control terms to increase the control robustness or to reduce the amplitude of ensuing oscillations.

8.5 Nonlinear Control

It is possible to use small nonlinear control terms to further enhance the stability of a system going through a period doubling bifurcation. We first introduce the following hypothesis:

Hypothesis 1 *Eq. (13) has a period-1 orbit at $x^*(\rho^*)$ where $x^*(\rho^*)$ is the fixed point at the critical parameter value ρ^*. Furthermore, the linearization of (13) at $x^*(\rho^*)$ possesses a simple eigenvalue $\lambda_1(\rho)$ with $\lambda_1(\rho^*) = -1$ and $\lambda_1'(\rho^*) \neq 0$, where $\lambda_1'(\cdot)$ is the derivative of $\lambda_1(\cdot)$ with respect to ρ.*

This hypothesis can be easily verified for TCP-RED map given by (13). Now we recall the nonlinear control theorem given in [1] for local control of period doubling bifurcation.

Theorem 4. *Under hypothesis 1 and for $l \cdot b \neq 0$, i.e., when the critical eigenvalue is controllable for linearized system, there is a feedback $u(x_k)$ with $u(x_k - x^*(\rho^*) = 0) = 0$, i.e., feedback control vanishes at the fixed point, which solves the local period doubling bifurcation control problem. Moreover, this can be accomplished with third order terms in $u(x_k)$, leaving the critical eigenvalue unaffected.*

Above theorem suggests a cubic control by itself can stabilize the system or a mixed control with linear terms can be used to enhance the stability of bifurcation in an extended parameter domain. This allows us to consider different functional forms for the control in (36). All these forms have been shown to enhance the stability of the fixed point, thus delaying the system bifurcations [1]. It is also important that only the error terms $x_k - x^*(\rho^*)$ from the nominal operating point is used to preserve the original operating point.

$$u_k = k_c {y_k}^3 \qquad \text{Cubic Control Law}$$
$$u_k = k_l y_k + k_c {y_k}^3 \quad \text{Mixed Control Law}$$

The stability analysis done in [1] also suggests that k_l and k_c be based on the computation of l and b. Clearly, we do not need a quadratic control due to critical eigendirection being linearly controllable. Cubic control can be used to change the nature of emerging period doubling orbit in the presence of uncertainty. According to the theoretical results in [1] it is possible to enhance the nonlinear stability terms by using just the cubic control terms. It is shown that stability coefficient β_2, which decides the nature of bifurcation in the absence of any control, equals

$$\beta_2 = -2\left(\frac{1}{2}\left(\frac{\partial^2 f}{\partial {q_k^n}^2}\right)^2 + \frac{1}{3}\left(\frac{\partial^3 f}{\partial {q_k^n}^3}\right)\right) . \tag{45}$$

This coefficient β_2, when evaluated for the linearized system, decides if the bifurcation will be super ($\beta_2 < 0$) or subcritical ($\beta_2 > 0$) [13]. With the cubic control terms β_2 is changed by the following value:

$$\Delta = -4C_u(r, r, r)lb \tag{46}$$

where $C_u(r, r, r)$ can be assigned any real value by an appropriate choice of cubic feedback to stabilize the ensuing bifurcation.

Along with the linear feedback term there is a sound reason to use small cubic terms in order to stabilize. The theorem from [1] supports this idea due to the fact that by using cubic term it is possible to stabilize the bifurcations with changes in parameters. There are indeed several reasons for using nonlinear feedback controls.

The effect of linear feedback control designed to stabilize the linearized version of the original system is difficult to determine. More precisely, when the bifurcation reappear at a different value of the bifurcation parameter, for instance, when the feedback control gain is small, the stability of this new bifurcation is not easily determined. Hence, using only a linear stabilizing feedback may be unacceptable if the goal is to stabilize a bifurcation and not merely to stabilize an equilibrium point for a fixed parameter value. In addition, in some cases a linear feedback that locally stabilizes an equilibrium point may result in globally unbounded behavior, whereas nonlinear feedback exists which stabilizes the equilibrium both locally and globally [25].

8.6 Numerical Example

In this section we study the effect of washout filter-aided control on RED by numerical simulations.

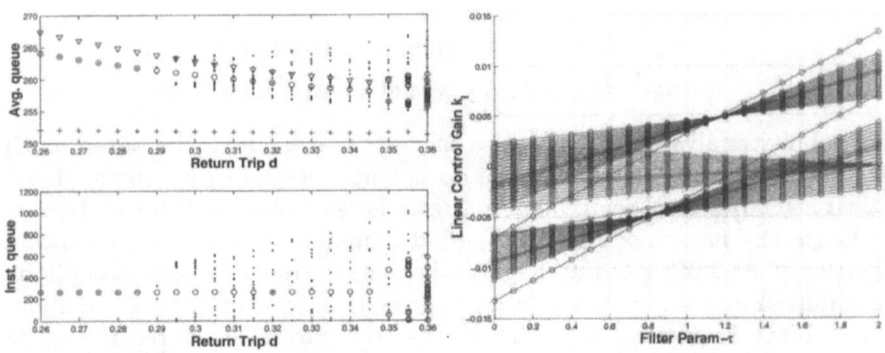

(a) (b)

Fig. 8: (a) Bifurcation diagram with and without control with respect to round trip delay d (with p_{max} modulation). Bifurcation diagrams in '.' and 'o' are plotted without and with control, respectively. (b) Allowed (τ, k_l) region lies below the line with 'Δ' for stability.

Fig. 8 plots the bifurcation diagram with respect to d and the stability region of (τ, k_l). Here we modulate p_{max} for feedback control, and only linear feedback control is used. The values of parameters used in the numerical example are as follows:

$$q_{max} = 747,\ q_{min} = 249,\ c{=}40 \text{ Mbps},\ K = \sqrt{3/2},$$
$$B = 3,735,\ w = 2^{-6},\ M = 4 \text{ kbits},\ N = 129,$$
$$k_l = -15/b,\ \tau = 0.2,\ d = \text{bifurcation parameter}$$

As shown in the fig 8(a) the washout filter-aided control delays the bifurcation. However, once the bifurcation takes place with feedback control, the system becomes even more unstable than the system without feedback control. This demonstrates the need for nonlinear feedback control as explained in the previous subsection.

References

1. E. H. Abed, H. O. Wang, and R. C. Chen. Stabilization of period doubling bifurcations and implications for control of chaos. *Physica D*, 70:154–164, 1994.
2. K. T. Alligood, T. D. Sauer, and J. A. Yorke. *Chaos: an introduction to dynamcal systems.* Springer-Verlag, New York, 1996.
3. S. Athuraliya, S. Low, V. H. Li, and Q. Yin. REM: active queue management. *IEEE Network*, 15:48–53, May/June 2001.
4. F. Baccelli, D. R. McDonald, and J. Reynier. A mean-field model for multiple TCP connections through a buffer implementing RED. TR-4449, INRIA, April 2002.
5. S. Banerjee, P. Ranjan, and C. Grebogi. Bifurcations in two-dimensional piecewise smooth maps - theory and applications in switching circuits. *IEEE Trans. on Circuits and Systems–I: Fundamental Theory and Applications*, 47(5):633–643, 2000.
6. M. D. Bernardo, M. I. Feigin, S. J. Hogan, and M. E. Homer. Local analysis of *c*-bifurcations in *n*-dimensional piecewise smooth dynamical systems. *Chaos Solitons and Fractals*, 10(11):1881–1908, 1999.
7. L.S. Brakmo and L.L. Peterson. TCP Vegas: end to end congestion avoidance on a global Internet. *IEEE Journal on Selected Areas in Communications*, 13(8):1465–80, October 1995.
8. V. Firoiu and M. Borden. A study of active queue management for congestion control. Proc. IEEE INFOCOM, 2000.
9. S. Floyd. TCP and explicit congestion notification. *ACM Computer Communication Review*, 24:10–23, October 1994.
10. S. Floyd, R. Gummadi, and S. Shenker. Adaptive RED: an algorithm for increasing the robustness of RED's Active Queue Management. Available at http://www.icir.org/floyd, August 2001.
11. S. Floyd and V. Jacobson. Random early detection gateways for congestion avoidance. *IEEE/ACM Transactions on Networking*, 1(4):397–413, Auguest 1993.
12. R. J. Gibbens and F. Kelly. Resource pricing and the evolution of congestion control. *available at http://www.statslab.cam.ac.uk/˜frank*, June 1998.
13. J. Guckenheimer and P. Holmes. *Nonlinear Oscillations, Dynamical Systems, and Bifurcations of Vector Fields.* Springer-Verlag, NY, 1983.
14. J. P. Hespanha, S. Bohacek, K. Obraczka, and J. Lee. Hybrid modeling of TCP congestion control. *Lecture notes in computer science*, (2034):291–304, 2001.

15. C. V. Hollot, V. Misra, D. Towsley, and W. Gong. A control theoretic analysis of RED. Proc. IEEE INFOCOM, 2001.
16. E. A. Jackson. *Perspectives of Nonlinear Dynamics.* Cambridge University Press, NY, 1991.
17. V. Jacobson. Congestion avoidance and control . *Computer Communication Review*, 18(4):314–29, August 1988.
18. Thomas Kailath. *Linear Systems.* Prentice Hall, Englewood CLiffs, NJ, 1980.
19. R. J. La, P. Ranjan, and E. H. Abed. Analysis of Adaptive Random Early Detection (ARED). In *Proc. of 18th ITC*, September 2003.
20. T. Li and J. A. Yorke. Period three implies chaos. *Americal mathematical monthly*, 82(10):985–992, December 1975.
21. D. P. Lindroff. *Theory of Sampled-data Control Systems.* Wiley, New York, 1965.
22. S. H. Low, F. Paganini, J. Wang, S. Adlakha, and J. C. Doyle. Dynamics of TCP/RED and a scalable control. In *Proc. IEEE INFOCOM*, 2002.
23. M. Mathis, J. Semke, J. Mahdavi, and T. Ott. The macroscopic behavior of the TCP congestion avoidance algorithm. *Computer communication review*, 27(3), 1997.
24. M. May and B. Lyles J. Bolot, C. Diot. Reasons not to deploy red. In *Proc. IWQoS'97*, 1997.
25. F. C. Moon and R. H. Rand. Parametric stiffness control of flexible structures. *NASA Conf Proc.: Workshop on Identification and Control of Flexible Space Structures*, pages 329–338, May 1984.
26. H. E. Nusse and J. A. Yorke. Border collision bifurcations including "period two to period three" for piecewise smooth maps. *Physica D*, 57(11):39–57, 1992.
27. E. Ott, C. Grebogi, and J. A. Yorke. Controlling chaos. *Physical Review Letters*, 64(11):1196–1199, 1990.
28. J. Padhye, V. Firoiu, D. Towsley, and J. Kurose. Modeling TCP Reno performance: a simple model and its empirical validation. *IEEE/ACM Transactions on Networking*, 8(2):133–45, April 2000.
29. K. Pyragas. Continuous control of chaos by self-controlling feedback. *Physics Letters A*, 170(6):421–428, 1992.
30. P. Ranjan and E. H. Abed. Nonlinear analysis and control of TCP-RED in a simple network model. In *Proc. ACC*, 2002.
31. P. Ranjan, E. H. Abed, and R. J. La. Nonlinear instabilities in TCP-RED. In *Proc. of the IEEE INFOCOM*, 2002.
32. P. Ranjan, E. H. Abed, and R. J. La. Trade-offs in rate control with communication delay. available at http://www.eng.umd.edu/~hyongla, September 2003.
33. A. N. Sharkovsky, S. F. Kolyada, A. G. Sivak, and V. V. Fedorenko. *Dynamics of one Dimensional maps.* Kluwer Academic, 1997.
34. P. Tinnakornsrisuphap and A. Makowski. Limit behavior of ECN/RED gateways under a large number of TCP flows. In *Proc. of the IEEE INFOCOM*, March 2003.

Synchronization in Complex Networks

Ljupco Kocarev[1] and Gábor Vattay[2]

[1] Institute for Nonlinear Science, University of California, San Diego, La Jolla, CA, USA; lkocarev@ucsd.edu

[2] Department of Physics of Complex Systems of the Eotvos Lorand University, Budapest, Hungary; vattay@gawain.elte.hu

1 Introduction

The study of complex systems pervades all of science, from cell biology to ecology, from computer science to meteorology. A paradigm of a complex system is a network [1] where complexity may come from different sources: topological structure, network evolution, connection and node diversity, and/or dynamical evolution. Examples of networks include food webs [2, 3], electrical power grids, cellular and metabolic networks, the World-Wide Web [4], the Internet backbone [5], neural networks, and co-authorship and citation networks of scientists. These networks consist of nodes which are interconnected by a mesh of links. The macroscopic behavior of a network is determined by both the dynamical rules governing the nodes and the flow occurring along the links.

Real networks of interacting dynamical systems – be they neurons, power stations or lasers – are complex. Many real-world networks are small-world [6] and/or scale-free networks [7]. The presence of a power-law connectivity distribution, for example, makes the Internet a scale-free network. The research on complex networks has been focused so far on the their topological structure [8]. However, most networks offer support for various dynamical processes. In this paper we propose to study one aspect of dynamical processes in non-trivial complex network topologies, namely their synchronization behaviors.

The general question of network synchronizability, for many aspects, is still an open and outstanding research problem [9]. There are, in general, two classes of results which give criteria under which a network of oscillators synchronizes. The first class of results uses Lyapunovs direct method by constructing a Lyapunov function which decreases along trajectories and gives analytical criteria for local or global synchronization. For example, in [10], the authors gave sufficient conditions for an array of linearly coupled systems to synchronize. A typical result states that the array will synchronize if the nonzero eigenvalues of the coupling matrix have real parts that are negative enough. The work in [10] has been extended and generalized in [11, 12, 13, 14, 15].

The second class of results uses linearized equations around the synchronization manifold and computes numerically the Lyapunov exponents of the variational equations. In this context, an important contribution has been given by Pecora and Carroll in [16], where, for a network of coupled chaotic oscillators, they derived the so-called *Master Stability Equation* (MSE), and introduced the corresponding *Master Stability Function* (MSF). Consequently, the stability analysis of the synchronous manifold [16] for the network under consideration can be decomposed in two sub-problems. The first sub-problem consists of deriving the MSF for the network nodes, *i.e.* to study in which region, of the complex plane the MSE admits a negative largest Lyapunov exponent (LE). The second sub-problem is to verify whether the eigenvalues of the so-called *connectivity matrix* [17] of the network, apart from the zero-eigenvalue, lie in the synchronization region(s) (see also [16, 17, 18]). This approach is particularly relevant because the MSE depends only on the nodes local dynamics and on the *coupling matrix* [17]. It turns out that the mathematical problem has the same dimension as the single network node. For example, when considering a network of coupled Rössler systems [19], the master stability equation has dimension three.

Recently, the synchronization phenomenon in scale-free dynamical networks has been investigated in [20, 21]. Robustness and/or fragileness of the networks' synchrony is discussed. Networks' synchronization and desynchronization processes in a scale-free network are illustrated by a prototype composing of Henon maps. A new general method to determine global stability of total synchronization in networks with different topologies is proposed in [22, 23]. This method combines the Lyapunov function approach with graph theoretical reasoning. In particular, the method is applied to the study of synchronization in rings of $2K$-nearest neighbor coupled oscillators. This method is extended to the blinking model of small-world networks where, in addition to the fixed $2K$-nearest neighbor interactions, all the remaining links are rapidly switched on and off independently of each other.

In this work, at first as a motivation, we study synchronization properties of a simple TCP/IP network. Recently, a new class of models has been introduced, which appears to be the most promising approach for scalable and accurate performance analysis of large TCP/IP networks. This new approach, often called 'fluid models', adopts an abstract deterministic description of the average network dynamics through a set of ordinary differential equations. For a large class of systems (described with a set of ordinary differential equations) the synchronization region may have following forms: an interval $(\beta_1, +\infty)$, an interval (β_1, α_1), or an empty set. Since large TCP/IP networks have typically complex topologies (power-law, scale-fee and/or hybrid models), we study synchronization in complex networks topologies. Section 2 is devoted to the analysis of synchronization properties of classical and power-law random graph models. We prove that random graphs networks synchronize. In section 3 we study synchronization properties of hybrid networks. We prove that although local graph networks do not synchronize for large N, adding only a small number of global edges makes these hybrid networks to synchronize. We close our paper with conclusion.

1.1 Synchronization in TCP/IP Networks

Traditional approaches to performance evaluation of telecommunication networks in general, and of packet networks in particular, have normally relied on attempts to describe as closely as possible the dynamics of network elements over a discrete state-space. Discrete models are a natural choice, since the operations of traffic sources, switches, and protocols, are normally governed by finite-state machines, whose dynamics determine the IP network performance. However, discrete models, requiring the description of the dynamics of the different network elements over their discrete state spaces, suffer from limited scalability, thus allowing only the performance analysis of rather small networking setups. This is the reason why only toy topologies are normally considered in IP network performance studies, and models almost invariably concentrate on a very limited subset of the network protocol stack.

A new class of semi-analytical models has recently been introduced, which appears to be the most promising approach for scalable and accurate performance analysis of large IP networks. This new approach, often called 'fluid models', adopts an abstract deterministic description of the average network dynamics through a set of ordinary differential equations [24, 25, 26, 27, 28], thus neglecting the short term, packet-by-packet description of the stochastic network dynamics. The resulting set of differential equations is then solved numerically, obtaining estimates of the time-dependent network behavior.

The most attractive property of fluid models resides in the fact that their complexity (i.e., the number of differential equations to be solved) is independent of the number of TCP flows and of link capacities, when considering traffic scenarios comprising only long-lived TCP flows (commonly called 'elephants'). In addition, fluid models have been recently proved to capture the limiting behavior of TCP elephants in single bottleneck topologies when the number of TCP flows grows very large [26, 29, 30, 31].

In [26, 27] a fluid model is proposed to describe the dynamics of the average window for TCP elephants traversing a network of drop-tail routers. The behavior of such a network is pulsing: congestion epochs in which some buffers are overloaded (and overflow) are interleaved to periods of time in which no buffer is overloaded, and no loss is experienced, due to the fact that previous losses forced TCP sources to reduce their sending rate. In such a setup, a careful analysis of the average TCP window dynamics at congestion epochs is necessary, whereas sources can be simply assumed to increase their rate at constant speed between congestion epochs. This behavior allows the development of fluid equations and an efficient methodology to solve them. Ingenious queueing theory arguments are exploited to evaluate the loss probability during congestion epochs, and to study the synchronization effect among sources that share the same bottleneck link. In this fluid model, the TCP algorithm is approximated by the continuous time Additive Increase–Multiplicative Decrease (AIMD) model. We give a short introduction to the AIMD model next. A more detailed description of AIMD model can be found in [26, 27].

The traffic load offered by a TCP is controlled by the so called *congestion window* variable. The congestion window limits the maximal number of data packets sent by the TCP during a round-trip time period. After a short transient period (*slow start*), the development of the congestion window of the ith TCP $w^{(i)}$, between two consecutive packet loss events, can be approximated by the following differential equation:

$$\frac{dw^{(i)}}{dt} = \frac{1}{T^{(i)}_{\mathrm{RTT}}(\mathbf{w})}, \tag{1}$$

where $T^{(i)}_{\mathrm{RTT}}$ denotes the round-trip time of the ith TCP connection. Note, that (1) is applicable only if the packet loss ratio is modest ($< 1 - 2\%$).

In the Internet, the packets might be buffered at the routers, which can affect the round-trip time, encountered by the packets. That is why the round-trip time can depend on the congestion windows ($\mathbf{w}$). If the buffering delays are negligible, the round-trip times can be considered to be constants, and (1) can be solved easily:

$$w^{(i)}(t) = w^{(i)}(0) + t/T^{(i)}_{\mathrm{RTT}} \tag{2}$$

(Additive Increase).

By Little's law, the stationary mean of the throughput $\bar{X}^{(i)}$ can be expressed by the mean of the congestion window $\bar{w}^{(i)}$:

$$P\bar{w}^{(i)} = T^{(i)}_{\mathrm{RTT}}\bar{X}^{(i)}, \tag{3}$$

where the coefficients are the packet size P in bits and the round-trip time $T^{(i)}_{\mathrm{RTT}}$. A heuristic, but reasonable assumption of the AIMD model is that Little's law would apply to instantaneous values as well: $Pw^{(i)}(t) = T^{(i)}_{\mathrm{RTT}}X^{(i)}(t)$.

Eqs. (1) and (2) are valid only between packet loss events. Let us denote the set of TCPs which share the link $e \in E$ by I_e, the nth packet loss event by T^e_n, the throughput of TCP $i \in I_e$ at T^e_n by $X^{(i)}_n = X^{(i)}(T^e_n)$ and the nth inter-loss period by $\tau^e_n = T^e_n - T^e_{n-1}$. Using the fact that losses occur at link $e \in E$ as soon as the link capacity is reached:

$$\sum_{i \in I_e} \left(X^{(i)}_n + \frac{P}{T^{(i)\,2}_{\mathrm{RTT}}} \tau^e_{n+1} \right) = C_e. \tag{4}$$

¿From (4), the "virtual" inter-congestion time τ^e_{n+1} can be expressed:

$$\tau^e_{n+1} = \frac{C_e - \sum_{i \in I_e} X^{(i)}_n}{\sum_{i \in I_e} P/T^{(i)\,2}_{\mathrm{RTT}}}. \tag{5}$$

Eq. (5) is only "virtual" inter-congestion time, since only the minimum of τ_{n+1}^{e} is realized at network level:

$$\tau_{n+1} = \min_{e \in E} \tau_{n+1}^{e}. \tag{6}$$

At a congestion event, some of the TCPs, which share the congested link(s), experience packet losses. The packet losses are modeled by the 0, 1 valued stationary stochastic processes $\xi_n^{(i)} \equiv \xi^{(i)}$. The mean of process $\xi^{(i)}$ is called *synchronization rate* of session i, $p^{(i)} = \mathbb{E}\,\xi^{(i)}$. Here the homogeneous situation is considered, i.e. $p^{(i)} = p$.

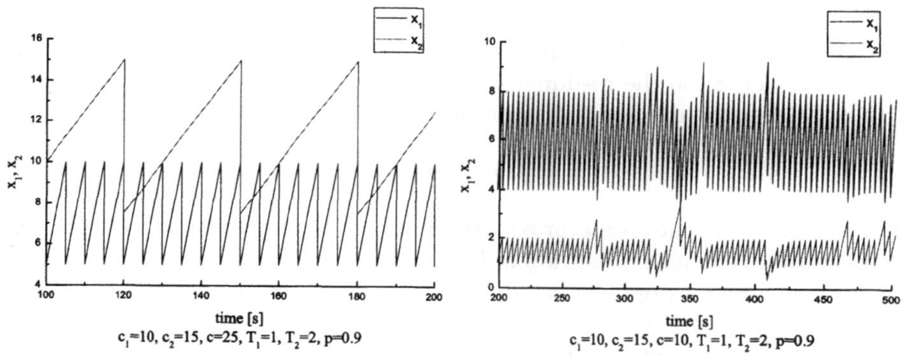

Fig. 1: The throughputs as function of the time. On the left: The TCPs oscillate independently, on the right: high synchronization rate

According to the AIMD model, the ith TCP, which utilizes the congested link, losses a data packet, if the random value of $\xi^{(i)}$ is 1 at a congestion epoch. Since every TCP halves its congestion window at a packet loss event (Multiplicative Decrease), the evolution of the throughput can be given by:

$$X_{n+1}^{(i)} = \left(1 - \chi_{I_B}(i)\,(1-\beta)\,\xi_{n+1}^{(i)}\right)\left(X_n^{(i)} + \frac{P}{T_{\mathrm{RTT}}^{(i)\,2}}\tau_{n+1}\right), \tag{7}$$

where I_B is the set of TCPs, which share the bottleneck link(s),

$$\chi_{I_B}(i) = \begin{cases} 1, & \text{if } i \in I_B \\ 0, & \text{if } i \notin I_B \end{cases}$$

is the characteristic function of set I_B, and $\beta = 1/2$.

We now consider a simple network, consists of two TCP flows which share a common link with capacity c. In this configuration the computer A sends data to the computer B on the first TCP route. The computer C sends data to the computer D on the second TCP route. Between A and B there are two links, with capacity c_1, which are used only by the first TCP. Between C and D there are two links, with capacity c_2, which are used only by the second TCP. Finally, there is a common link, with capacity C, which is shared by first and second TCP.

Figure 1 shows the result of a numerical simulation, for which we assume the following simple scenario: the routers have no buffer capacity, there is no delay, and the RTT times are assumed constant over the time. In this figure, x_1 is the first TCP throughput and x_2 is the second TCP throughput. One can see clearly the synchronization effect on the right figure.

1.2 Synchronization in Networks

Let us consider a network comprising N identical nodes, each being a (chaotic) oscillator. Let $\boldsymbol{x}_i$ be the m-dimensional vector of dynamical variables for the i-th node. Let the dynamics of each node be described by:

$$\dot{\boldsymbol{x}} = \boldsymbol{f}(\boldsymbol{x}) - \sigma \boldsymbol{L} \otimes \boldsymbol{H}\boldsymbol{x}, \tag{8}$$

where $\boldsymbol{f}$ describes the oscillator equations, σ is the overall strength of coupling, and the $N \times N$ matrix $\boldsymbol{L}$ is the Laplacian matrix representing the connection topology of the network: $l_{ij} = -1$ if nodes i and j are connected, $l_{ii} = k_i$ if node i is connected to k_i other nodes, and $l_{ij} = 0$ otherwise. The coupling matrix $\boldsymbol{H} = (h_{ij})$ contains the information about which variables are utilized in the coupling and is defined as $h_{ii} = 1$, if the i-th component is coupled, and $h_{ii} = 0$, otherwise.

The matrix $\boldsymbol{L}$, which will be our main concern, is positive semi-definite and symmetric. Its smallest eigenvalue is $\gamma_0 = 0$. Denote by γ_k the k-th smallest eigenvalue of $\boldsymbol{L}$, respecting the multiplicities, $k = l, 2, \ldots, N$. In particular, γ_N, is the maximal eigenvalue of $\boldsymbol{L}$. In this case MSF depends only on one parameter, α. We denote by $\mathcal{S} \subseteq I\!R$ the region where the master stability function (MSF) is negative and call it *synchronization region*. For the system (8), the synchronization region $\mathcal{S}$ may have one of the following forms: $\mathcal{S} = \emptyset$, $\mathcal{S} = (\beta_1, \infty)$, and $\mathcal{S} = (\beta_1, \alpha_1)$.

In the remaining of this manuscript we will focus on the second and third scenarios since they are generic for a typical dynamical system. It is easy to see that for the second scenario the condition of stable synchronous state is $\sigma\gamma_2 > \beta_1$, while for the third this condition reads $\gamma_N/\gamma_2 <\mid \alpha_1/\beta_1 \mid$. Therefore, for a large class of (chaotic) oscillators there exist two classes of networks; class-A: networks for which the condition of stable synchronous state is $\sigma\gamma_2 > a$, and class-B: networks for which this condition reads $\gamma_N/\gamma_2 < b$, where $a = \beta_1$ and $b =\mid \alpha_1/\beta_1 \mid$ are constant that depend on $\boldsymbol{f}$, the synchronous state $\boldsymbol{x}_1 = \boldsymbol{x}_2 = \ldots = \boldsymbol{x}_N$ and the matrix $\boldsymbol{H}$, but not on the Laplacian matrix $\boldsymbol{L}$. For typical oscillators $b > 1$.

2 Synchronization in random graphs

2.1 Preliminaries

A *graph* is an ordered pair of disjoint sets (V, E) such that E is a subset of the set of unordered pairs of V. The set V is the set of *vertices* and E is the set of *edges.* If G is a graph then $V = V(G)$ is the vertex set of G and $E = E(G)$ is the edge set. The edge $\{x, y\}$ is said to *join* the vertices x and y and is denoted by xy. Thus xy and yx means exactly the same edge; the vertices x and y are the *endvertices* of this edge. If $xy \in E(G)$ then x and y are *adjacent* or *neighboring* vertices of G and the vertices x and y are *incident* with edge xy.

The *order* of G is the number of vertices; it is denoted by $|G|$, where $|\cdot|$ denotes the number of elements (cardinality) of a set. The *size* is the number of edges; it is denoted by $e(G)$. We write G_N for an *arbitrary graph of order* N. Similarly $G(N, m)$ denotes an *arbitrary graph of order* N *and size* m.

The set of vertices adjacent to a vertex $x \in G$ is denoted by $\Gamma(x)$. The *degree* of x is $d(x) = |\Gamma(x)|$. The *minimum degree* of the vertices of a graph G is denoted by $\delta(G)$ and the *maximum degree* by $\Delta(G)$. If $\delta(G) = \Delta(G) = k$, that is every vertex of G has degree k then G is said to be *k-regular* graph. If $V(G) = \{x_1, x_2, \ldots, x_N\}$, then $\delta(G) = d(x_1) \leq d(x_2) \leq \ldots \leq d(x_N) = \Delta(G)$ is a *degree sequence* of G. The average degree or simply degree of a graph is $d(G) = \sum_i d(x_i)/N = 2e(G)/|G|$. The degree distribution $p_d(k)$ denotes the fraction of vertices that have degree equal to k.

The size of a graph of order N is at least 0 and most $N(N-1)/2$. Clearly for every m, $0 \leq m \leq N(N-1)/2$, there is a graph $G(N, m)$. A graph of order N and size $N(N-1)/2$ is called a *complete n-graph* and is denoted by K_N. A *path* is a graph P of the form:

$$V(P) = \{x_0, x_1, \ldots, x_l\}, E(P) = \{x_0x_1, x_1x_2, \ldots, x_{l-1}x_l\}.$$

This path is usually denoted by $x_0x_1 \ldots x_l$. The vertices x_0 and x_l are *endvertices* of P and $l = e(P)$ is the *length* of P. We say that P is a path from x_0 to x_l or an $x_0 - x_l$ path.

A *walk* W in G is an altering sequence os vertices and edges, say x_0, α_1, $x_1, \alpha_2, \ldots$, α_l, x_l, where $\alpha_i = x_{i-1}x_i$, $1 \leq i \leq l$. For simplicity we write $W = x_0x_1 \ldots x_l$. Note that a path is a walk with distinct vertices. If a walk $W = x_0x_1 \ldots x_l$ is such that $l \geq 3$, $x_0 = x_l$, and the vertices x_i, $0 < i < l$, are distinct from each other and x_0 then W is said to be a *cycle*. The symbol P_l denotes an arbitrary path of length l and C_l denotes a cycle of length l.

Given vertices x, y, their *distance* $d(x, y)$ is the minimum length of an $x-y$ path. If there is no $x - y$ path then $d(x, y) = \infty$. A graph is *connected* if for every pair $\{x, y\}$ of distinct vertices there is a path from x to y. The *diameter* of the graph G is $\operatorname{diam}(G) = \max_{x,y} d(x, y)$. The *radius* of the graph G is $\operatorname{rad}(G) = \min_x \max_y d(x, y)$.

There are several ways to associate a matrix to a graph. The usual adjacency matrix $\boldsymbol{A}$ associated with a (simple) graph has eigenvalues quite sensitive to the maximum degree (which is a local property). The combinatorial Laplacian $\boldsymbol{L} = \mathrm{d} - \boldsymbol{A}$ with D denoting the diagonal degree matrix is a major tool for enumerating spanning trees and has numerous applications [32]. Another matrix associated with a graph is the (normalized) Laplacian $\tilde{\boldsymbol{L}} = \mathrm{i} - \mathrm{d}^{-1/2}\boldsymbol{A}\mathrm{d}^{-1/2}$ which controls the expansion/isoperimetrical properties (which are global) and essentially determines the mixing rate of a random walk on the graph [33]. The traditional random matrices and random graphs are regular or almost regular so the spectra of all the above three matrices are basically the same (with possibly a scaling factor or a linear shift). However, for graphs with power law distribution, the above three matrices can have very different distributions [34].

Recall $\gamma_1 = 0 \le \gamma_2 \le \ldots \le \gamma_N$, repeated according to their multiplicities, are eigenvalues of the matrix $\boldsymbol{L}$. These eigenvalues are called *Laplace eigenvalues* of the graph G. Laplace eigenvalues of the complete graph K_N are $\gamma_1(K_N) = 0$ and $\gamma_k(K_N) = N$ for $2 \le k \le N$. The Laplace eigenvalues of the N-cycle C_N are the numbers

$$2 - 2\cos\left(\frac{2k\pi}{N}\right), \qquad k = 0, \ldots, N-1.$$

It is easy to see that 0 is always an eigenvalue of $\boldsymbol{L}$, and than $(1, 1, \ldots, 1)^T$ is the corresponding eigenvector. More precisely, we have the following description of the multiplicity of 0 as an eigenvalue of $\boldsymbol{L}$.

Theorem 1. *The multiplicity of 0 as an eigenvalue of* $\boldsymbol{L}$ *is equal to the number of connected components of* G.

This implies if $\gamma_2 > 0$ then the graph is connected. The following inequalities hold

Theorem 2.

$$\gamma_2(G) \le \frac{N}{N-1}\delta(G) \le \frac{N}{N-1}\Delta(G) \le \gamma_N(G) \le 2\Delta(G). \tag{9}$$

The proof of the above two theorems can be found, for example, in [35, 36].

2.2 Synchronization in Classical Random Networks

We turn now to random graphs. The primary model for the classical random graphs is the Erdös-Rényi model $\mathcal{G}_q$, in which edge is independently chosen with the probability q for some given $q > 0$. Let $G(N, q)$ be a random graph on N vertices.

For the model of a random graph we take a sequence of probability spaces $(\Gamma(N, q))_N$, where q is a real number between 0 and 1, and N is an integer.

We shall assume that q is fixed, but in general it may depend on N. The probability space $\Gamma(N, q)$ consists of all labelled simple graphs on N vertices, and an edge between an arbitrary pair of vertices appears with probability q, i.e. $\Gamma(N, q)$ has 2^M elements, where $M = N(N-1)/2$, and each graph in $\Gamma(N, q)$ with m edges has the probability equal to $q^m(1-q)^{M-m}$. By $P_{N,q}(X)$ we will denote the probability of an event $X \subseteq \Gamma(N, q)$ in the probability space $\Gamma(N, q)$. Let $\rho(G)$ mean that the graph G has the property ρ. We say that almost every graph has property ρ (or ρ happens *asymptotically almost surely* (a.a.s)), if

$$\lim_{N \to \infty} P_{N,q}\{G \in \Gamma(N, q) \mid \rho(G)\} = 1.$$

Theorem 3. *Let $G(N, q)$ be a random graph on N vertices. For sufficiently large N, the class-A network $G(N, q)$ almost surely synchronize for arbitrary small coupling σ. For sufficiently large N, almost every class-B network $G(N, q)$ with $b > 1$ is synchronizable.*

Proof. The proof of the theorem follows from the following result [37]. Let q be a fixed real number between 0 and 1. For almost every graph and every $\varepsilon > 0$

$$qN - \sqrt{(2+\varepsilon)pqN \log N} < \gamma_2(G) < qN - \sqrt{(2-\varepsilon)pqN \log N}, \qquad (10)$$

and

$$qN + \sqrt{(2-\varepsilon)pqN \log N} < \gamma_N(G) < qN + \sqrt{(2+\varepsilon)pqN \log N}. \qquad (11)$$

Therefore, for large N, $\gamma_2 \approx N$, while γ_N/γ_2 approaches 1. Now, for class-A networks the condition for synchronization reads $\sigma > a/N$ and σ can be chosen arbitrary small. For class-B networks with $b > 1$, since γ_N/γ_2 approaches 1, when $N \to \infty$, it follows that the network almost surely synchronizes.

2.3 Synchronization in Power-Law Networks

There are several approaches for studying power-law graphs. In the the first approach, one constructs power-law graphs with prescribed degree sequence. Bender and Canfield [38] introduced a model, so called configuration model, to construct a random graph with a prescribed degree sequence. This model was refined by Bollobas [36]. Recently, Molloy and Reed [39, 40] used the configuration model to show that if some conditions are satisfied, then the graph almost surely has a giant component. The advantage of the configuration model is to generate graphs exactly with the prescribed degrees. However, there are several disadvantages of the configuration model. The analysis of the configuration model is much more complicated due to the dependency of the edges. A random graph from the configuration model is in fact a multi-graph instead of a simple graph. The probability of having multiple edges increases rapidly when the degrees increase.

Another line of approach is evolution models, in which one generates a vertex/edge at a time, starting from a node or a small graph. We briefly mention several such evolution models. Barabasi and Albert [7] describe the following graph evolution process. Starting with a small initial graph, at each time step they add a new node and an edge between the new node and each of m random nodes in the existing graph, where m is a parameter of the model. The random nodes are not chosen uniformly. Instead, the probability of picking a node is weighted according to its existing degree (the edges are assumed to be undirected). Using heuristic analysis with the assumption that the discrete degree distribution is differentiable, they derive a power law for the degree distribution with a power of 3, regardless of m. A power law with power 3 for the degree distribution of this model was independently derived and proved by Bollobas et al. [41]. Kumar at el. [42] proposed three evolution models: "linear growth copying", "exponential growth copying", and "linear growth variants". Aiello et al. described a general random graph evolution process in [43] for generating directed power law graphs with given expected in-degrees and out-degrees. Recently, Cooper and Frieze [44] independently analyzed the above evolution of adding either new vertices or new edges and derived power law degree distribution for vertices of small degrees.

In this section we consider a random model introduced recently by Chung and Lu [45], which produces graphs with a given expected degree sequence. Therefore, this model does not produce the graph with exact given degree sequence. Instead, it yields a random graph with given expected degree sequence.

We consider the following class of random graphs with a given expected degree sequence

$$w = (w_1, w_2, \ldots, w_n).$$

The vertex v_i is assigned vertex weight w_i. The edges are chosen independently and randomly according to the vertex weights as follows. The probability p_{ij} that there is an edge between v_i and v_j is proportional to the product $w_i w_j$ where i and j are not required to be distinct. There are possible loops at v_i with probability proportional to w_i^2 , i.e.,

$$p_{ij} = \frac{w_i w_j}{\sum_k w_k}, \tag{12}$$

and we assume $\max_i w_i^2 < \sum_k w_k$. This assumption ensures that $p_{ij} \leq 1$ for all i and j. We denote a random graph with a given expected degree sequence w by $G(w)$. For example, a typical random graph $G(N, q)$ (see the previous section) on N vertices and edge density q is just a random graph with expected degree sequence $(qN, qN, \ldots, qN)$. The random graph $G(w)$ is different from the random graphs with an exact degree sequence such as the configuration model. We will use d_i to denote the actual degree of v_i in a random graph G in $G(w)$, where the weight w_i denotes the expected degree. The following proposition is proven in [45].

Proposition 1. *With probability* $1 - 2/N$, *all vertices* v_i *satisfy*

$$2\sqrt{w_i \log N} \leq d_{v_i} - w_i \leq \frac{2}{3}\log N + \sqrt{\left(\frac{2}{3}\log N\right)^2 + 4w_i \log N}. \tag{13}$$

Now we give some definitions. The expected average degree d of a random graph G in $G(w)$ is defined to be

$$d = \frac{1}{N}\sum w_i. \tag{14}$$

For a subset S of vertices, the volume of S, denoted by $\mathrm{Vol}(S)$, is the sum of expected degrees in S:

$$\mathrm{Vol}(S) = \sum_{v_i \in S} w_i.$$

In particular, the volume $\mathrm{Vol}(G)$ of $G(w)$ is just $\mathrm{Vol}(G) = \sum_i w_i = Nd$.

If a graph strictly follows the power law, then the average degree as well as its connectivity will be completely determined by the exponent of the power law (see [46]). However, for most realistic graphs, the power law holds only for a certain range of degrees, namely, for the degrees which not too small and not too large. We will consider the following model with the consideration that most examples of massive graphs satisfying power law have exponent $\beta > 2$.

In this paper we consider the model $M(N, \beta, d, m)$, where N is the number of vertices, $\beta > 2$ is the power of the power law, d is the expected average degree, and m is is the expected maximum degree (or an upper bound for the range of degrees that obey the power law), such that $m^2 = o(Nd)$. We assume that the i-th vertex v_i has expected degree

$$w_i = ci^{-\frac{1}{\beta-1}},$$

for $i_0 \leq i < N + i_0$. Here c depends on the average degree d and i_0 depends on the maximum expected degree m. It is easy to compute that the number of vertices of expected degree between k and $k+1$ is of order $c'k^{-\beta}(\beta - 1)$ as required by the power law. To determine c, we consider

$$\mathrm{Vol}(G) = \sum_i w_i \approx c\frac{\beta-1}{\beta-2}N^{1-\frac{1}{\beta-1}}.$$

Since $Nd \approx \mathrm{Vol}(G)$, we have

$$c = \frac{\beta-2}{\beta-1}dN^{\frac{1}{\beta-1}}.$$

From

$$m = ci_0^{-\frac{1}{\beta-1}},$$

it follows

$$i_0 = N \left[\frac{d}{m} \frac{(\beta-2)}{(\beta-1)} \right]^{\beta-1} .$$

Let k be the expected minimum degree. Then

$$k = \frac{\beta-2}{\beta-1} d \left[1 + \left(\frac{d}{m} \frac{(\beta-2)}{(\beta-1)} \right)^{\beta-1} \right]^{-\frac{1}{\beta-1}} ,$$

which can be written as

$$k \approx d \left[1 + \left(\frac{d}{m} \right)^{\beta-1} \right]^{-\frac{1}{\beta-1}} .$$

Assuming further that $d \ll m$, we have $k \approx d$.

Theorem 4. *Let $M(N, \beta, d, m)$ be a random power-law graph on N vertices. For sufficiently large N, if d grows with N, then the class-A network $M(N, \beta, d, m)$ almost surely synchronize for arbitrary small coupling σ. Otherwise, if $d < \infty$, then a class-A network synchronizes only if $\sigma > a/d$. For sufficiently large N, almost every class-B network $M(N, \beta, d, m)$ is synchronizable only if $\lim_{N\to\infty} m/d < b$, assuming that d grows with N. Otherwise, if $d < \infty$, then a class-B network does not synchronize.*

Proof. ¿From equations (9) and (13) it follows that for large N $\gamma_N(M) \approx \Delta \approx m$. We now show that $\gamma_2(M) \approx d$.

For the considered model d can be in any range greater than 1: it does not have to grow with N [47]. It is proven in [48] that the function $\gamma_2(G)$ is non-decreasing for graphs with the same set of vertices. i.e. $\gamma_2(G_1) \leq \gamma_2(G_2)$ if $G_1 \subseteq G_2$ and G_1, G_2 have the same set of vertices. Let G_2 be our $M(N, \beta, d, m)$ random graph and G_1 be a d-regular random graph which has the same set of vertices as G_2. Then $\gamma_2(M) \geq \gamma_2(G_1)$.

According to [50] the isoperimetric number of almost every d-regular random graph satisfies $i \geq d/2 - \sqrt{k \ln 2}$ for large enough d. According to [49], γ_2 satisfies $i \leq \sqrt{\gamma_2(2d - \gamma_2)}$. Therefore, we have

$$\gamma_2(G_1) \geq d - \sqrt{\frac{3}{4} d^2 - d(\ln 2 - \sqrt{d \ln 2})}.$$

¿From equations (9) and (13) it follows that for large N, $\gamma_2(M) < d$. Combining the last two inequalities, we find that $\gamma_2(M) \approx d$.

If d grows as $N \to \infty$, we conclude that the class-A network $M(N, \beta, d, m)$ almost surely synchronize for arbitrary small coupling σ. On the other hand, if $d < \infty$, then a class-A network synchronizes only if $\sigma > a/d$. ¿From $\gamma_N/\gamma_2 \approx m/d$, we see that for sufficiently large N, almost every class-B network $M(N, \beta, d, m)$ is synchronizable only if $\lim_{N\to\infty} m/d < b$.

Note, however, that if σ_1 and σ_2 are the values of the coupling parameter for which synchronization is achieved in models $G(N,q)$ and $M(N,\beta,d,m)$, respectively, then $\sigma_1 < \sigma_2$. In addition, note also that the limes $\lim_{N\to\infty} m/d$ can have any values from the set $[1,\infty)$. Therefore, random model $G(N,q)$ synchronies better than the model $M(N,\beta,d,m)$.

3 Synchronization in Hybrid Networks

It has been observed that many realistic networks possess the so-called small world phenomenon, with two distinguishing properties: small distance between any pair of nodes, and the clustering effect that two nodes are more likely to be adjacent if they share a neighbor. In this section, we consider a hybrid graph model proposed by Chung and Lu [51], which has both aspects of the small world phenomenon. Roughly speaking, a hybrid graph is a union of a global graph (consisting of "long edges" providing small distances) and a local graph (consisting of "short edges" respecting local connections).

Examples of local graphs include paths and cycles. More generally, we define a local graph as follows: consider a lattice graph where the vertices are in a d-dimensional lattice where each vertex is a d-dimensional vector in the hypercube $[0,r]^d$ with integer entries. Suppose each vertex is connected to its nearest neighbors. This graph, also known as the grid graph, has diameter D, which as a function of the number of vertices N, and has maximum vertex degree $\Delta = 2d$.

Theorem 5. *When $N \to \infty$ local graphs for both class-A and class-B oscillators do not synchronize.*

Proof. It is know that, see for example [12],

$$\gamma_2 < \frac{2d\ln(N-1)}{2(D-2)-\ln(N-1)},$$

if $2(D-2)-\ln(N-1) > 0$. Therefore, $\gamma_2 \to 0$ as $N \to \infty$ for the grid graphs. This is also true when the vertices are connected to neighbors in an arbitrary local neighborhood. On the other hand, $\gamma_N \le 2\Delta(G) = 4d$. Therefore, $\gamma_N/\gamma_2 \to \infty$ as $N \to \infty$.

A hybrid graph consists of two parts: a global graph and a local graph. The edge set of the hybrid graph is a disjoint union of the edge set of the global graph G and that of the local graph L. We consider two cases: classical random graph model $G(N,q)$ described in the section III.B and power-law random graph model $M(N,\beta,d,m)$ described in the section III.C. For local graph L we consider the grid graph, although other choices are also possible. For example, Chung and Lu use two parameters to describe the local connectivity. For any fixed two integers $k \ge 2$ and $l \ge 2$, a graph L is called "locally (k,l)-connected" if for any edge uv, there are at least l edge-disjoint paths with length at most k joining from u to v (including the edge uv). For example, the grid graph $C_n \square C_n$ is locally $(3,3)$-connected as well as locally $(5,9)$-connected. For any two points u and v, the probability of choosing an edge between u and v is denoted by $p(u,v)$, defined as follows: $p(u,v) = 1$ if uv is an edge of L, $p(u,v) = q$ for a classical random graph, and $p(u,v) = w_u w_v \rho$ for a power-law graph.

Let now consider a hybrid network for which equation of the motion can be written as:

$$\dot{\boldsymbol{x}} = \boldsymbol{f}(\boldsymbol{x}) - \sigma(\boldsymbol{L}_L + \boldsymbol{L}_G) \otimes \boldsymbol{H}\boldsymbol{x}, \tag{15}$$

where $\boldsymbol{L}_L$ is the matrix describing the local graph L, and $\boldsymbol{L}_G$ is the coupling matric of the global graph G. Let $N_{total} = N(N-1)/2$ be the total number of edges (links) in a network with N nodes and N_L be the total number of local edges. Then $N_G = N_{total} - N_L$ is the number of all possible global edges. Let pN_G, where $0 \leq p \leq 1$, be a number of global edges.

Theorem 6. *Assume N is large enough and let G be a global graph (classical random graph model or power-law model). Then for class-A networks, given a, there exist a number p, such that $\sigma_c(p) \ll \sigma_c(0)$, where $\sigma_c(p) = a/\gamma_2(p)$, $\gamma_2(p)$ is the second eigenvalue of the matrix $\boldsymbol{L}_L + \boldsymbol{L}_G$, and $\gamma_2(0)$ is the second eigenvalue of the matrix $\boldsymbol{L}$. For class-B networks, given $b > 1$, there exist a number p, such that $\gamma_N(p)/\gamma_2(p) < b$, where $\gamma_2(p)$ and $\gamma_N(p)$ are the second and the N-th eigenvalue, respectively, of the matrix $\boldsymbol{L}_L + \boldsymbol{L}_G$.*

Proof. Since for $p = 1$, the matrix $\boldsymbol{L}_L + \boldsymbol{L}_G$ is fully connected, it follows that $\gamma_i(1) = N$, $i \geq 1$; hence $\gamma_2(1) = N$. On the other hand, we know that $\gamma_2(0)$ is a small number. It is proven in [48] that the function $\gamma_2(G)$ is non-decreasing for graphs with the same set of vertices. i.e. $\gamma_2(G_1) \leq \gamma_2(G_2)$ if $G_1 \subseteq G_2$ and G_1, G_2 have the same set of vertices. Therefore, $\gamma_2(p)$ is non-decreasing function of p. Similar argument holds for class-B networks, for which $\gamma_N(p)/\gamma_2(p)$ is non-decreasing function of p. Thus, for both classes of networks (class-A and class-B), there exists a critical value of p, p_c, such that for $p > p_c$, almost all networks (15) are synchronizable.

We now present an example. Let the local graph L be a circle and $N = 128$. It is easy to compute that $\gamma_2 = 0.00241$ and $\gamma_N/\gamma_2 = 1659.75$.

Assume that the global graph is a classical random graph model. Consider first class-A oscillators for which $a = 1$ and $\sigma \leq 10$. Since $\gamma_2 = 0.00241$, the local network L of 128 oscillators does not synchronize. The dependence of γ_2 on p is shown in Figure 1. Since $\sigma\gamma_2 > a$, it follows that the hybrid graph $L + G$ synchronizes if $\gamma_2 > a/\sigma = 0.1$. ¿From Fig. 1 one easily finds that $\gamma_2 > 0.1$ already for $p = 0.005$. We consider now a network of class-B oscillators for which $b = 40$. Since $\gamma_N/\gamma_2 = 1659.75$, the local network L does not synchronize. The dependence of γ_N/γ_2 on p is shown in Figure 2. Since the condition for synchronization is $\gamma_N/\gamma_2 < b$, it follows that the hybrid graph $L + G$ synchronizes for $p = 0.01$. Therefore, adding only a small number of global edges makes the oscillators synchronize.

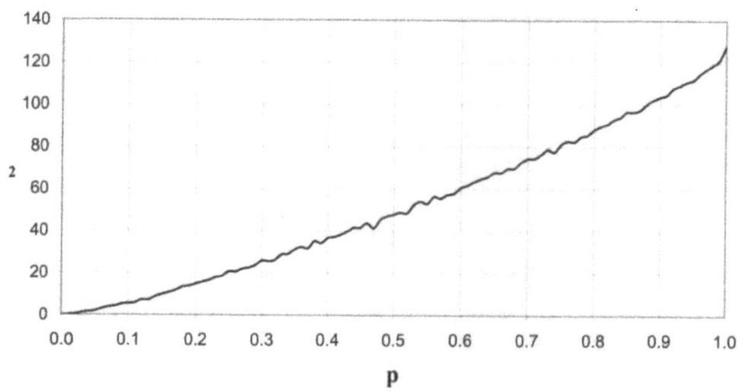

Fig. 2: γ_2 versus p for the hybrid model with $N = 128$, in which the local graph is a circle and the global graph is a classical random graph model.

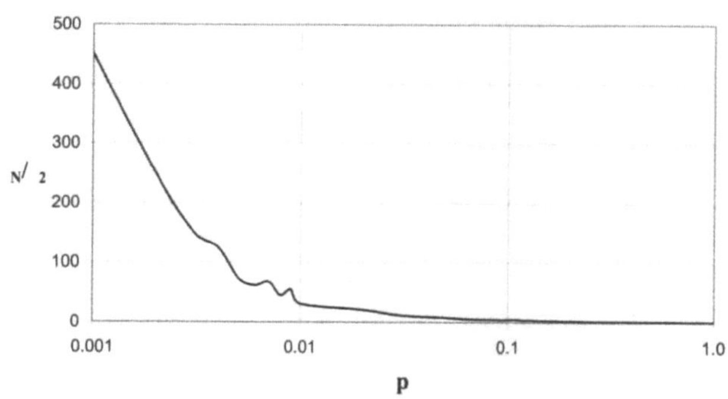

Fig. 3: γ_2/γ_N versus p for the hybrid model with $N = 128$, in which the local graph is a circle and the global graph is a classical random graph model.

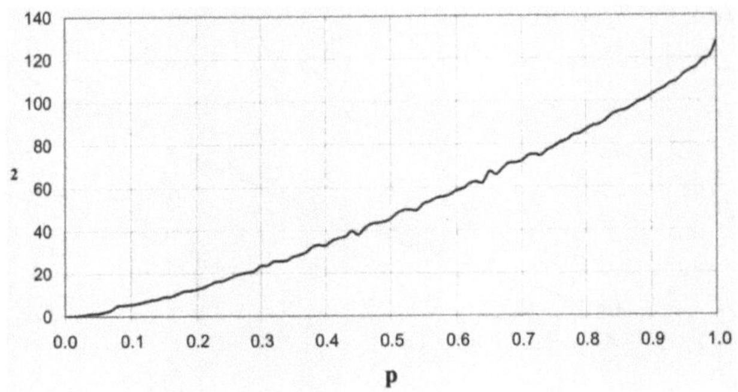

Fig. 4: γ_2 versus p for the hybrid model with $N = 128$ and $m = 5$, in which the local graph is a circle and the global graph is a power-law graph model.

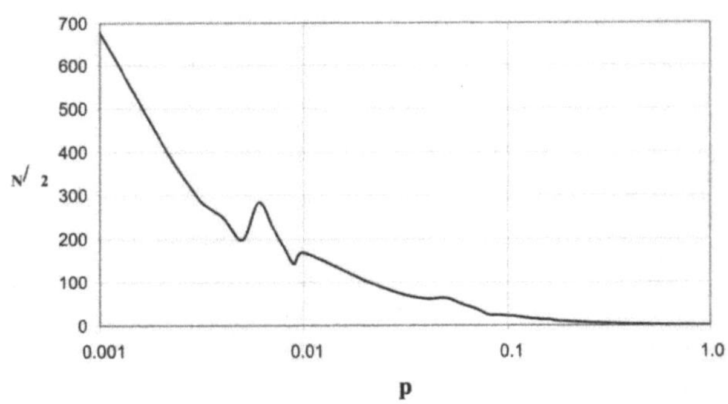

Fig. 5: γ_2/γ_N versus p for the hybrid model with $N = 128$ and $m = 5$, in which the local graph is a circle and the global graph is a power-law graph model.

Assume now that the global graph is a power-law graph model. Numerically we consider the model generated in the following way. First, we choose m nodes at random from all N nodes with equal probabilities and assign them to be centers. Second, we add links (global edges) by connecting one node chosen at random from all N nodes to another node randomly chosen from the m centers. Third, when all centers are fully connected with other nodes, we start uniformly to add links between the rest of the nodes. The dependence of of γ_2 and γ_N/γ_2 on p for such model is shown in Figure 3 and Figure 4, respectively for $m = 5$. From these figures and our numerical experiments, we may conclude: (i) $\gamma_N(p)$ increases reaching the maximum values N for smaller value of m; thus, γ_N reaches the value N in the fastest way for $m = 1$, and (ii) γ_2 is not effected by m. Therefore, the random model with m centers only influences synchronization property of class-B networks: if one adds global edges using the model with centers, the network becomes more difficult to synchronize. Thus, for example, class-B network with $b = 40$ will synchronize for $p = 0.08 > 0.01$.

4 Conclusion

In this chapter we have studied synchronization in networks with different topologies. We summarize the main conclusions of this chapter as follows:

- For a large class of oscillators there exist two classes of networks; class-A: networks for which the condition of stable synchronous state is $\sigma\gamma_2 > a$, and class-B: networks for which this condition reads $\gamma_N/\gamma_2 < b$, where a and b are constant that depend on local dynamics, synchronous state and the coupling matrix, but not on the Laplacian matrix of the graph describing the topology of the network.
- Let $G(N, q)$ be a classic random graph (Erdös-Rényi model) on N vertices. We proved that for sufficiently large N, the class-A network $G(N, q)$ almost surely synchronize for arbitrary small coupling σ. For sufficiently large N, almost every class-B network $G(N, q)$ with $b > 1$ is synchronizable.
- Let $M(N, \beta, d, m)$ be a random power-law graph on N vertices. We proved that for sufficiently large N, if d grows with N, then the class-A network $M(N, \beta, d, m)$ almost surely synchronize for arbitrary small coupling σ. Otherwise, if $d < \infty$, then a class-A network synchronizes only of $\sigma > a/d$. For sufficiently large N, almost every class-B network $M(N, \beta, d, m)$ is synchronizable only of $\lim_{N\to\infty} m/d < b$, assuming that d grows with N. Otherwise, if $d < \infty$, then a class-B network does not synchronize.
- Hybrid graphs are synchronizable, while oscillators in local graphs do not synchronize.

Acknowledgment

We are grateful to F. Chung, C.W. Wu, B. Mohar, A. Fekete, K. Diriczi, and M. Sterjev for stimulating discussions. This research is supported in part by the NSF and Hungarian Scientific Research Fund OTKA T03793.

References

1. A.-L. Barabasi: *Linked: The New Science of Networks* (Perseus Publishing, Cambridge, Massachusetts, 2002)
2. J. E. Cohen, F. Briand, C. M. Newman: *Community Food Webs: Data and Theory* (Springer, Berlin, 1990)
3. R. J. Williams, N. D. Martinez: "Simple rules yield complex food webs", Nature **404**, 180 (2000)
4. A. Broder et al: "Graph structure in the web", Comput. Netw. **33**, 309 (2000)
5. M. Faloutsos, P. Faloutsos, C. Faloutsos: "On power-law relationships of the internet topology", Comp. Comm. Rev. **29**, 251 (1999)
6. D. J. Watts, S. H. Strogatz: "Collective dynamics of 'small-world' networks", Nature **393**, 440 (1998)
7. A.-L. Barabasi, R. Albert: "Emergence of scaling in random networks", Science **286**, 509 (1999).
8. R. Albert, A.-L. Barabasi: "Statistical mechanics of complex networks", Rev. Mod. Phys. **74**, 47 (2002).
9. S. Strogatz, *Sync: The Emerging Science of Spontaneous Order* (Hyperion, New York, 2003).
10. C. W. Wu, L. O. Chua: "Synchronization in an array of linearly coupled dynamical systems," IEEE Trans. Circuits Syst. I, **42**, 430 (1995)
11. C. W. Wu: Synchronization in arrays of coupled chaotic circuits and systems: Theory and applications. In: *Controlling Chaos and Bifurcations in Engineering Systems*, ed by G. Chen (Boca Raton, FL: CRC Press, 1999) chapter 23
12. C. W. Wu: ' "Synchronization in Arrays of Coupled Nonlinear Systems: Passivity, Circle Criterion, and Observer Design," IEEE Trans. Circuits Syst. I, **48**, 1257 (2001)
13. C. W. Wu: *Synchronization in Coupled Chaotic Circuits and Systems* (Singapore: World Scientific, 2002)
14. C. W. Wu: "Synchronization in Coupled Arrays of Chaotic Oscillators With Nonreciprocal Coupling," IEEE Trans. Circuits Syst. I, **50**, 294 (2003)
15. C. W. Wu: "Perturbation of coupling matrices and its effect on the synchronizability in arrays of coupled chaotic systems," Physics Letters A **319**, 495 (2003)
16. L. Pecora, T. Carroll: "Master stability functions for synchronized coupled systems," Physical Review Letters, **80**, 2109 (1998)
17. K.S. Fink, G. Johnson, T. Carroll, D. Mar, L. Pecora: "Three coupled oscillators as a universal probe of synchronization stability in coupled oscillator arrays," Phys. Rev. E, **61**, 5080 (2000)
18. M. Barahona, L. M. Pecora: "Synchronization in small-world systems," Physical Review Letters, **89**, 054101 (2002)

19. E. Ott: *Chaos in Dynamical Systems* (Cambridge University Press, Cambridge, 1993)
20. X. F. Wang, G. Chen: "Synchronization in Scale-Free Dynamical Networks: Robustness and Fragility," IEEE Trans. Circuits Syst. I, **49**, 54 (2002)
21. X. Li, G. Chen: "Synchronization and Desynchronization of Complex Dynamical Networks: An Engineering Viewpoint," IEEE Trans. Circuits Syst. I, **50**, 1381 (2003)
22. V. N. Belykh, I. V. Belykh, M. Hasler: Connection graph stability method for synchronized coupled chaotic systems. Physica D, in press, 2004.
23. I. V. Belykh, V. N. Belykh, M. Hasler: Blinking model and synchronization in small-world networks with a time-varying coupling. Physica D, in press, 2004.
24. S. Misra, W. B. Gong, D. Towsley: Fluid-Based Analysis of a Network of AQM Routers Supporting TCP Flows with an Application to RED. ACM SIGCOMM 2000, Stockholm, Sweden (2000)
25. C. V. Hollot, V. Misra, D. Towsley, W. B. Gong: On the Design of Improved Controllers for AQM Routers Supporting TCP Flows. IEEE Infocom 2001, Anchorage, Alaska, USA (2001)
26. F. Baccelli, D. Hong: Interaction of TCP Flows as Billiards. IEEE Infocom 2003, San Francisco, CA, USA (2003)
27. F. Baccelli, D. Hong: Flow Level Simulation of Large IP Networks. IEEE Infocom 2003, San Francisco, CA, USA (2003)
28. Y. Liu, F. Lo Presti, V. Misra, D. Towsley: Fluid Models and Solutions for Large-Scale IP Networks. ACM Sigmetrics 2003, San Diego, CA, USA (2003)
29. S. Deb, S. Shakkottai, R. Srikant: Stability and Convergence of TCP-like Congestion Controllers in a Many-Flows Regime. IEEE Infocom 2003, San Francisco, CA, USA (2003)
30. S. Shakkottai, R. Srikant: How Good are Deterministic Fluid Models of Internet Congestion Control? IEEE INFOCOM 2002, New York, USA (2002)
31. P. Tinnakornsrisuphap, A. Makowski: Limit Behavior of ECN/RED Gateways Under a Large Number of TCP Flows. IEEE Infocom 2003, San Francisco, CA, USA (2003)
32. N. Biggs: *Algebraic Graph Theory* (Cambridge University Press, 1993)
33. F. Chung: *Spectral Graph Theory*, (AMS Publications, 1997)
34. F. Chung, L. Lu, V. Vu: Proceedings of National Academy of Sciences, **100**, 6313 (2003)
35. B. Bollobás: *Graph Theory: An Introductory Course* (Springer-Verlag, New York, 1979).
36. B. Bollobás: *Random Graphs* 2nd edition, (Cambridge University Press, Cambridge, 2001)
37. M. Juvan, B. Mohar: "Laplace eigenvalues and bandwidth-type invariants of graphs," J. Graph Theory, **17**, 393 (1993)
38. E. A. Bender, E. R. Canfield: "The asymptotic number of labelled graphs with given degree sequences," J. Combinat. Theory (A), **24**, 296 (1978)
39. M. Molloy, B. Reed: A critical point for random graphs with a given degree sequence. Random Structures and Algorithms, **6**, 161 (1995)
40. M. Molloy, B. Reed: The size of the giant component of a random graph with a given degree sequence, Combin. Probab. Comput. **7**, 295 (1998)
41. B. Bollobas, O. Riordan, J. Spencer, G. Tusnady: "The Degree Sequence of a Scale-Free Random Graph Process," Random Structures and Algorithms, **18**, 279 (2001)

42. R. Kumar, P. Raghavan, S. Rajagopalan, D. Sivakumar, A. Tomkins, E. Upfal: Stochastic models for the Web graph. Proceedings of the 41st Annual Symposium on Foundations of Computer Science (2000), pp 57–65
43. W. Aiello, F. Chung, L. Lu: Random evolution in massive graphs. In *Handbook on Massive Data Sets* eds. J. Abello et al., (Kluwer Academic Publishers, 2002), pp 97–122
44. C. Cooper, A. Frieze: A general model of web graphs. Proceedings of ESA (2001), pp 500–511
45. F. Chung and L. Lu: "Connected components in random graphs with given expected degree sequences," Annals of Combinatorisc **6**, 125 (2001)
46. W. Aiello, F. Chung, L. Lu: A random graph model for massive graphs. Proceedings of the Thirty-Second Annual ACM Symposium on Theory of Computing, (2000), pp 171–180
47. F. Chung, personal communication.
48. M. Fiedler: "Algebraic connectivity of graphs," Czech. Math. J., **23**, 298 (1973)
49. B. Mohar: "Isoperimetric numbers of graphs," J. Combinatorial Theory, Series B, **47**, 274 (1989)
50. B. Bollobas: "The isoperimetric number of random regular graphs," Euro. J. Combinatorics, **9**, 241 (1988)
51. F. Chung, L. Lu: The small world phenomenon in hybrid power law graphs. Preprint, UCSD, (2003)

Dynamic Complexity in the Internet Traffic

Misako Takayasu

Department of Computational Intelligence and Systems Science, Tokyo Institute of Technology, 4259 Nagatsuta-cho, Midori-ku, Yokohama 226-8502, Japan
takayasu@dis.titech.ac.jp

The Internet is one of the most enormous autonomously distributed network systems created by human being. There is no central control in the Internet system, However, the Internet protocols enable information transmission in the widely opened and flexible network.

The Internet is found to have a similar universal macroscopic behavior to those of natural materials, a dynamic phase transition. In this chapter, we report our observation and simulation results showing the evidence for the phase transition in the Internet between congested and non-congested phases accompanying critical fluctuations at the phase transition point.

The input mean flow density into the Internet is regarded as the control parameter of this phase transition. Phenomenal results show power-law distribution of congestion duration time, divergence of correlation length of jams, and discontinuity in the differential coefficient of the probability of jam occurrence at the critical point. Also the $1/f$ power spectrum is confirmed both by observation and by numerical simulations at the critical point. Our observation results show that the transmission efficiency become the highest at the critical point where fluctuations become largest. By introducing a minor revision to transmission protocols it is possible to improve the traffic efficiency drastically. The physics viewpoint of introducing phase transition asserts new kind of system controlling ideas to the prospective advanced information network.

1 Introduction

Our daily life is supported by massive traffic systems, for examples, urban automobile traffics [1, 2, 3, 4, 5, 6], dynamical transport systems of information transmission, and more over, the world wide currency traffics in the financial market of recent works on econophysics [7, 8, 9, 10].

Although each use of such traffic systems is based on human intention, massive traffics cancel individuality, and their statistics are likely to follow universal laws similar to those established for natural systems [11, 12, 13]. In this chapter we focus on the Internet traffic.

In the early stage of the Internet the traffics were rather sparse and congestion was not an attractive issue until 1990's [14, 15]. As the increase of both number of users and transporting file-sizes, information congestion became a hot topic in the latest 15 years [16, 17, 18, 19, 20, 21, 22, 23].

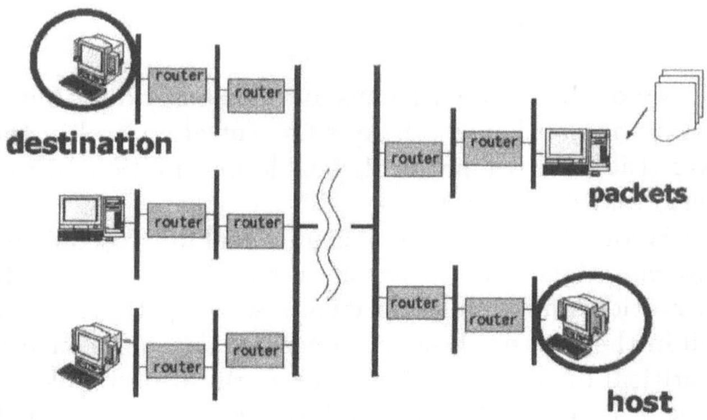

Fig. 1: The structure of the Internet

The Internet is physically consisted of three typical components as schematically shown in Fig.1: The first elements are computers at the peripheral parts of the network. They are called in various ways such as user computers, hosts, senders or destinations". These are the computers using at our office and home transferring e-mails or seeking for profitable sites at World Wide Web. The second elements are called routers, gateways or nodes. They are the internal computers of the network, which have function of relay stations of packet transportation. The last components are called cables or links which connect these peripheral and internal computers.

Information in the Internet is transported by the basic protocols called TCP/IP which ensures information delivery in the autonomously distributed computer networks[24]. The information emitted from a host computer is divided into a series of numbered packets. A typical packet size is about 1 Kilo-byte and it has header information of both addresses of the destination and the sender. Neighboring routers periodically exchange information of neighboring connections called the routing tables in order to update their circumstance of connection. When a packet comes into a router from a link the header information is retrieved by the router, and an adequate output link is chosen by referring the routing table, which is estimated as the closest to the destination. After all packets are reached at the destination computer, the information is re-constructed from packets to the original form.

Though the bandwidth of links and the performance of routers are revised by replacing them with new elements, the number of users and the size of transferring data are increasing even more rapidly. As a result, we often encounter congestion, or traffic jam of packets. We experience such congestion when accessing to a popular website. Sometimes it takes quite longer time to open a window, which implies that the packets encountered congestion.

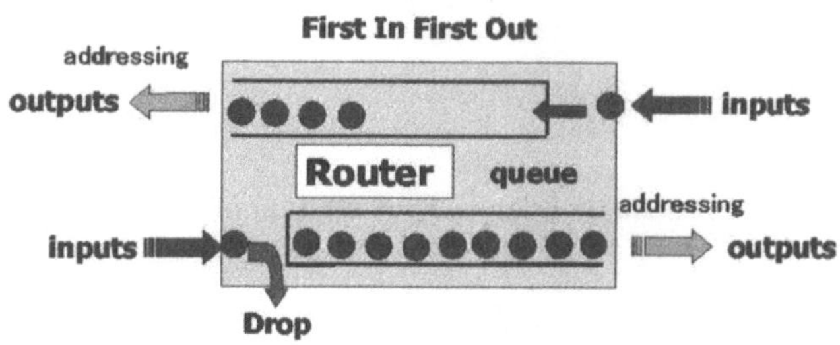

Fig. 2: Schematic view of a router

Let us give more precise vision of what is happening in the internet at the congestion period: As we mentioned above a router is a relay station and it autonomously decides the way of sending a packet following the "First-In-First-Out" rule. When a router receives a packet while previous packet is in the process, the latter packet goes into a memory queue waiting for process. In the case that many packets almost simultaneously arrive at a router, the queue length of packets becomes longer, and as a result, a packet spends non-negligible time at the router. Summation of these staying time at each router along the path reflects the congestion level of the packets on the way to the destination. If all memory buffers are occupied at a router, the information of spilled out packets are simply discarded as schematically shown in Fig.2. We call this feature as packet loss. In such a case the lost packets are re-transmitted from the host computer following the TCP as will be described in detail in section 5. To be more precise another type of congestion can be found in Ethernet cables by collision of packets[25, 26].

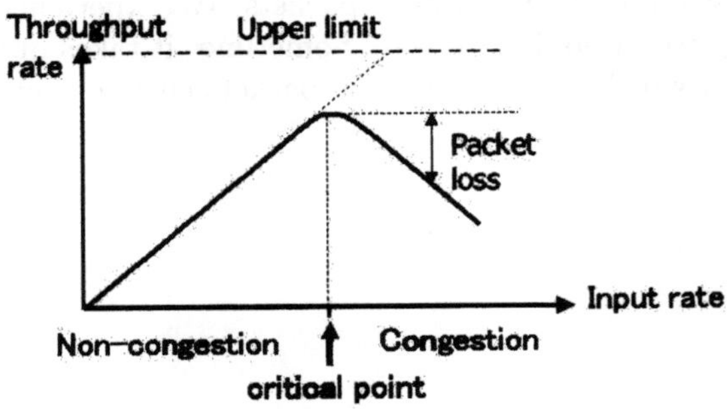

Fig. 3: Schematic input-output relation of the Internet

The general property of traffic congestion can be characterized by the nonlinear input-output relation as schematically shown in Fig.3. In the case input rate of packets is small, the output rate at the destination host is nearly equal to the input rate. This state is called the non-congestion phase. For higher input density the output rate decays due to packet re-transmission caused by packet loss. This state is called the congested phase. It is interesting that the maximum output rate is attained at the threshold of these two states where packet loss rate appears. The input rate and output rate are nearly equal at the threshold, however, the network is very unstable and sensitive to a perturbation and a jam occurs easily. As a result we can observe fractal properties such as the $1/f$ power spectrum and power law distributions at the threshold input rate as described in the following sections.

2 Statistical Model of Simple Queue

Basically, the critical phenomena is essential in any kind of traffic systems with queues at the buffers following "First-in-First-out" rule. Even the simplest theoretical model of a single queue shows phase transition behaviors [23]. The simplest queue is composed of three parts: the input gate, the buffer and the output gate. Let us consider an imaginary buffer that can store infinite number of packets. In each discrete time step, a random natural number of packets, $I(t)$, are enqueued into the buffer. We consider the case that the output flow, $F(t)$, has a maximum value F_{max} in each time step reflecting the router's processing speed.

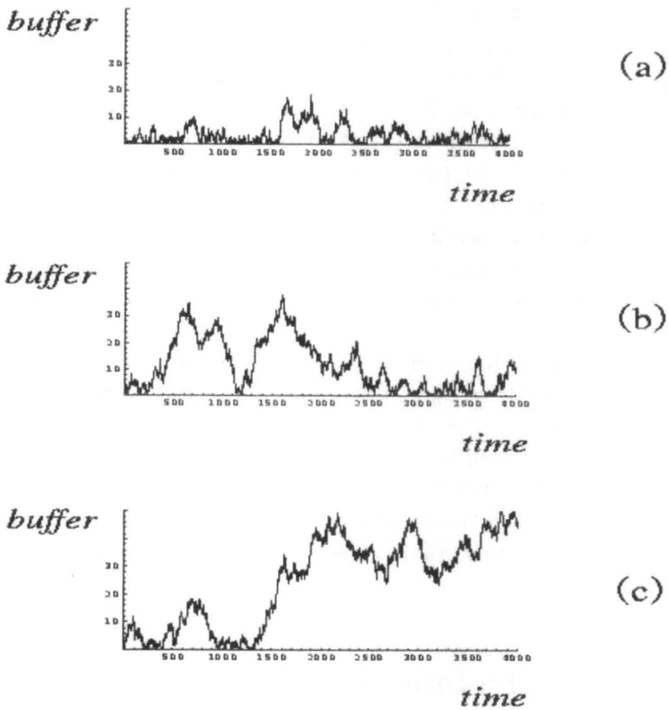

Fig. 4: Examples of fluctuation of a buffer queue length for (a) small mean input, (b) critical mean input, and (c) large mean input

The output flow dynamics from the buffer is given as follows;

$$F(t) = F_{max}, \quad \text{if} \quad B(t-1) + I(t) \geq F_{max},$$
$$F(t) = B(t-1) + I(t), \quad \text{if} \quad B(t-1) + I(t) < F_{max},$$

where $B(t)$ represents the number of packets in the buffer which takes a non-negative numbers as defined by

$$B(t) = max\{B(t-1) + I(t) - F_{max},\ 0\}. \tag{1}$$

This simple buffer dynamics shows a critical behavior similar to the mean field percolation model qualitetively[27]. Fig.4 shows examples of queue length sequences for three different conditions; (a) $\langle I(t)\rangle < F_{max}$, (b) $\langle I(t)\rangle = F_{max}$, and (c) $\langle I(t)\rangle > F_{max}$.

In the case of $\langle I(t)\rangle = F_{max}$, the injection flow and output flow balance. Under such circumstance the queue length $B(t)$ fluctuates randomly according to the time similar to the 1-dimension Brownian motion with reflection wall at $B(t) = 0$. For the case $\langle I(t)\rangle < F_{max}$, only a small positive deviation near $B(t) = 0$ can be observed in $B(t)$ as if there are attractive forces to the $B(t) = 0$. On the other hand, $B(t)$ shows a trend to go infinity as $t \to \infty$ for the case of $\langle I(t)\rangle > F_{max}$. If we focus on the probability P_∞ of having infinite duration time of maximum flow $F(t) = F_{max}$, we can observe finite probability of P_∞ for $\langle I(t)\rangle > F_{max}$. And at $\langle I(t)\rangle = F_{max}$, the cumulative probability of duration time, $P(\geq t)$, of taking $F(t) = F_{max}$, which follows the same distribution as duration time of taking $B(t) \geq F_{max}$, becomes $P(\geq t) \propto t^{-1/2}$ since it corresponds to the distribution of first recurrence time in random walk model [28]. As expected by the random work theory the power-spectra of the time sequence for the queue length $B(t)$ and output flow $F(t)$ follow f^{-2} and $f^{-1/2}$, respectively.

In case that the buffer has a finite capacity, we confirm that the power exponent of $P(\geq t)$ for $\langle I(t)\rangle = F_{max}$ does not change but it has an exponential cut off at a finite time scale and the power-spectrum becomes a white noise for smaller wave numbers. Though these qualitative behaviors are simillar to the observation results of real traffic as we show in the next section, $1/f$ power spectrum is not observed in this simple model, which is often observed in real Internet traffic. The dynamics of $1/f$ power spectrum is obtained from more heterogeneous complexity.

3 Observation of the Internet Flow

From 1997 to 1998 we connected a monitor computer to a link (Ethernet, 10Mbps) between two routers of Wide Area Network (WAN) and we observed size and time stamp of all packets going through the cable, both upstream and downstream by using "tcpdump" command. We recorded 9 sets of 4 hours information density fluctuation, with a bin size of 0.1 seconds[29, 30].

An example of flow density fluctuations is shown in Fig.5. We notice that spike-like fluctuations always exist in short time scale and there are large density modulations in long time scale. The data does not look stationary for the whole scale of 4 hours, however, we find quicker decay around several hundred second at maximum in locally observed auto-correlation function. Therefore, we can treat the data as quasi-stationary in the time scale of several hundred seconds and divide the data into subsets consist of several hundred seconds time series. In the following analysis we set a subset bin size as 500 seconds determined by our data in the above way.

The characteristic of a subset is classified as following three typical cases: Non-congested case, as we see in subset (a) in Fig.5, flow density is rela-

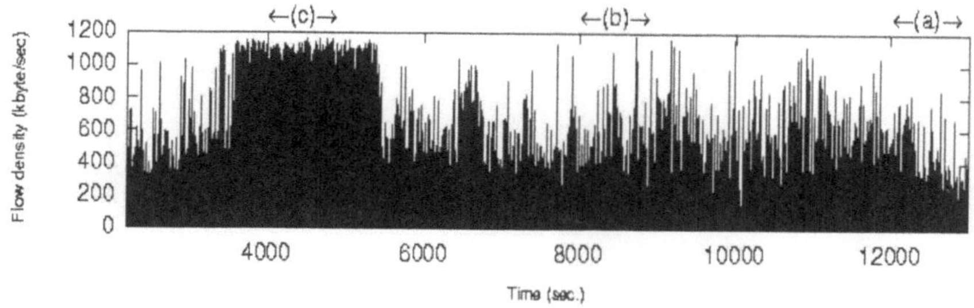

Fig. 5: An examples of real traffic flow for succesive 4 hours

tively low, and there are high density pulses appearing rather randomly. The fluctuations can be modeled nicely by the classical traffic model based on a Poissonian white noise.

Intermediate case, as we see in subset (b) in Fig.5, flow density is intermediate, and there are both large and small fluctuations. The estimated correlation time scale is much longer than the former case and we can confirm critical behaviors as discussed in detail in the following.

Congested case, as we see in subset (c) in Fig.5, flow density is fluctuating randomly near the bandwidth (1200 Kilo byte per second). In our observation, we cannot detect lost packets directly, however, it is reasonable to expect that there is much information loss occurred in period (c). The fluctuation around the mean flow density can be modeled also by a white noise.

The corresponding probability densities of flow density for these three cases are plotted in Fig.6. In the non-congested subset (a), the probability density has a large peak at a very low flow density, and the distribution decay sharply at larger flow density as shown by dotted line in Fig.6. In the congested subset (c), the probability distribution has a high peak at a large flow density a little below the physical upper limit. The interesting distribution is observed at the intermediate case (b), which gives a nearly flat wide distribution.

The fact that the widening of the distribution at the subset (b) is the key ingredient consistent with the existence of phase transition, which is very different from simple shift of the peak position in the distribution according to the increment of the mean flow density. This broadening of the probability density function can be characterized by observing the half-value width of the probability density as a function of the mean flow density of each subset of size 500 seconds box. In Fig.7, the horizontal axis shows the mean flow density of subset, vertical axis shows the half-value width of the probability density. The half-value width of the distribution increases sharply untill a certain mean flow density value. The peak of the half-value width become maximum at the mean flow density of subset (b).

It is well established in the study of physics that the autocorrelation length takes the largest value at the critical point of a phase transition. This general

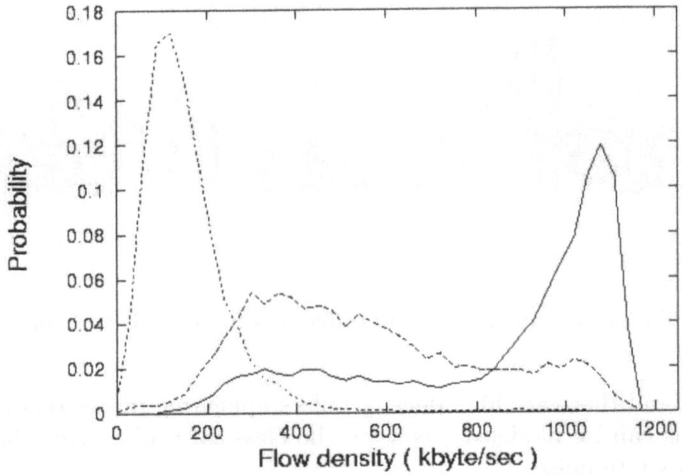

Fig. 6: Probability densities of flow density for the congested case (a)(dotted line), the intermediate case (b)(dashed line),and the congested case (c)(solid line).

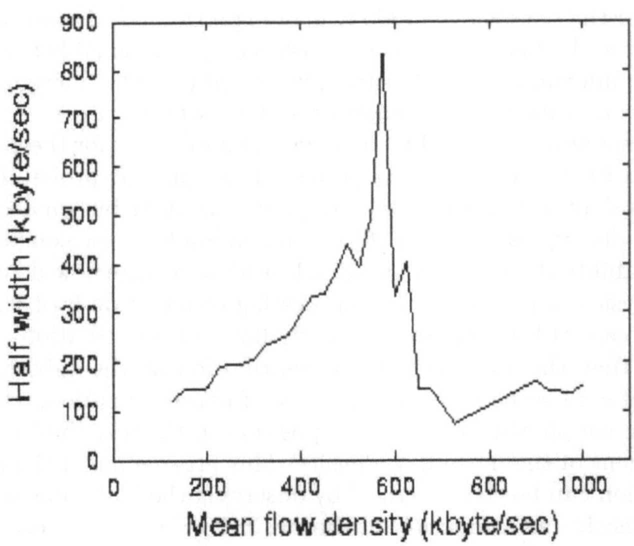

Fig. 7: Half-value width of the probability density as a function of the mean flow density

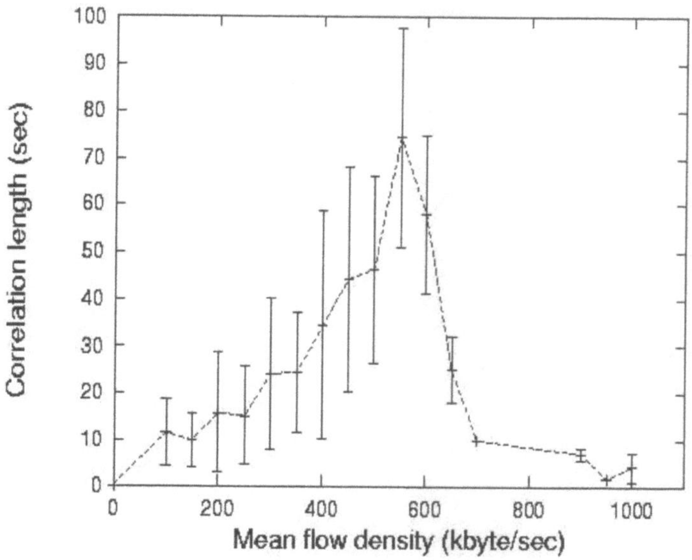

Fig. 8: Correlation length as a function of the mean flow density

property is also confirmed in the congested subsets of the Internet as shown in Fig.8. The autocorrelation function of a subset generally shows slow decay for a short time scale and it decays exponentially from certain time scale. We estimate the correlation length defined by the starting time scale of observing exponential decay in the autocorrelation. As expected the correlation length takes large values for intermediate flow densities and becomes maximum at the critical flow density which we estimated by the largest half-width of the probability density.

Apparently, this peak point gives the critical flow density of the phase transition of the Internet traffic. The above two analysis can be applied generally for estimating the critical flow density from given flow density fluctuation data. The critical mean flow density is about the half of the bandwidth in this case.

It is also well-known that some physical quantities follow power law distributions at the critical point in a general phase transition of the second order, or a continuous phase transition. In order to extract a power-law nature at the critical point, we introduce a quantity called the congestion duration time by defining a threshold value[31]. When the flow density takes a value higher than a threshold value, we regard that packets are caught in congestion at that time. We set the threshold value to be (*mean value*)+2(*mean deviation*). Then most of the peak positions of the flow density distributions are smaller than this threshold value when the mean density is below the critical point.

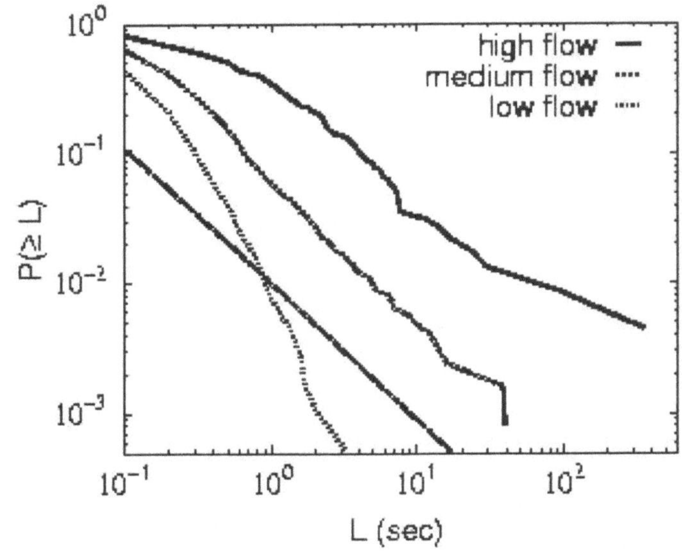

Fig. 9: The cumulative distribution of congestion duration time in log-log scale. The non-congested case (a)(dotted line), the intermediate case(dashed line), and the congested case (c)(solid line)

We observe successive time that the flow density keeps exceeding the threshold value for L seconds, and we plot the cumulative distribution of L, $P(\geq L)$, for each subset. Here, we plotted three lines corresponding to (a), (b) and (c) in Fig.9.

In the case of (a), a quick exponential decay is observed showing that congestion occurs independently like a Poisson process and the congestion state does not exist for long time. For the subset whose mean flow density is nearly critical, the case of (b), we get a power-law distribution with exponent close to -1. This is another evidence of the critical behaviors and it is known that this exponent is consistent with the so-called $1/f$ fluctuations [19]. In the congested box of (c), the probability of finding large jams becomes higher as expected naturally and a plateau appears for larger value of L. It should be noted that each curve in Fig.9 is obtained from 5000 data points in each subset without taking any ensemble average. Namely, the statistics of each subset is very stable with this bin size.

In order to confirm the critical behaviors more clearly we show an example of information flow fluctuation near the critical flow density in Fig.10. An enlargement of a part of the fluctuations gives statistically self similar fluctuations as the original scale, and a more enlargement gives also self similar

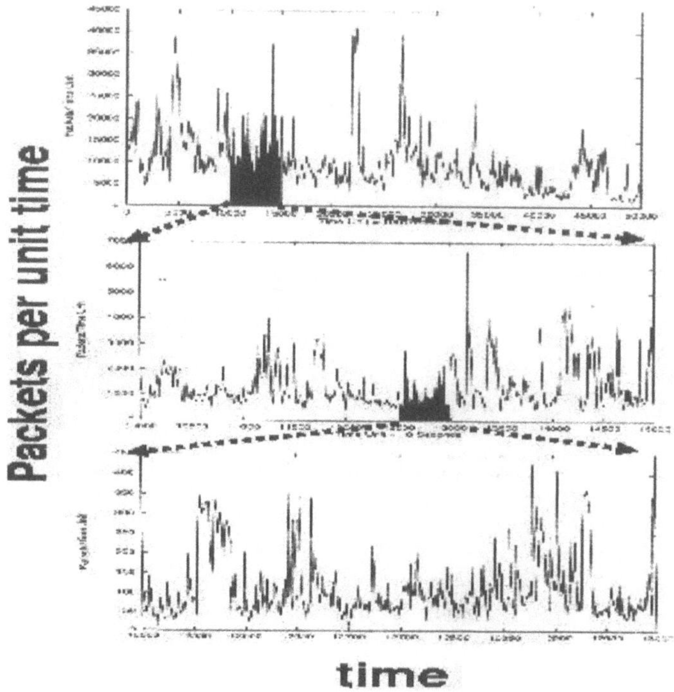

Fig. 10: Self-similarity of critical flow fluctuation

fluctuations. Here, the bottom figure is 100 times enlarged from the top figure. This is the fractal property at the critical point.

This scale-invariant property can also be confirmed by observing the power spectrum of the fluctuations. As shown in Fig.11 the power spectrum is approximated by a straight line with slope -1 in the log-log plot. Namely, the Internet traffic follows the $1/f$ fluctuations at the critical point. The time range holding the $1/f$ spectrum is from a second up to a few hours.

In the present case the mean flow density for each subset of time plays an important role in characterizing the condition of subsets. Therefore let us call the mean flow density of a subset as a control parameter of the system, and the non-congested, intermediate, and congested subsets as non-congested phase, critical point, and congested phase, respectively by the common phase transition terminology.

In the physics systems accompanying static phase transition, the control parameter can be really controlable like temperature in spin systems, or probability of existing a particle in percolation, however, the mean flow density in the Internet is not a fixed parameter since it strongly depends on the massive

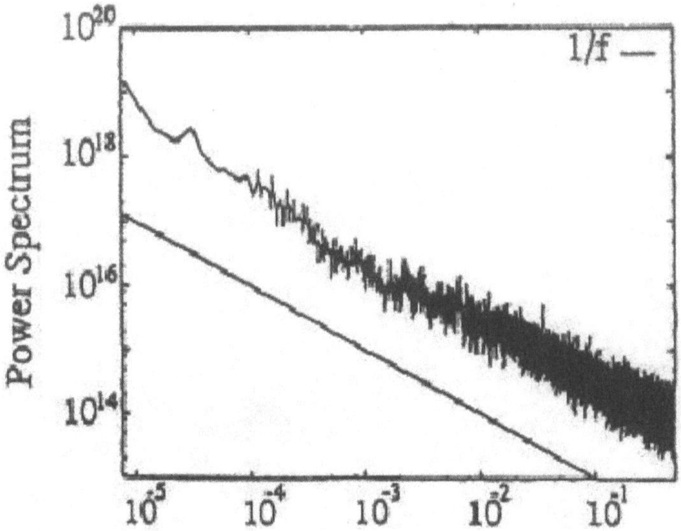

Fig. 11: Power spectrum of the critical flow

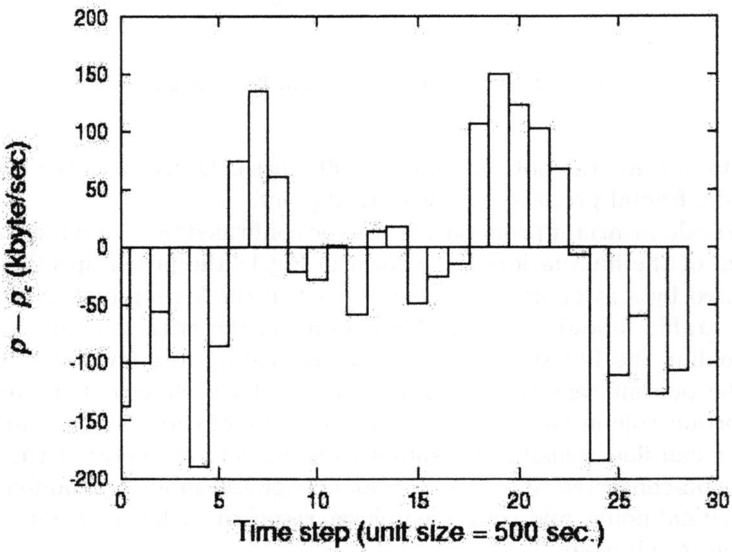

Fig. 12: Fluctuations of control parameter arround the critical density. ρ denotes the mean flow density and ρ_c denotes the critical value. Zero line parallel to the hrizontal axis shows the critical value of 580 kilo-byte/sec estimated from our data.

users attitude. This type of phase transition is dynamical and the control parameter fluctuates rather spontaneously from time to time. In Fig.12 we show temporal fluctuation of control parameter around critical points. Horizontal axis shows sequence of subsets(time) and vertical axis shows the mean flow density of subsets around the critical pointρ_c.

In order to observe the whole view of phase transition both below and above the critical point, it is important to avoid choosing a link which is continuously too sparse or too crowded. Also we checked the traffic properties by the data obtained from several different links and confirmed that all the results are consistent with the above description, however, the estimated critical mean flow density is quite different. The critical mean flow density strongly depends on the neighboring physical circumstance. In our observation, the critical flow density is always around $50 - 60$ % of the minimum bandwidth of neighboring links. More observation is needed to clarify the relation between the critical flow density and the maximum flow densities of the local environments.

It is now evident that all results from the Internet flow observation are consistent with the theory of the phase transition accompanying critical behaviors between two phases of non-congested and congested traffics. The whole behaviors of this phase transition look similar to those of ordinary second order phase transitions in physical systems in equilibrium, however, there is one novel difference that the present system is essentially dynamical and it is not stationary in long time sacle.

4 Ping Experiment and Visualization of the Internet Traffics

So far we have confirmed temporal critical behaviors observed at a link in the Internet. In this section, we pay attention to the development of spatial congestion such as contagion of congestion in the network.

A global congestion property can be observed indirectly by using the *ping* command in the UNIX operating system [19, 32]. This command is designed for confirmation of whether the destination computer is active or not. Namely, if a host sends a *ping* packet (ICMP echo request packet) to a destination, the destination computer immediately sends back a reply packet(ICMP echo reply packet) to the host like an echo.

A *ping* packet has the size of 64 bytes including the information of a departure time stamp. By comparing the arrival time and the departure time stamp the round trip time (RTT) can be calculated. From the round trip times of the echo packet we estimate the level of congestion along the path. If the path is crowded then a packet spends more time in buffers along the path so that RTT quantitatively reflects the state of congestion along the Internet path.

We performed an experiment of sending *pings* periodically (evey 0.1 seconds for 8 hours) to a destination host and observed the time series of fluctuations of the round trip time that is expected to be proportional to the level of congestion. In this experiment we choose a destination carefully since a packet route to a destination in the Internet is not always stable and symmetric [33]. In order to avoid such problem and to make sure that *ping* packets follow the same route during the round trip we check the packet route by using the UNIX command *traceroute* for every 10 minutes. *traceroute* memorizes a list of IP address along the path. The lost packet at a router during the trip is measured as a maximum RTT.

We observed time-series of round-trip-time along a path in the Internet, and we found that the traffic degree of a path is well characterized by the three typical state, non-congested state, congested state and critical state. Examples of round-trip-time series obtained from the *ping* experiment is shown in Fig.13.

Top figure of Fig.13 shows a time series of RTT obtained from a path at non-congested state. Middle figure of Fig.13 shows a series of RTT obtained from a path at critical state. Bottom figure of Fig.13 shows a series of RTT obtained from a path at congested state.

The corresponding power spectra for Fig.13 are shown in Fig.14. If traffic in a path is in non-congested state and congested state (top and bottom figure, respectively) the power spectra of the fluctuation follow white noise in low frequency range. When traffic in a path is at the critical state a clear $1/f$ power spectrum can be observed, as shown in middle figure of Fig.14.

In order to analyze the congestion duration time, L, of RTT sequence, we set a threshold RTT as (*mean value*)+2(*mean deviation*) as we did in the analysis of packet flow densitiy. Then, we observe the duration time lengths L of congestion which exceed the threshold value in RTT data and obtain the cumulative distribution $P(\geq L)$ of congestion duration time as shown by cross in Fig.15 for non-congested state, and Fig.16 for congested state. One can explain Fig.15 type of exponential decay by considering Poisson process but Fig.16 can never been explained in the same manner.

Let us introduce a simple theoretical explanation for both types of interval distributions by introducing a phase transition model to the Internet traffic. The dashed lines in Fig.15 and Fig.16, which fit nicely to the observation results, are explained by the following theory.

We consider two macroscopic stochastic features in congestion process of network in a discrete time step:

1. Congestion propagates from one router to adjacent routers by a probability p(reproduction process).
2. Congestion at a router is spontaneously eased by a probability q (annihilation process).

As we regard a congested router as a particle and a sparse router as an empty site, such a stochastic model on a lattice is well-known under the name of *Contact Process*(CP) [34]. If the reproduction rate p is too large compared with the annihilation rate q, the system converges to a trivial fixed point in which the system is occupied with particles. This state is called the survival state. On the contrary, if the annihilation rate q is high compared with the reproduction rate p, the system converges to a condition in which all sites are empty. This state is called the extinction phase. It is known that a phase transition occurs at a condition between two phases of extinction and survival with the ratio p/q as a control parameter.

Let us define W_∞ as a probability that offsprings starting from a single particle survives after infinite time. It is known that $W_\infty = 0$ below the critical point, and W_∞ takes finite value for the case of above the critical point. The CP is the simplest non-trivial stochastic model showing the extinction-survival phase transition, and it is known to have wide universality class, which means many models having different evolution rules converge to the CP by repeating coarse-graining operations of the renormalization [35].

By considering a mean-field approximation of CP the probability distribution of duration time of congestion is estimated as follows. Let us consider a survival probability of a particle's offsprings after L time steps be $W(\geq L)$. Then $W(\geq t)$ and $W(\geq t+1)$ are related as follows:

$$W(\geq t+1) = (1-p-q)W(\geq t) + p[1-(1-W(\geq t))^2], \tag{2}$$

where the first term of right-hand side shows the probability that a particle does not reproduce nor annihilate in the first time step and will survive for L time steps. The second term denotes the case that the particle reproduces another particle in the first time step and either of the particles exists for L time steps. Approximating the equation by taking the continuum limit, Eq.(2) can be solved as

$$W(\geq t) \propto \frac{\delta}{1-e^{-p\delta(t+t_0)}}, \qquad \delta = 1 - \frac{q}{p}, \tag{3}$$

where t_0 is a constant. At the critical point, $\delta = 0$, $W(\geq t)$ becomes a power law as follows:

$$W(\geq t) \propto \frac{1}{t}. \tag{4}$$

$W(\geq t)$is to be compared with the distribution of congestion duration time. We can reproduce our observation results by choosing adequate parameters of t_0 and δ in Eq.(3). The dotted line in Fig.15 is fitted by parameters below the critical point($\delta < 0$), and that of Fig.16 is fitted by parameters above the critical point($\delta > 0$). Since the congestion duration time follows a power-law as $W(\geq t) \propto 1/t$ at the critical point, corresponding power-spectrum can be numerically calculated as $1/f$ noise as already seen in middle figure in Fig.14.

We extend the *ping* experiment to visualize the space-time behavior of congestion along a path of 2 end hosts [36]. We prepare two end hosts at a

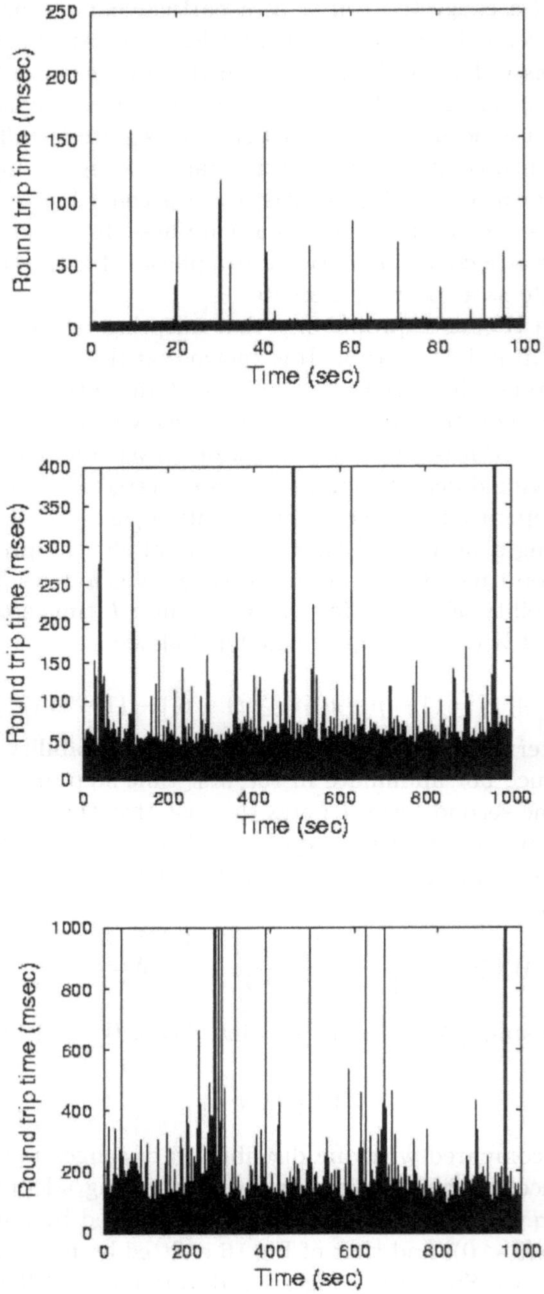

Fig. 13: Fluctuations of round trip time by *ping* experiment for non-congested state, for critical state, and for congested state, from the top to the bottom.

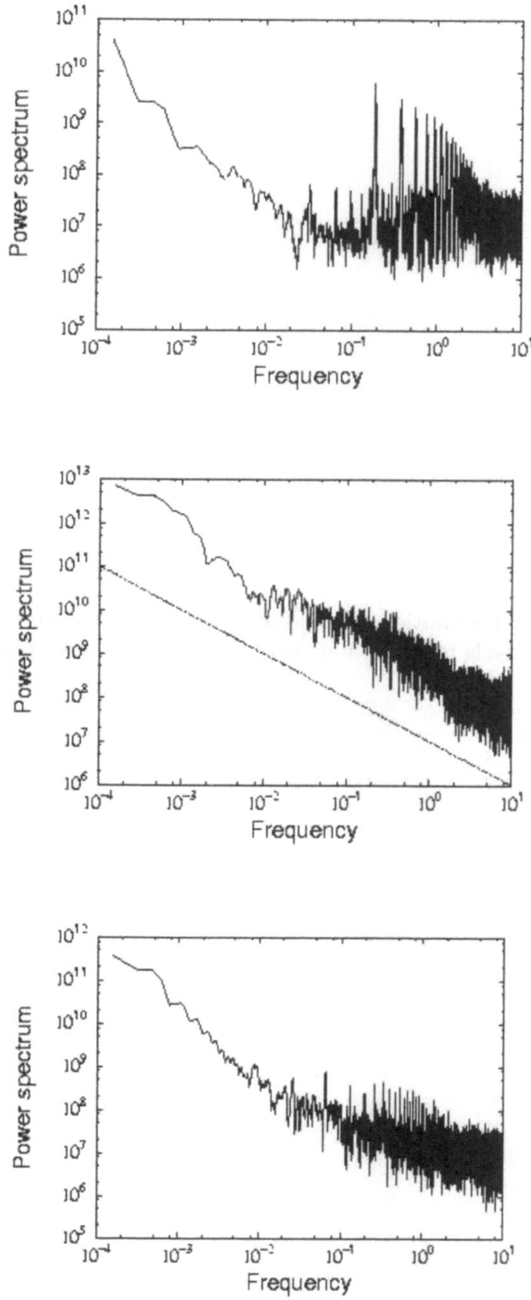

Fig. 14: Power Spectrum for non-congested state, critical state, and congested state, from the top to the bottom

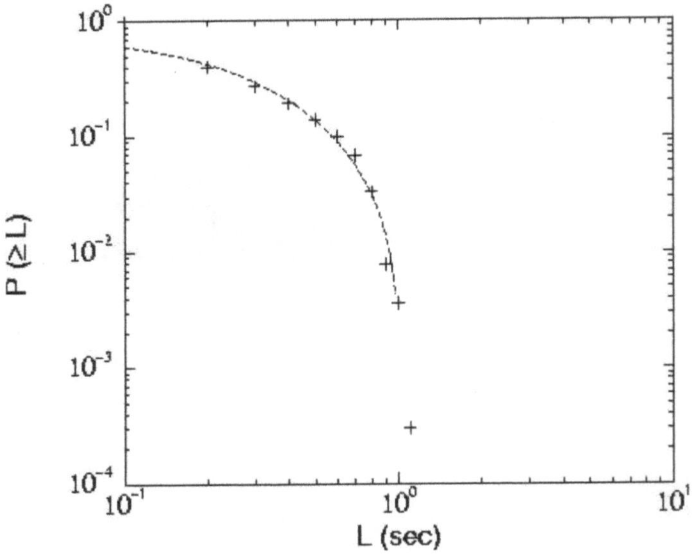

Fig. 15: Cumulative distribution of congestion duration time for the non-congested state: Real data analysis in cross, and theoritical analysis in dashed line.

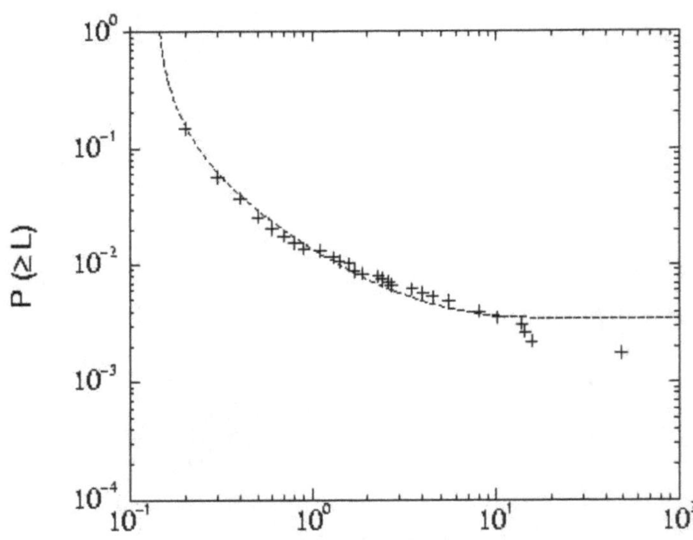

Fig. 16: Cumulative distribution of congestion duration time for cogested state: Real data analysis in cross, and theoritical analysis in dashed line.

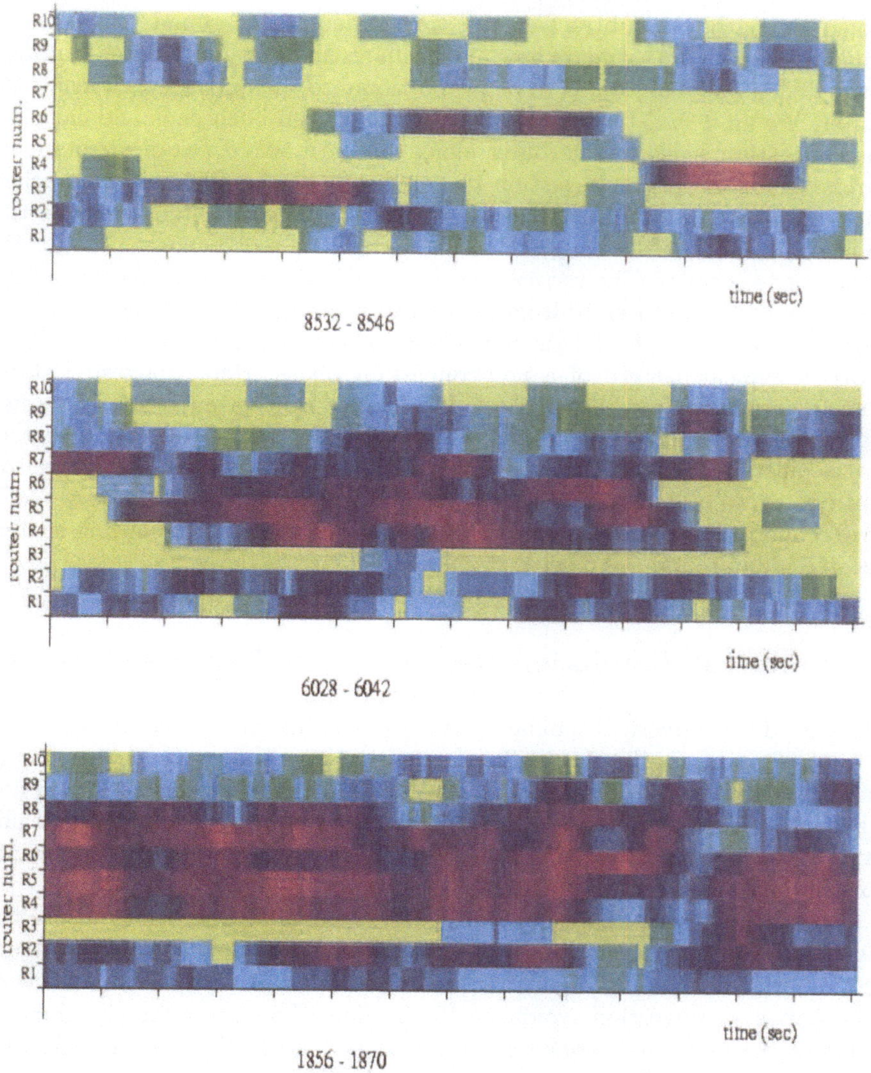

Fig. 17: Spatial-temporal pattern of Internet congestion for a non-congested path, a critical path, and a congested path, from top to bottom

distance. From both ends we send ping packets simultaneously to all routers along the path. By subtracting the time difference of round trip times of adjacent routers, we estimate how much time is spent at each router, and that estimated time is considered to be proportional to the queue length inside the router. We emit *ping* packets every 0.1 second from both ends and calculate the congestion level of each router along the path. We check our experiment carefully not to create congestion by comparing the number of *ping* packets with that of background packet flows. Fig.17 represents the result of this tomography of Internet congestion. The horizontal axis denotes time for the period of 14 seconds. The vertical axis is the router numbers along the path in order; router 1 to router 10 from bottom to the top. The darkness shows the degree of congestion; from light congestion to heavy congestion. The top figure shows a typical pattern of a non-congestion phase. Here, congested states appear and disappear rather randomly among routers and each congested state continues only a few seconds. In the critical case the congestion pattern makes clusters in the space-time configuration. By watching carefully we can find contagion of congestion among neighboring routers with time scale of a few seconds. Correlation coefficient of congestion among neighboring routers become high around the critical state[36].

5 Numerical Simulation Results of the Internet Traffics

We started the numerical simulation study of the Internet packet transmission in 1996[37], since in the real traffic observation we are not able to control the important parameter of input rate. As shown in the following we obtain non trivial critical behaviors due to the TCP feedback controls and all basic traffic can be realized in a very simple network topology with independent white noise inputs [38, 39].

The TCP involves following 3 basic flow control algorithms to insure the high reliability in packet transmission; window-based flow control for a connection, re-transmission algorithm, and transmission rate controls such as slow start and congestion avoidance. These algorithms are complicated in details but basically they work for accelerate and brake the transmission rate according to the traffic flow of the path in the Internet.

Let us explain the essence of standard TCP protocol and how it works: When a host computer sends information, the information is embedded in a series of numbered packets then it establishes a connection between a destination host. At first the origin host sends one packet. When the traffic is not congested and the packet reaches to the destination without loss, the destination sends an Acknowledgement packet(ACK packet in short) requesting the next sequence number(packet number 2) to the sender. If the sender receives an ACK packet in certain waiting time, then the origin sends two packets next. By receiving two ACK packets requesting sequence number 3 and 4 successfully within a waiting time, then the sender sends four packets, then eight packets, sixteen packets, and so on. So, in each time the sender doubles the number of sending packets(window size) until the number reaches the given maximum number. The brake works when congestion is detected. If a packet is lost by an overflow at a router, then the window size is set to 1. And if more packet loss occurs, the host computer refrains from sending a packet for a period. In this way the traffic flow density controlled by TCP tends to be around the critical point.

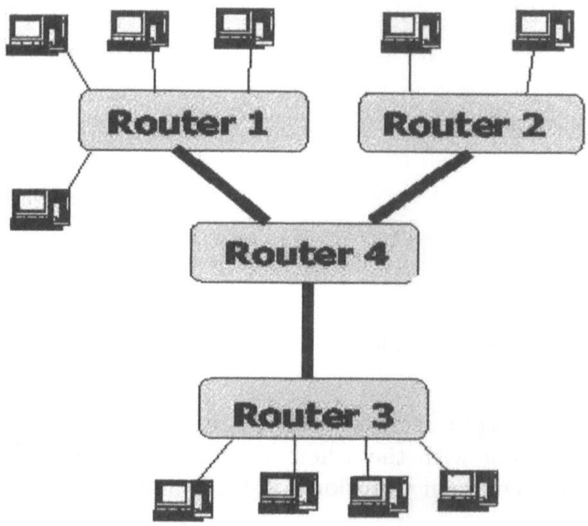

Fig. 18: Network structure of the numerical simulation

The packet motion following TCP is investigated by numerical simulations using a free software called ns-2[40]. This simulator is widely used for the study of Internet since it reproduces the details of the protocols. We assume a simple network structure consisted of 4 routers as schematically shown in Fig.18. Hosts are connected to the peripheral routers and they are assumed to make a connection with a given probability to a randomly chosen destination following the TCP.

In the simulation we give the rate for creating a new connection to a destination as a control parameter, and in each connection, 100 data packets are sent. The queue length of each router is set to be 100 in the simulation.

Fig.19 shows the input-output relation obtained by our numerical simulation. Input rate is given by the number of connection per second, and output is given by the number of finished connections during the simulation. The output increases linearly for small input, and it becomes maximal at the critical

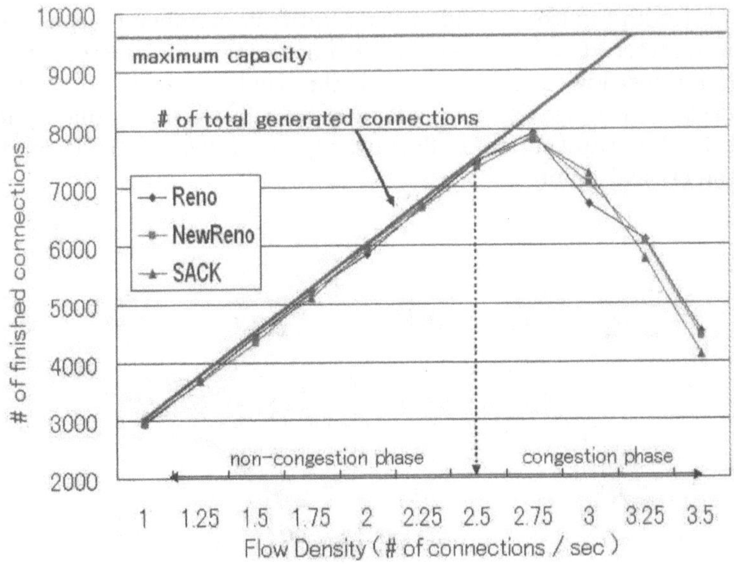

Fig. 19: Input-output relation by the TCP simulation

input rate, and the output decreases rapidly for larger input rate. This result is in perfect agreement with the schematic figure shown in Fig.3. Here, we plotted results of 3 common variations of TCP, and the results make no big difference.

The basic congestion properties of the simulation are the same as we observe in the real system. At the critical point we can confirm the $1/f$ power spectrum, the flat broad probability distribution of flow density, and a power-law distribution of congestion duration time, etc. In Fig.20 we show the distribution of congestion duration time obtained by TCP simulation. We confirm the feature for the distribution is consistent with the real traffic. For the low density input case, jam duration time follows exponential function, and, for the high input rate, we obtain a function with plateau in long duration time range. At the critical input rate, we confirm a straight line with slope -1 in the log-log plot, which is also consistent with the observation results of real traffic. Also, by observing the packet loss rate, we confirm sudden increase around the critical point likewise the probability for the infinite size of cluster in percolation. On the other hand, the packet loss rate is almost zero for below the critical point. From the characteristic of phase transition, we confirm that the critical point is most efficient in packet transmission since the output becomes maximum value and the packet loss rate is small, statistically.

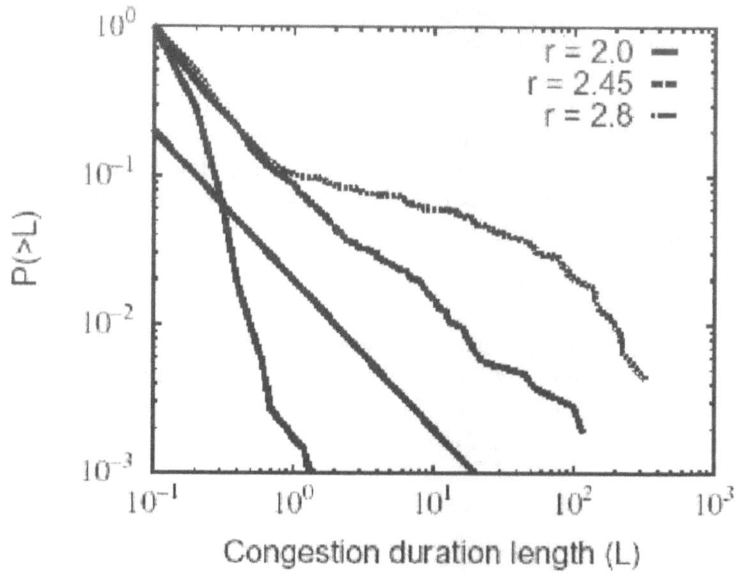

Fig. 20: Cumulative distribution of congestion duration time by the TCP simulation

We can deduce the following results from this simplified numerical simulation which can reproduce almost all basic phase transition properties as demonstrated.

1. The complicated network topology in the real system is not essential for the basic statistical properties of congestion.
2. Human activity is not relevant in our simulation as hosts make connections with randomly chosen destinations.
3. File size distributions are not playing an important role as we found that realistic phase transition properties can be obtained even with a fixed file size.

We showed the theoretical result in section 2 that the simplest single queue model without TCP can show a similar congestion phase transition, however, the power spectrum of output flow follows not $1/f$ but $1/f^{0.5}$ at the critical point [23]. The exponent of the congestion duration time distribution also follows a power law with exponent -0.5, not -1.

From theoretical viewpoints the $1/f$ spectrum has been analyzed in various ways. A macroscopic model of congestion(CT) can explain the $1/f$ behaviors at the critical point as we showed in previous section[19]. Also a microscopic model of competitive output on an Etheret cable proves that an exponential increase of waiting time at packet-losses is responsible for the $1/f$ spectrum [25, 26]. Both of these models show phase transition behaviors, and the $1/f$

spectrum appears only at the critical point. There are models that produce critical fluctuations automatically.

A general model of random time sequence called the Self-Modulation Process(SMP) model is also known to produce random fluctuations with the $1/f$ power spectrum [41, 42]. This model is based on a general moving average and its application to information traffics is now being developed. Consider a set of points whose t-th interval is denoted by $X(t)$. Then the basic formulation of SMP is shown as follows:

$$X(t+1) = \gamma(t)X(t) + R(t), \tag{5}$$

where $\gamma(t)$ denotes a random value satisfying $\langle\gamma(t)\rangle = 1$, and $R(t)$ denotes a small random value, $R(t) \approx 0$. By solving the equation, we obtain the cumulative distribution for $X(t)$, $Prob(\geq X)$, and the power spectrum for the fluctuation of $X(t)$, $S(f)$, as follows:

$$Prob(\geq X) \propto \frac{1}{X}, \tag{6}$$

$$S(f) \propto \frac{1}{f}. \tag{7}$$

Applying this model to the real traffic, δt is recommended to be set around the TCP re-transmission timer, and $X(t)$ can be considered as the congestion duration time. The auto-correlation for $X(t+1)/X(t)$ is expected to be zero, following the model[42].

6 An Attempt to Revise TCP : Introduction of LEAP

Here we clarify the reason for rapid decrease of the output in the congested phase. In Fig.21 the vertical axis shows the averaged number of required packets in order to send the given 100 data packets in a connection as a function of the input rate. From this figure we find that the number of extra or wasted packets increases rapidly for input rate larger than the critical point. There are two types of wasted packets. One is the case that data packets really dropped from the router on the way to the destination. In this case re-transmission is needed to reconstruct the data at the destination. The other type is the case that data packets reached successfully to the destination, however, the TCP misunderstood them to be lost and re-transmited the packets. This situation occures at the sender host by timeout of ACK packet's arrival or loss of ACK packet on the return path. These wasted packets are called **duplicate packets** shown in the lower part in Fig.21.

In order to avoid sending duplicate packets, we revise the TCP in the following way. When the host does not receive ACK packet within the waiting time, the host sends the next data packet assuming that the first data packet has reached the destination. Therefore, a duplicate packet will be re-sent only

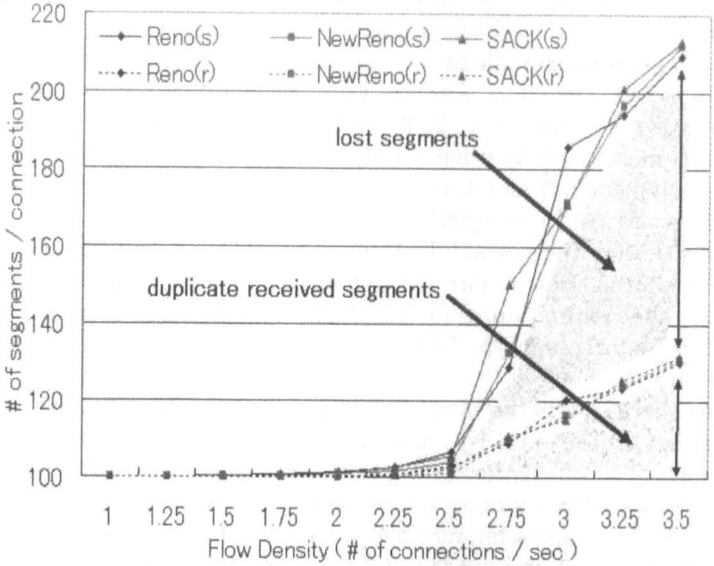

Fig. 21: Number of transmitting packets to send 100 data packets by the TCP simulation

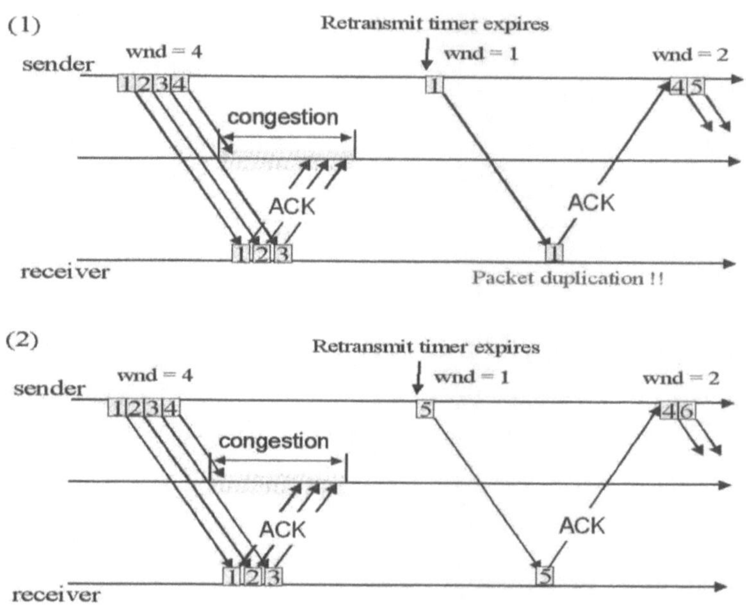

Fig. 22: Comparison of packet transmission rules for (1)TCP and (2) LEAP

if there is an ACK packet requesting the sequential number. This modification of TCP is named LEAP as the host leaps retransmitting the duplicate packets.

Fig.22(2) gives an example how this modified rule, LEAP, works comparing with nomal TCP, Fig.22(1). Assume that the window size is 4 and a host sends data packets number 1, 2, 3, and 4. If the packets 1,2 and 3 reach the destination successfully, then the destination computer sends ACKs with requesting numbers 2, 3 and 4, respectively. We also assume that all of these ACK packets are lost by congestion.

In this situation the normal TCP, shown in (1), re-sends the data packet 1 when the waiting time expires. As soon as the data packet 1 reaches the destination, the destination sends an ACK with number 4 since it already have received data 1, 2 and 3. Then normal TCP sends two succesive packets number 4 and 5.

On the other hand, in the same situation, LEAP works differently, as shown in (2). Instead of resending 1 by timeout of the waiting time, the host sends the data packet 5. When the destination computer receives the data packet 5, then it replies ACK number 4, since data packet 4 is still missing. If the host recieved ACK requesting number 4, then the host sends packets number 4 and 6. Namely, LEAP re-sends a data packet only when it is requested by an ACK.

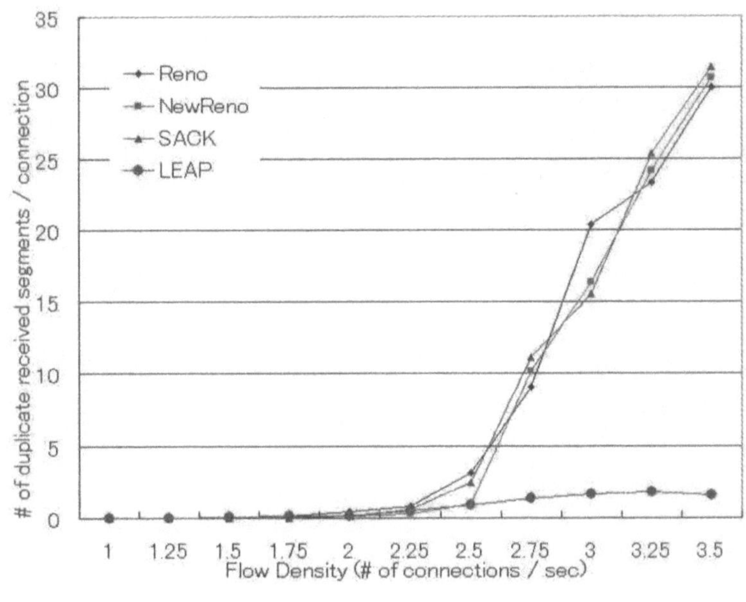

Fig. 23: Number of duplicate recieved packets for sending 100 packets by LEAP compared with normal TCP

In this way we can avoide sending the duplicate packets. Fig.23 demonstrates a drastic decrease of the number of duplicated packets by LEAP. The number of duplicate packets can not be zero, because there still exist cases that data packets are involved in heavy congestion and those packets spent considerably long time at queues of routers, but did not discarded. This problem occurs by an inappropriate way of defining the waiting time. The duplicate packet number will become zero if the waiting time estimation can be revised.

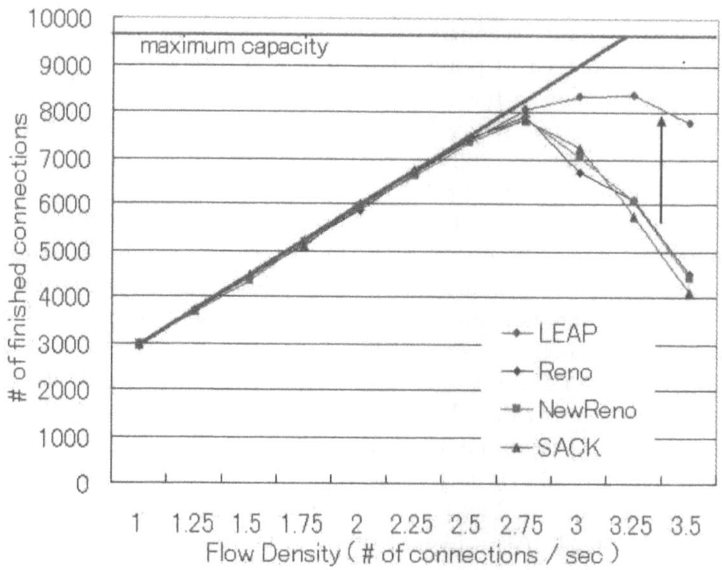

Fig. 24: Input-output relation by LEAP

Although our new protocol LEAP is just a very minor modification of ordinary TCP, the result is very much promising. Fig.24 shows the input-output relation by LEAP. As known from this figure the critical flow density shifts at about 20% larger input density, and the peak output increases about 5%. In the heavily congested state we confirmed the output flow increases about 80% compared to the normal TCP cases. Traffic control of congested phase is generally not easy, but this result shows a surprising success of totally new way of controlling traffics by shifting the critical point to the higher density.

The biggest merit of the LEAP is that the modification of TCP is quite small and implementation to a real system is rather easy. We have started applying LEAP to real systems and we will inform our research results in advance.

7 Discussions

Historically self-similar properties of information traffics were firstly discovered by the pioneering work done by Leland et al in 1994 [17]. They observed the packet flow density fluctuation in an Ethernet of Local Area Network (LAN) and found that the flow density fluctuation has a self-similar nature such as the $1/f$ fluctuations. The reason for such fractal fluctuation has been considered to be due to the human activities of using the Internet [21].

On the other hands, in our view of phase transition, their results can be understood as a special case of our observation that the local traffic of the observing point is adequately congested and the mean flow densities are always fluctuating near the critical point. The origins of the critical behaviors and $1/f$ power spectrum in traffic density flow are in various source of network dynamics in different time scale; TCP feedback control[38, 39], Ethernet collision avoidance dynamics[25, 26], and congestion contageon among neighboring routers[19, 23]. Our simulation results of using ns-2 advocate that human behaviors at the ends and complex network topology are not relevant to the critical behaviors since the qualitative characteristics are realized by simple toplogy of network and random inputs. As discussed in the preceding sections the most efficient condition for the Internet is at the critical point between congested and non-congested states. The best control is not to keep the system beyond the critical point since the packet drop rate suddenly increases and throughput of the traffic decreases rapidly above the critical point. We call this new type of system control as the *Critical Flow Control.*

The new method LEAP is an end-control that end-users or peripheral computers are the targets. There are other approaches, such as router control by tuning queue length. Also, like automobile with computerized-navigation, it might be possible to avoid congestion by using dynamical intelligent routing tables that controls packet motion in the real time.

Therefore, traffic control of heavily congested phase is a challenging task not only for the Internet, but also for general traffic systems. For future study of information network, an interesting aspect is that the $1/f$ power spectrum is observed at the critical point with the highest performance. It is well known that the brain waves and heart beat intervals are generally showing the $1/f$ fluctuations. This may imply that information traffic control of the brain is ideally designed and the *Critical Flow Control* might already been installed.

The author would like to thank Mr.Takamitsu Sato, Dr. Alex Tretyakov, Mr. Yoshiteru Takeshima, Dr. Kensuke Fukuda, Dr. Yoshifumi Nishida, Dr.Yousuke Tamura for collabolating on a part of the study during some period of the 10 years research on this topics. Mr.Takamitsu Sato started mesuring network congestion by using *ping* command in 1994 independent of Csabai's work[32]. Dr. Kensuke Fukuda extended the *ping* experiment to observe the spacial-temporal evolution of congestion as showed in the later half of section 4. Also, he devoted in simulation of Internet by using ns-2. Mr. Yoshiteru Takeshima developed the simulation of collision on an Ethernet cable and found a phase transition between random output phase and oscillated phase. Dr.Yousuke Tamura formulated the LEAP and it is a successful application of controllig the critical point. And she would like to acknowledge Dr.Hideki Takayasu for collabolation and encouragement during the whole period of this research.

References

1. T. Musha and H. Higuchi: Jpn. J. Appl. Phys. **15**, 1271 (1976)
2. K. Nagel and M. Schrekenberg: J. Phys. I France, **2**, 2221 (1992)
3. M. Takayasu and H. Takayasu: Fractals **4**, 860 (1993)
4. D.E. Wolf, M. Schreckenberg, A. Bachem(editors) *Traffic and Granular Flow*, (Springer, Singapore 1996)
5. M. Schreckenberg, D.E. Wolf, M. Schreckenberg(editors) *Traffic and Granular Flow'97*, (Springer, Singapore 1998)
6. M. Fukui, Y.Sugiyama, M. Schreckenberg, D. E. Wolf(editors) : *Traffic and Granular Flow'01*, (Springer, Singapore 2002)
7. H.Takayasu(editor): *Empirical Science of Financial Fluctuations: The Advent of Econophysics*, (Springer, Tokyo 2002).
8. H.Takayasu(editor): *The Application of Econophysics*, (Springer, Tokyo 2003).
9. Rosario N. Mantegna and H. Eugene Stanley: *An Intrduction to Econophysics: Correlations and Complexity in Finance*, (Cambridge University Press, Cmbridge 2000).
10. Didier Sornette: *Why Stock Markets Crash: Critical Events in Complexity Financial Systems*, (Princeton University Press, New Jersey 2003)
11. B. B. Mandelbrot: *The Fractal Geometry of Nature*, (W. H. Freedman and Co., New York, 1983)
12. H. E. Stanley: *Introduction to Phase Transition and Critical Phenomena*, (Oxford University Press, New York, 1971)
13. Per Bak: *How Nature Works*, (Springer-Verlag, New York, 1996)
14. L. Kleinlock: *Queueing Systems*, (John Wiley and Sons, Singapore 1976)
15. R. Jain and S.Routhier: IEEE Journal on Selected Area in Communications, **4(6)** (1986)
16. W. E. Leland, M. S. Taqqu, W. Willinger and D. V. Wilson: IEEE/ACM Trans. Networking, **2**, **NO.1** (1991)
17. W. E. Leland, D. V. Willson: In *Proceedings of INFOCOM '91*,IEEE (1994) pp1360–1366
18. V. Paxon, and S. Floyd: IEEE/ACM Trans. Networking **3** 226 (1995)
19. M. Takayasu, H. Takayasu and T. Sato: Physica A **233**, 824 (1996)
20. W. Willinger, M. S. Taqqu, R. Sherman: IEEE/ACM Trans. Networking **5(1)**, 71 (1997)
21. M. E. Crovella, A. Bestavros: IEEE/ACM Trans. Networking **5(6)**, 835 (1997)
22. B. A. Huberman, and R. M. Lukose: Science **277** (1997), 535
23. M. Takayasu, A.Yu. Tretyakov, K. Fukuda, and H.Takayasu: Phase Transition and $1/f$ Niose inthe Internet Packet Trasport. In: *Traffic and Granular Flow'97*, ed by M. Schreckenberg, D.E. Wolf (Springer, Singapore 1998) pp 57–67
24. D.E. Comer: *Internetworking with TCP/IP*, vol 3, (Plentice-Hall, Inc., New Jersey 1995)

25. Y. Takeshima, K. Fukuda, H. Takayasu, and M. Takayasu: *An Analysis on the An Analysis onthe Critical Phenomena in CSMA/CD Network Traffic Model by Computer Simulation*, In The Institute of Electronics; Information and Comunication Engineers Transactions B, **J84-B**, 840 (2001) (in Japanese)
26. K. Fukuda, H. Takayasu and M. Takayasu: Physica A **287**, 289 (2000)
27. Dietrich Stauffer, A. Aharony: *Introduction to Percolation Theory*, reviced 2nd edn. (Taylor and Francis, London, 1994)
28. W. Feller: *An Introduction to Probability Theory and Its Applications*, 2nd edn. Vol.II (Willey, New York, 1971)
29. M. Takayasu, K. Fukuda, and H. Takayasu: Physica A **274**, 30 (1999)
30. M. Takayasu, H. Takayasu, and K. Fukuda: Physica A **277**, 248 (2000)
31. M. Takayasu: Physica A **197**, 371 (1993)
32. I. Csabai: Journal of Physics A:Math.Gen. **27** 414 (1994)
33. V. Paxon: IEEE/ACM Trans. Networking **5** 601 (1997)
34. T. M. Liggett: *Interacting Particle System*, (Springer, New York, 1985)
35. H. Takayasu, N. Inui, and A. Y. Tretyakov: Phys. Rev. E **49** 1070 (1994)
36. K. Fukuda, H. Takayasu, and M. Takayasu: Fractals **7**, 23 (1999)
37. A. Yu. Tretyakov, H. Takayasu and M. Takayasu: Physica A **253**, 315 (1998)
38. K. Fukuda, M. Takayasu, and H. Takayasu: *Communication Networks and distributed systems: Modelings and simulation conference (CNDS 2001), Society for Computer Simulations* (2002) pp 55.
39. K.Fukuda, M.Takayasu, and H.Takayasu: Analysis of Minimal model of Internet Traffic. In: *Traffic and Granular Flow'01*, ed by M. Fukui, Y.Sugiyama, M. Schreckenberg, D. E. Wolf (Springer, Singapore 2002) pp 389–400
40. http://www-mash.cs.berkeley.edu/ns
41. H. Takayasu, A. Sato and M. Takayasu: Phys. Rev. Lett., **76**, 966 (1997)
42. M. Takayasu and H. Takayasu, Physica A **324**, 101 (2003)
43. Y. Tamura, M. Takayasu, and H. Takayasu: J. Information Processing Society of Japan.,**45(6)**, 1672 (2004)

Index

Understanding Complex Systems

Edited by J.A. Scott Kelso

McDaniel, R.R. Jr.; Driebe, D.J. (Eds.)
Uncertainty and Surprise in Complex Systems: Questions on Working with the Unexpected
211 p. 2005 [3-540-23773-9]

Kerner, B.S.
The Physics of Traffic: Empirical Freeway Pattern Features, Engineering Applications, and Theory
682 p. 2005 [3-540-20716-3]

Kleidon, A.; Lorenz, R.D. (Eds.)
Non-equilibrium Thermodynamics and the Production of Entropy: Life, Earth, and Beyond
260 p. 2005 [3-540-22495-5]

Jirsa, V.A.; Kelso, J.A.Scott. (Eds.)
Coordination Dynamics: Issues and Trends
272 p. 2004 [3-540-20323-0]

MIX
Papier aus verantwortungsvollen Quellen
Paper from responsible sources
FSC® C105338
www.fsc.org

If you have any concerns about our products,
you can contact us on
ProductSafety@springernature.com

In case Publisher is established outside the EU,
the EU authorized representative is:
Springer Nature Customer Service Center GmbH
Europaplatz 3, 69115 Heidelberg, Germany

Printed by Libri Plureos GmbH
in Hamburg, Germany